OXFORD ENGINEERING SCIENCE SERIES

THE OXFORD ENGINEERING SCIENCE SERIES

1 D. R. RHODES: *Synthesis of planar antenna sources*
2 L. C. WOODS: *Thermodynamics of fluid systems*
3 R. N. FRANKLIN: *Plasma phenomena in gas discharges*
4 G. A. BIRD: *Molecular gas dynamics*
5 H.-G. UNGER: *Planar optical waveguides and fibres*
6 J. HEYMAN: *Equilibrium of shell structures*
7 K. H. HUNT: *Kinematic geometry of mechanisms*
8 D. S. JONES: *Methods in electromagnetic wave propagation* (*Two volumes*)
9 W. D. MARSH: *Economics of electric utility power generation*
10 P. HAGEDORN: *Non-linear oscillations* (*Second edition*)
11 R. HILL: *Mathematical theory of plasticity*
12 D. J. DAWE: *Matrix and finite element displacement analysis of structures*
13 N. W. MURRAY: *Introduction to the theory of thin-walled structures*
14 R. I. TANNER: *Engineering rheology*
15 M. F. KANNINEN and C. H. POPELAR: *Advanced fracture mechanics*
16 R. H. T. BATES and M. J. McDONNELL: *Image restoration and reconstruction*
18 K. HUSEYIN: *Multiple-parameter stability theory and its applications*
19 R. N. BRACEWELL: *The Hartley transform*
20 J. WESSON: *Tokamaks*
21 P. B. WHALLEY: *Boiling, condensation, and gas-liquid flow*
22 C. SAMSON, M. LeBORGNE, and B. ESPIAU: *Robot control: the task function approach*
23 H. J. RAMM: *Fluid dynamics for the study of transonic flow*
24 R. R. A. SYMS: *Practical volume holography*
25 W. D. McCOMB: *The physics of fluid turbulence*
26 Z. P. BAZANT and L. CEDOLIN: *Stability of structures: principles of elastic, inelastic, and damage theories*
27 J. D. THORNTON: *Science and practice of liquid–liquid extraction* (*Two volumes*)
28 J. VAN BLADEL: *Singular electromagnetic fields and sources*

Singular Electromagnetic Fields and Sources

J. VAN BLADEL

Laboratorium voor Elektromagnetisme en Acustica
Rijksuniversiteit Gent

CLARENDON PRESS · OXFORD

1991

Oxford University Press, Walton Street, Oxford OX2 6DP

Oxford New York Toronto
Delhi Bombay Calcutta Madras Karachi
Petaling Jaya Singapore Hong Kong Tokyo
Nairobi Dar es Salaam Cape Town
Melbourne Auckland

and associated companies in
Berlin Ibadan

Oxford is a trade mark of Oxford University Press

Published in the United States
by Oxford University Press, New York

British Library Cataloguing in Publication Data
Bladel, J. van 1922–
Singular electromagnetic fields and sources.
1. Electromagnetic fields
I. Title
530.141
ISBN 0-19-856200-4

Library of Congress Cataloging in Publication Data
Bladel, J. van (Jean), 1922–
Singular electromagnetic fields and sources/J. van Bladel.
p. cm. — (Oxford engineering science series)
Includes bibliographical references.
1. Electromagnetic fields—Mathematics. I. Title. II. Series.
QC665.E4B574 1991 530.1'41—dc20 90-7724
ISBN 0-19-856200-4

Typeset by Macmillan India Ltd., Bangalore 25
Printed in Great Britain by
St Edmundsbury Press Ltd
Bury St Edmunds, Suffolk

PREFACE

The present monograph is meant to complement the several excellent texts on electromagnetic theory which are available today. The contents should be of interest to graduate students in electrical engineering and physics, as well as to practising electromagneticists in industrial and academic laboratories. The overall purpose of the book is to discuss, in more detail than would a typical general treatise, the various 'infinities' which occur in electromagnetic fields and sources. To achieve this goal, the text has been divided into three parts.

The first discusses the 'distributional' representation of strongly concentrated charges and currents. It is well-known that a point charge at r_0 can be represented by a volume density $\rho = q\delta|\boldsymbol{r} - \boldsymbol{r}_0|$, i.e. the first term in a *multipole* expansion for ρ. More general sources—scalar or vectorial—require additional terms. These involve derivatives of δ-functions, and are discussed at length in Chapter 2. The analysis there is based on an elementary presentation of Schwartz' Theory of Distributions, given in Chapter 1. In that chapter, as in the rest of the book, the approach is unashamedly that of the 'applied mathematician'.

Multipole expansions can be written for magnetic currents $\boldsymbol{K}$ as well as electric currents $\boldsymbol{J}$. It is well-known that electric currents can be replaced by equivalent magnetic currents, and vice versa. Several sections in Chapter 2 are devoted to an extensive study of these equivalences, particularly with respect to sources which are concentrated on a *surface*.

The second part of the book analyses the *fields* associated with concentrated sources. In the case of a static point charge, potential and electric field have singularities of the order of $1/|\boldsymbol{r} - \boldsymbol{r}_0|$ and $1/|\boldsymbol{r} - \boldsymbol{r}_0|^2$, respectively. When the source is time harmonic, however, stronger singularities occur. For the electric Green's dyadic, for example, they are of the order of $1/|\boldsymbol{r} - \boldsymbol{r}_0|^3$. The way to handle such singularities has generated an abundant—and often controversial—literature. Chapter 3 surveys the various possible approaches, but ultimately puts the accent on the distributional and modal aspects of the theory.

The third part of the book is devoted to an analysis of field behaviour near *geometrical* singularities such as sharp edges, tips of cones, and vertices of sectors. The mathematics involved are quite elementary, and the emphasis is laid on the presentation of numerical data useful to the practising electromagneticist.

The scope of the monograph is seen to be quite modest. It is clear that additional topics could have been included, e.g. a treatment of Green's dyadics in non-homogeneous media (particularly layered ones), or a discussion of field behaviour near foci and caustics. These topics were deliberately left aside to safeguard the compact character of the book.

Many authors mentioned in the text took the trouble to read the paragraphs in which their work was quoted, and to suggest improvements and additions. These distinguished colleagues cannot be thanked individually, given their number. An exception must be made for Professor J. Boersma, whose extensive and critical remarks considerably increased the mathematical accuracy of many a section, particularly in Chapters 3 and 5.

Any formal qualities present in the text should be credited to the author's daughter Viveca, who applied her literary talents to a thorough criticism of the style of the original manuscript.

Finally, the author wishes to acknowledge the support given by his colleague and friend Professor P. E. Lagasse, and the competence with which two devoted secretaries, Mrs Buysse and Mrs Naessens, struggled with figures and equations.

Ghent J.V.B.
October 1990

CONTENTS

LIST OF SYMBOLS

General notation

Standard notations are used for:

—the electromagnetic fields $\boldsymbol{e}$, $\boldsymbol{h}$, $\boldsymbol{d}$, $\boldsymbol{b}$;

—coordinates such as polar coordinates (r, φ, z) and spherical coordinates (R, φ, θ);

—special functions, such as Bessel and Hankel functions, and Legendre, Gegenbauer, and Chebychev polynomials;

The imaginary symbol is j, and the harmonic time factor is $e^{j\omega t}$.

Capitals are used to represent complex phasors (e.g. $\boldsymbol{E}$ for $\boldsymbol{e}$).

The Naperian logarithm is denoted by $\log_e$.

A_i denotes an incident quantity

A_{sc} denotes a scattered quantity

$o(1/x)$ and $O(1/x)$: a function $f(x)$ is $o(1/x)$ or $O(1/x)$ depending on whether $\lim_{x \to \infty} x f(x)$ is zero or different from zero (but finite).

$\doteqdot$ means proportional to;

$\approx$ means almost equal to;

$\underset{x}{\sim}$ means asymptotically equal to, for $x \to \infty$.

Symbols

$\boldsymbol{a}$ = magnetic vector potential (see Section 2.6)

$\boldsymbol{c}$ = electric vector potential (see Section 2.6)

$\boldsymbol{c}_s$ = strength of a double layer of current (see Section 1.12)

$\boldsymbol{e}_c$ = "cavity" electric field (see Section 3.18)

$\boldsymbol{e}_m, \boldsymbol{f}_m, \boldsymbol{g}_m, \boldsymbol{h}_m$ = cavity eigenvectors (see Appendix C)

f = frequency

h_i = metrical coefficient (see Section 1.5)

$\boldsymbol{j}$ = volume density of electric current

$\boldsymbol{j}_s$ = surface density of electric current

$k = \omega/c = 2\pi/\lambda$ = wavenumber in vacuum

$\boldsymbol{k}$ = volume density of magnetic current

$\boldsymbol{k}_s$ = surface density of magnetic current

p = acoustic pressure

$\boldsymbol{p}_e$ = electric dipole moment (see Section 2.1)

$\boldsymbol{p}_m$ = magnetic dipole moment (see Section 2.3)

q = electric charge

$\mathbf{q}_e$ = electric quadrupole dyadic (see Section 2.1)
$\mathbf{q}_m$ = magnetic quadrupole dyadic (see Section 2.5)
$\boldsymbol{r} = x\boldsymbol{u}_x + y\boldsymbol{u}_y + z\boldsymbol{u}_z$ = radius vector from the origin
$\boldsymbol{u}_a$ = unit vector in the direction in which parameter a is measured
$\boldsymbol{v}$ = velocity
$G(\boldsymbol{r}|\boldsymbol{r}')$ = a Green's function
$\mathbf{G}(\boldsymbol{r}|\boldsymbol{r}')$ = a Green's dyadic (see Appendix B)
$\mathbf{G}_e(\boldsymbol{r}|\boldsymbol{r}')$ = electric Green's dyadic (see Section 3.14)
$\mathbf{G}_m(\boldsymbol{r}|\boldsymbol{r}')$ = mixed Green's dyadic (see Section 3.12)
$\mathbf{G}_h(\boldsymbol{r}|\boldsymbol{r}')$ = magnetic Green's dyadic (see Section 3.19)
$\mathbf{I}_{xy} = \boldsymbol{u}_x\boldsymbol{u}_x + \boldsymbol{u}_y\boldsymbol{u}_y$ = identity dyadic in the (x, y) plane
$\mathbf{L}_{V^*}$ = depolarization dyadic relative to a volume V^* (see Section 3.10)
$R_c = \sqrt{(\mu_0/\varepsilon_0)}$ = characteristic resistance of vacuum
R_i = principal radius of curvature (see Section 1.5)
tr = trace of a dyadic (see Appendix B)
$\mathrm{Y}(x)$ = unit, step, or Heaviside function (see Section 1.3)
$Y_s(\boldsymbol{r})$ = three-dimensional Heaviside function (see Section 1.10)
γ_m, γ_n = propagation (or damping) constant (see Section 3.24)
$\delta(x)$ = one-dimensional delta-function (see Section 1.1)
$\delta^{(m)}(x)$ = m^{th} derivative of the delta-function (see Section 1.8)
$\delta(\boldsymbol{r})$ = three-dimensional delta-function (see Section 1.4)
δ_s = Dirac function on a surface (see Section 1.5)
δ_c = Dirac function on a curve (see Section 1.5)
δ_{ik} = Kronecker's delta ($\delta_{ii} = 1$; $\delta_{ik} = 0$ for $i \neq k$)
$\boldsymbol{\delta}_l$ = longitudinal Dirac dyadic (see Section 3.20)
$\boldsymbol{\delta}_t$ = transverse Dirac dyadic (see Section 3.20)
ε_r = dielectric constant (dimensionless)
$\varepsilon_0 = 4\pi\, 10^{-7}$ (F m^{-1})
$\varepsilon = \varepsilon_0\varepsilon_r$
μ_r = magnetic permeability (dimensionless)
$\mu_0 = 10^{-9}/36\pi$ (H m^{-1})
$\mu = \mu_r\mu_0$
$\omega = 2\pi f$ = angular frequency
π = a Hertz potential (see Section 3.18)
ρ = volume density of electric charge (in C m^{-3})
ρ_s = surface density of electric charge (in C m^{-2})
ρ_c = electric charge density on a curve C (in C m^{-1})
ρ_L = electric charge density on a straight line (in C m^{-1})
$\boldsymbol{\rho} = x\boldsymbol{u}_x + y\boldsymbol{u}_y$ = radius vector from the origin, in the (x, y) plane
σ = conductivity

τ = dipole-layer density on a surface (see Section 1.12)
ϕ_m = Dirichlet eigenfunction (see Appendices C and D)
ψ_n = Neumann eigenfunction (see Appendices C and D)
Ω = a solid angle
$\mathscr{D}$ = space of test functions (see Section 1.2)
$\mathscr{D}'$ = space of distributions (see Section 1.3)
$\mathscr{E}$ = an energy per unit length (see Section 4.2)

Operators

$\mathrm{div}_s \boldsymbol{j}_s$ = surface divergence of a tangential vector (see Section 1.15)
$\mathrm{grad}_s A$ = surface gradient of a scalar function (see Section 2.13)

1

DELTA-FUNCTIONS AND DISTRIBUTIONS

The delta-function and its derivatives are frequently encountered in the technical literature. The function was first conceived as a tool which, if properly handled, could lead to useful results in a particularly concise way. Its popularity is now justified by solid mathematical arguments, developed over the years by authors such as Sobolev, Bochner, Mikusinski, and Schwartz. In the following pages we give the essentials of the Schwartz approach (distribution theory). The level of treatment is purely utilitarian. Rigorous exposés, together with descriptions of the historical evolution of the theory, may be found in the numerous texts quoted in the bibliography.

1.1 The δ-function

The idea of the δ-function is quite old, and dates back at least to the times of Kirchoff and Heaviside (van der Pol *et al.* 1951). In the early days of quantum mechanics, Dirac put the accent on the following properties of the function:

$$\int_{-\infty}^{\infty} \delta(x)\mathrm{d}x = 1, \qquad \delta(x) = 0 \quad \text{for } x \neq 0. \tag{1.1}$$

The notation $\delta(x)$ was inspired by δ_{ik}, the Kronecker delta, equal to 0 for $i \neq k$, and to 1 for $i = k$. Clearly, $\delta(x)$ must be 'infinite' at $x = 0$ if the integral in (1.1) is to be unity. Dirac recognized from the start that $\delta(x)$ was not a function of x in the usual mathematical sense, but something more general which he called an 'improper' function. Its use, therefore, had to be confined to certain simple expressions, and subjected to careful codification. One of the expressions put forward by Dirac was the 'sifting' property

$$\int_{-\infty}^{\infty} f(x)\delta(x)\mathrm{d}x = f(0). \tag{1.2}$$

This relationship can serve to define the delta function, not by its value at each point of the x axis, but by the ensemble of its scalar products with suitably chosen 'test' functions $f(x)$.

It is clear that the infinitely-peaked delta function can be interpreted intuitively as a strongly concentrated forcing function. The function may represent, for example, the force density produced by a unit force acting on a one-dimensional mechanical structure, e.g. a flexible string. This point of view leads to the concept of $\delta(x)$ being the limit of a function which becomes more and more concentrated in the vicinity of $x = 0$, whereas its integral from $-\infty$ to $+\infty$ remains equal to one. Some of the limit functions which behave in that manner are shown in Fig. 1.1. The first one is the rectangular pulse, which becomes 'needle-like' at high values of n (Fig. 1.1a). The other ones are (de Jager 1969; Bass 1971)

$$\lim_{n\to\infty} \frac{n}{\sqrt{\pi}} e^{-n^2x^2} \qquad \text{(shown in Fig. 1.1b),} \qquad (1.3)$$

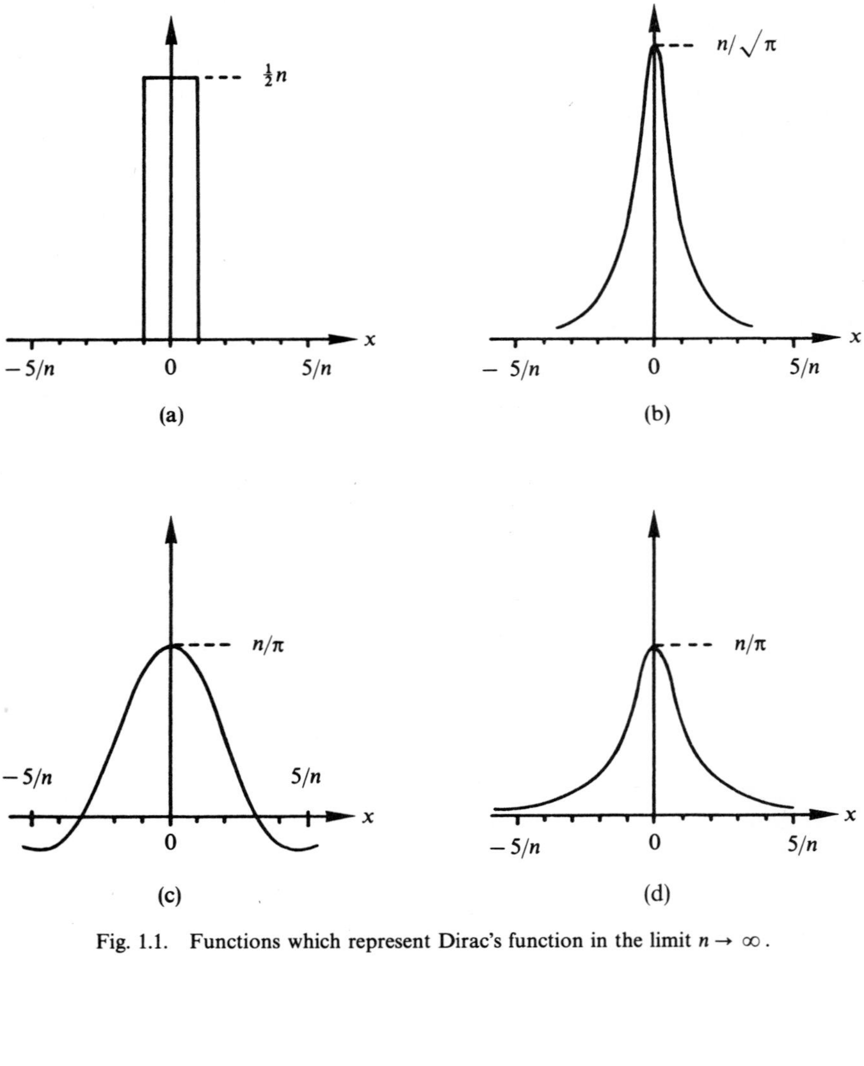

Fig. 1.1. Functions which represent Dirac's function in the limit $n \to \infty$.

$$\lim_{n\to\infty} \frac{\sin nx}{\pi x} \qquad \text{(shown in Fig. 1.1c)}, \tag{1.4}$$

$$\lim_{n\to\infty} \frac{n}{\pi(1+n^2x^2)} = \lim_{n\to\infty}\left(-\frac{n}{\pi}\mathrm{Im}\frac{1}{nx+\mathrm{j}}\right) \qquad \text{(shown in Fig. 1.1d)}. \tag{1.5}$$

1.2 Test functions and distributions

The notion of distribution is obtained by generalizing the idea embodied in (1.2), namely that a function is defined by the totality of its scalar products with reference functions termed *test functions.* The test functions used in the Schwartz theory are complex continuous functions $\phi(\boldsymbol{r})$ endowed with continuous derivatives of all orders. Such functions are often termed 'infinitely smooth'. They must vanish outside some finite domain, which may be different for each ϕ. They form a space $\mathscr{D}$. The smallest closed set which contains the set of points for which $\phi(\boldsymbol{r}) \neq 0$ is the *support* of ϕ. A typical one-dimensional test function is

$$\phi(x) = \begin{cases} \exp \dfrac{|ab|}{(x-a)(x-b)} & \text{for } x \text{ in } (a, b), \\ 0 & \text{for } x \text{ outside } (a, b). \end{cases} \tag{1.6}$$

The support of this function is the interval $[a, b]$. At the points $x = a$ and $x = b$, all derivatives vanish, and the graph of the function has a contact of infinite order with the x axis. A particular case of (1.6) is

$$\phi(x) = \begin{cases} \exp \dfrac{-1}{1-x^2} & \text{for } |x| < 1, \\ 0 & \text{for } |x| \geqslant 1. \end{cases} \tag{1.7}$$

In n dimensions, with $R^2 = x_1^2 + \cdots + x_n^2$, we have

$$\phi(\boldsymbol{r}) = \begin{cases} \exp \dfrac{-1}{1-R^2} & \text{for } |\boldsymbol{r}| < 1, \\ 0 & \text{for } |\boldsymbol{r}| \geqslant 1. \end{cases} \tag{1.8}$$

A few counterexamples are worth mentioning: $\phi(x) = x^2$ (for all x) is *not* a test function because its support is not bounded. The same is true of $\phi(x) = \sin|x|$, which furthermore has no continuous derivative at the origin.

To introduce the concept of 'distribution', it is necessary to first define *convergence* in $\mathscr{D}$ (Schwartz 1965). A sequence of functions $\phi_m(x)$ belonging to $\mathscr{D}$ is said to converge to $\phi(x)$ for $m \to \infty$ if

(1) the supports of the ϕ_m are contained in the same closed domain, independently of m;
(2) the ϕ_m and their derivatives of all orders converge uniformly to ϕ and its corresponding derivatives.

The next step is to define a *linear functional* on $\mathscr{D}$. This is an operation which associates a complex number $t(\phi)$ with every ϕ belonging to $\mathscr{D}$, in such a way that

$$t(\phi_1 + \phi_2) = t(\phi_1) + t(\phi_2), \qquad t(\lambda\phi) = \lambda t(\phi), \tag{1.9}$$

where λ is a complex constant. The complex number $t(\phi)$ is often written in the form

$$t(\phi) = \langle t, \phi \rangle \tag{1.10}$$

The functional is *continuous* if, when ϕ_m converges to ϕ for $m \to \infty$, the complex numbers $t(\phi_m)$ converge to $t(\phi)$. *Distributions are continuous linear functionals* on $\mathscr{D}$. They form a vector space $\mathscr{D}'$. To clarify these concepts, assume that $\tau(x)$ is a locally integrable function (i.e. a function which is integrable over any compact set). Such a function generates a distribution by the operation (Schwartz 1965)

$$\tau(\phi) = \langle \tau, \phi \rangle \stackrel{\text{def}}{=} \int_{-\infty}^{\infty} \tau(x)\phi(x)\mathrm{d}x. \tag{1.11}$$

Many distributions cannot be written as an integral of that form, except in a formal way. For such cases the 'generating function' $\tau(x)$ becomes a symbolic function, and (1.11) only means that the integral, whenever it is encountered in an analytical development, may be replaced by the value $\tau(\phi)$. It should be noted, in this respect, that experiments do not yield instantaneous, punctual values of quantities such as a force or an electric field. Instead, they generate *integrated* outputs, i.e. averages over some non-vanishing intervals of time and space. The description of a quantity by scalar products of the form (1.11) is therefore quite acceptable from a physical point of view.

1.3 Simple examples

A first simple example is the integral of ϕ from 0 to ∞. This integral is a distribution, which may be written as

$$\langle \mathrm{Y}, \phi \rangle \stackrel{\text{def}}{=} \int_{0}^{\infty} \phi(x)\mathrm{d}x = \int_{-\infty}^{\infty} \mathrm{Y}(x)\phi(x)\mathrm{d}x. \tag{1.12}$$

The generating function is the Heaviside unit function $\mathrm{Y}(x)$, defined by the values

$$\mathrm{Y}(x) = \begin{cases} 0 & \text{for } x < 0, \\ 1 & \text{for } x \geqslant 0. \end{cases} \tag{1.13}$$

As a second example, we consider a function $f(x)$, possibly undefined at c and unbounded near c, but integrable in the intervals $(a, c - \varepsilon)$ and $(c + \eta, b)$, where ε and η are positive. If

$$I = \lim_{\substack{\varepsilon \to 0 \\ \eta \to 0}} \left(\int_a^{c-\varepsilon} f(x)\mathrm{d}x + \int_{c+\eta}^{b} f(x)\mathrm{d}x \right) \tag{1.14}$$

exists for ε and η approaching zero independently of each other, this limit is termed the integral of $f(x)$ from a to b. Sometimes the limit exists only for $\varepsilon = \eta$. In such a case, its value is the principal value of Cauchy, and one writes

$$\mathrm{PV} \int_a^b f(x)\mathrm{d}x = \lim_{\varepsilon \to 0} \left(\int_a^{c-\varepsilon} f(x)\mathrm{d}x + \int_{c+\varepsilon}^{b} f(x)\mathrm{d}x \right). \tag{1.15}$$

An example of such an integral is

$$\mathrm{PV} \int_{-a}^{a} \frac{\mathrm{d}x}{x} = \lim_{\varepsilon \to 0} \left(\int_{-a}^{-\varepsilon} \frac{\mathrm{d}x}{x} + \int_{\varepsilon}^{a} \frac{\mathrm{d}x}{x} \right)$$
$$= \lim_{\varepsilon \to 0} (\log \varepsilon - \log a + \log a - \log \varepsilon) = 0. \tag{1.16}$$

The function $1/x$ does not define a distribution since it is not integrable in the vicinity of $x = 0$. But a well-defined meaning may be attached to $\mathrm{PV}(1/x)$ by introducing the functional:

$$\langle \mathrm{PV}\,(1/x), \phi \rangle \stackrel{\text{def}}{=} \mathrm{PV} \int_{-\infty}^{\infty} \frac{\phi(x)}{x} \mathrm{d}x = \int_{-\infty}^{\infty} \mathrm{PV}\,(1/x)\,\phi(x)\mathrm{d}x. \tag{1.17}$$

The third example, of great importance in mathematical physics, is the original Dirac distribution, which associates the value $\phi(0)$ with any test function $\phi(x)$. Thus,

$$\langle \delta_0, \phi \rangle \stackrel{\text{def}}{=} \phi(0) = \int_{-\infty}^{\infty} \delta(x)\phi(x)\mathrm{d}x. \tag{1.18}$$

Similarly,

$$\langle \delta_{x_0}, \phi \rangle \stackrel{\text{def}}{=} \phi(x_0) = \int_{-\infty}^{\infty} \delta(x - x_0)\phi(x)\mathrm{d}x. \tag{1.19}$$

As an application

$$\int_{-\infty}^{\infty} x^m\, \delta(x)\phi(x)\mathrm{d}x = \int_{-\infty}^{\infty} \delta(x)\,[x^m \phi(x)]\,\mathrm{d}x = 0 \tag{1.20}$$

holds when m is a positive integer, in which case $x^m\phi(x)$ is a test function. Property (1.20) may therefore be written in symbolic form as

$$x^m\delta(x) = 0. \tag{1.21}$$

As mentioned above, $\delta(x)$ has no 'values' on the x axis, but the statement that the delta function $\delta(x)$ is zero in the vicinity of a point such as $x_0 = 1$ can be given a well-defined meaning by introducing the concept 'support of a distribution'. A distribution $\langle t, \phi \rangle$ is said to vanish in an interval Δ if, for every $\phi(x)$ which has its support in that interval, $\langle t, \phi \rangle = 0$. This clearly holds, in the case of $\delta(x)$, for intervals Δ which do not contain the origin. The support of t is what remains of the x axis when all the Δ intervals have been excluded. The support of δ_0 is therefore the point $x = 0$ (Schwartz 1965).

1.4 Three-dimensional delta-functions

The three-dimensional δ-function is defined by the sifting property

$$\langle \delta_0, \phi \rangle \stackrel{\text{def}}{=} \phi(0) = \iiint \delta(\boldsymbol{r})\phi(\boldsymbol{r})\,\mathrm{d}V; \tag{1.22}$$

here and in the future, the omission of the integration limits means that the integral is extended over all space.

In Cartesian coordinates, the volume element is $\mathrm{d}x\,\mathrm{d}y\,\mathrm{d}z$, and $\delta(\boldsymbol{r})$ can be written explicitly as

$$\delta(\boldsymbol{r}) = \delta(x)\,\delta(y)\,\delta(z). \tag{1.23}$$

In a more general coordinate system, the form of $\mathrm{d}V$ determines that of $\delta(\boldsymbol{r})$. Let (u, v, w) be a set of curvilinear coordinates. The volume element at a regular point is $J\,\mathrm{d}u\,\mathrm{d}v\,\mathrm{d}w$, where J denotes the Jacobian of the transformation from the (x, y, z) coordinates into the (u, v, w) coordinates. More explicitly:

$$J = \begin{vmatrix} \dfrac{\partial x}{\partial u} & \dfrac{\partial x}{\partial v} & \dfrac{\partial x}{\partial w} \\ \dfrac{\partial y}{\partial u} & \dfrac{\partial y}{\partial v} & \dfrac{\partial y}{\partial w} \\ \dfrac{\partial z}{\partial u} & \dfrac{\partial z}{\partial v} & \dfrac{\partial z}{\partial w} \end{vmatrix}. \tag{1.24}$$

The three-dimensional delta function can be expressed in terms of one-dimensional functions by the relationship

$$\delta(u - u_0, v - v_0, w - w_0) = \frac{\delta(u - u_0)\delta(v - v_0)\delta(w - w_0)}{J(x_0, y_0, z_0)}. \tag{1.25}$$

The singular points of the coordinate system are those at which the Jacobian vanishes. At such points, the transformation from (x, y, z) into (u, v, w) is no longer of the one-to-one type, and some of the (u, v, w) coordinates become ignorable, i.e. they need not be known to find the corresponding (x, y, z). Let J_k be the integral of J over the ignorable coordinates. Then δ is the product of the δ's relative to the nonignorable coordinates, divided by J_k. In cylindrical coordinates, for example, J is equal to r, and

$$\begin{aligned}\delta(\mathbf{r}-\mathbf{r}_0) &= \delta(r-r_0, \varphi-\varphi_0, z-z_0) \\ &= (1/r_0)\delta(r-r_0)\delta(\varphi-\varphi_0)\delta(z-z_0).\end{aligned} \tag{1.26}$$

Points on the z axis are singular, and φ is ignorable there. We therefore write

$$\delta(\mathbf{r}-\mathbf{r}_0) = \delta(r)\delta(z-z_0)\Big/\int_0^{2\pi} r\,\mathrm{d}\varphi = \frac{1}{2\pi r}\delta(r)\delta(z-z_0). \tag{1.27}$$

This representation is valid with the convention

$$\int_0^\infty \delta(r)\mathrm{d}r = 1. \tag{1.28}$$

If one chooses

$$\int_0^\infty \delta(r)\mathrm{d}r = \tfrac{1}{2}, \tag{1.29}$$

then the $1/2\pi$ factor in (1.27) should be replaced by $1/\pi$.

In spherical coordinates, J is equal to $R^2 \sin\theta$, and

$$\delta(\mathbf{r}-\mathbf{r}_0) = \delta(R-R_0)\delta(\varphi-\varphi_0)\delta(\theta-\theta_0)/R^2\sin\theta\,. \tag{1.30}$$

On the polar axis (where θ_0 is zero or π), the azimuth φ is ignorable, and

$$\delta(\mathbf{r}-\mathbf{r}_0) = \delta(R-R_0)\delta(\theta-\theta_0)/2\pi R^2\sin\theta\,. \tag{1.31}$$

At the origin, both φ and θ are ignorable, and

$$\delta(\mathbf{r}-\mathbf{r}_0) = \delta(R)/4\pi R^2\,. \tag{1.32}$$

This formula holds when $\delta(R)$ satisfies (1.28), with r replaced by R. If $\delta(R)$ is assumed to satisfy (1.29), then the factor $1/4\pi$ in (1.32) must be replaced by $1/2\pi$.

1.5 Delta-functions on lines and surfaces

The electric charge density ρ_s on a surface is a concentrated source; hence it should be possible to express its value in terms of some appropriate δ-function. A first method to achieve this goal is based on partial separation of variables in the (v_1, v_2, n) coordinate system (Fig. 1.2a). The n coordinate is

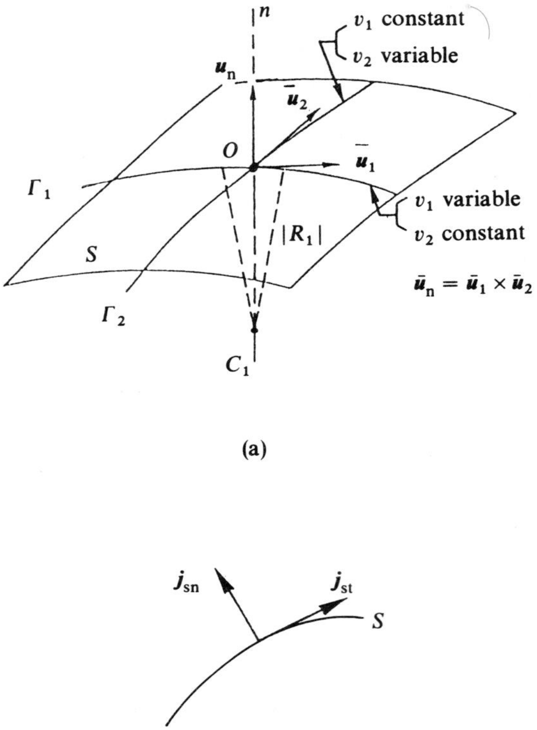

(a)

(b)

Fig. 1.2. Coordinates and currents on a surface.

measured along the normal, whereas the (v_1, v_2) coordinates fix the position of a point on S. The lines of constant v_1 (or v_2) are the (orthogonal) lines of curvature Γ_2 (or Γ_1). On a line such as Γ_1, the normals at consecutive points intersect at a common point C_1, termed the centre of curvature. Similar considerations hold for Γ_2. The distances from P to C_1 and C_2, counted positive in the direction of $\boldsymbol{u}_n$, are the two principal radii of curvature: R_1 and R_2. With this sign convention, they are negative at a point where the surface is convex. On a sphere of radius a, for example, $R_1 = R_2 = -a$.

An increase $(\mathrm{d}v_1, \mathrm{d}v_2, \mathrm{d}n)$ in the coordinates results in a displacement $\mathrm{d}l$ given by

$$\mathrm{d}l^2 = h_1^2 \mathrm{d}v_1^2 + h_2^2 \mathrm{d}v_2^2 + \mathrm{d}n^2. \tag{1.33}$$

The quantities h_1 and h_2 are the metrical coefficients.

To determine the distributional form of the surface charge density, we start from a strongly concentrated volume density ρ straddling the surface S. Let us assume that the law of variation of ρ with n is the same for all points of S. Under those circumstances, we write $\rho = g(v_1, v_2)f(n)$. When the concentration increases without limit, the volume charge goes over into a surface charge, and the 'profile function' $f(n)$ becomes a $\delta(n)$ function. We write

$$\rho = \rho_s(v_1, v_2)\delta(n). \tag{1.34}$$

Integrated over space, this expression yields

$$\iiint \rho \, dV = \iiint \rho_s(v_1, v_2)\delta(n) dS \, dn = \iint_S \rho_s(v_1, v_2) dS. \tag{1.35}$$

where $dS = h_1 h_2 \, dv_1 \, dv_2$. A representation such as (1.34) is particularly useful when separation of variables is applicable, and n is one of the coordinates. The correct distributional representation of ρ_s, however, is not based on $\delta(n)$, but on a symbolic (or generalized) function δ_s, defined by the functional

$$\langle \delta_s, \phi \rangle \stackrel{\text{def}}{=} \iint_S \phi(\boldsymbol{r}) dS = \iiint \delta_s \phi(\boldsymbol{r}) dV. \tag{1.36}$$

The meaning of this relationship is the usual one, that is whenever the volume integral is encountered in an analytical development, it may be replaced by the surface integral. The support of δ_s is the surface S. A distribution of surface charge density ρ_s may now be represented by the volume density

$$\rho = \rho_s(v_1, v_2)\delta_s. \tag{1.37}$$

The corresponding functional is (de Jager 1969, 1970)

$$\langle \rho_s \delta_s, \phi \rangle \stackrel{\text{def}}{=} \iint_S \rho_s \phi \, dS = \iiint \rho_s \delta_s \phi \, dV. \tag{1.38}$$

Similarly, a surface electric current may be written in the form (Fig. 1.2b)

$$\boldsymbol{j} = \boldsymbol{j}_s(v_1, v_2)\delta_s = \boldsymbol{j}_{st}(v_1, v_2)\delta_s + \boldsymbol{j}_{sn}(v_1, v_2)\delta_s. \tag{1.39}$$

The normal component of this current represents a surface distribution of elementary currents, oriented along the normal.

Finally, a distribution of linear charge density ρ_c on a curve C may be represented by the volume density

$$\rho = \rho_c \delta_c, \tag{1.40}$$

where

$$\langle \rho_c \delta_c, \phi \rangle \stackrel{\text{def}}{=} \int_C \rho_c \phi \, dC = \iiint \rho_c \delta_c \phi \, dV. \tag{1.41}$$

1.6 Multiplication of distributions

There is no natural way to define the product of two distributions. A locally integrable function, for example, generates a distribution; but the product of two such functions might not be locally integrable, and hence it might not generate a distribution. To illustrate the point, consider the function $f(x) = 1/\sqrt{|x|}$, which is locally integrable. Its square $f^2(x) = 1/|x|$, however, is not, and does *not* define a distribution. In general, the more f is irregular, the more g must be regular if the product fg is to have a meaning. Multiplication by an infinitely differentiable function $\alpha(x)$, however, is always meaningful. More precisely:

$$\langle \alpha t, \phi \rangle \overset{\text{def}}{=} \int_{-\infty}^{\infty} \alpha(x)t(x)\phi(x)\mathrm{d}x = \int_{-\infty}^{\infty} t(x)[\alpha(x)\phi(x)]\mathrm{d}x = \langle t, \alpha\phi \rangle. \tag{1.42}$$

This result is based on the fact that $\alpha\phi$ is a test function. An example of application of (1.42) is relationship (1.21). Another one is

$$\alpha(x)\delta(x - x_0) = \alpha(x_0)\delta(x - x_0). \tag{1.43}$$

The restriction to infinitely differentiable $\alpha(x)$ is not always necessary. The function $\alpha(x)\delta(x)$, for example, has a meaning, namely $\alpha(0)\delta(x)$, once $\alpha(x)$ is continuous at the origin.

It should be noted that multiplication of distributions, even when defined, is not necessarily associative. For example:

$$\left(\frac{1}{x}x\right)\delta(x) = \delta(x), \qquad \frac{1}{x}[x\delta(x)] = \frac{1}{x}0 = 0. \tag{1.44}$$

1.7 Change of variables

The operation 'change of variables' starts from a generating function $t(x)$, and introduces $t[f(x)]$ by means of the formula (Friedman 1969)

$$\int_{-\infty}^{\infty} t[f(x)]\phi(x)\mathrm{d}x = \int_{-\infty}^{\infty} t(y)\left(\frac{\mathrm{d}}{\mathrm{d}y}\int_{f(u)<y} \phi(u)\mathrm{d}u\right)\mathrm{d}y. \tag{1.45}$$

The right-hand side has a meaning, provided that the term in big parentheses is a test function. Let us investigate, for example, the properties of $\delta(\alpha x - \beta)$. From (1.45):

$$\int_{-\infty}^{\infty} \delta(\alpha x - \beta)\phi(x)\mathrm{d}x = \int_{-\infty}^{\infty} \delta(y)\left(\frac{\mathrm{d}}{\mathrm{d}y}\int_{\alpha u - \beta < y} \phi(u)\mathrm{d}u\right)\mathrm{d}y. \tag{1.46}$$

Assume first that $\alpha > 0$. The integral becomes

$$\int_{-\infty}^{\infty} \delta(y)\left(\frac{d}{dy}\int_{-\infty}^{(1/\alpha)(y+\beta)} \phi(u)du\right)dy = \frac{1}{\alpha}\phi\left(\frac{\beta}{\alpha}\right); \tag{1.47}$$

hence

$$\delta(\alpha x - \beta) = (1/\alpha)\delta(x - \beta/\alpha). \tag{1.48}$$

Consider now the case $\alpha < 0$. The integral takes the form

$$\int_{-\infty}^{\infty} \delta(y)\left(\frac{d}{dy}\int_{(1/\alpha)(y+\beta)}^{\infty} \phi(u)du\right)dy = \int_{-\infty}^{\infty} \delta(y)\left[-\frac{1}{\alpha}\phi\left(\frac{y+\beta}{\alpha}\right)\right]dy$$

$$= -\frac{1}{\alpha}\phi\left(\frac{\beta}{\alpha}\right); \tag{1.49}$$

hence

$$\delta(\alpha x - \beta) = -(1/\alpha)\delta(x - \beta/\alpha). \tag{1.50}$$

The two cases may be combined into a single formula:

$$\delta(\alpha x - \beta) = (1/|\alpha|)\delta(x - \beta/\alpha). \tag{1.51}$$

Similarly

$$\delta[(x-a)(x-b)] = (1/|b-a|)[\delta(x-a) + \delta(x-b)] \quad (a \neq b). \tag{1.52}$$

A particular application of (1.51) is

$$\delta(x) = \delta(-x). \tag{1.53}$$

The δ-function is therefore an 'even' function, a property which is in harmony with the profile of the curves shown in Fig. 1.1.

Additional useful formulas may be obtained from (1.45). For example:

$$\delta(x^2 - a^2) = (1/2|a|)[\delta(x-a) + \delta(x+a)]. \tag{1.54}$$

As a particular case:

$$|x|\delta(x^2) = \delta(x). \tag{1.55}$$

Consider further a function $f(x)$ which varies monotonically, vanishes at $x = x_0$, and satisfies $f'(x_0) \neq 0$. For such a function,

$$\delta[f(x)] = (1/|f'(x_0)|)\delta(x - x_0). \tag{1.56}$$

Result (1.51) follows directly from this formula. Finally, (1.45) also leads to

$$\langle t(x/a), \phi(x)\rangle = |a|\langle t(x), \phi(ax)\rangle. \tag{1.57}$$

1.8 The derivative of a distribution

The derivative of a distribution t is a new distribution t', defined by the functional

$$\langle t', \phi \rangle \stackrel{\text{def}}{=} -\left\langle t, \frac{d\phi}{dx} \right\rangle = -\int_{-\infty}^{\infty} t \frac{d\phi}{dx} dx = \int_{-\infty}^{\infty} \frac{dt}{dx} \phi \, dx. \tag{1.58}$$

Every distribution, therefore, has a derivative: a property which obviously has no analogue in the classical theory of functions.

One expects the generating function $t'(x)$ to coincide with the usual derivative when both t and dt/dx are continuous. That this is so may be shown by the following elementary integration:

$$-\int_{-\infty}^{\infty} t \frac{d\phi}{dx} dx = -\left[t(x)\phi(x)\right]_{-\infty}^{\infty} + \int_{-\infty}^{\infty} \frac{dt}{dx} \phi dx = \int_{-\infty}^{\infty} \frac{dt}{dx} \phi dx. \tag{1.59}$$

To obtain this result, we took into account that $t(x)$ is bounded, and that $\phi(x)$ vanishes at $x = \infty$ and $x = -\infty$.

As a second application, consider the automatic introduction of a δ-function into the derivative of a function $t(x)$ which suffers a jump discontinuity A at x_0 (Fig. 1.3). From (1.58), since t remains bounded in x_0,

$$-\int_{-\infty}^{\infty} t \frac{d\phi}{dx} dx = -\int_{-\infty}^{x_0^-} t \frac{d\phi}{dx} dx - \int_{x_0^+}^{\infty} t \frac{d\phi}{dx} dx \tag{1.60}$$

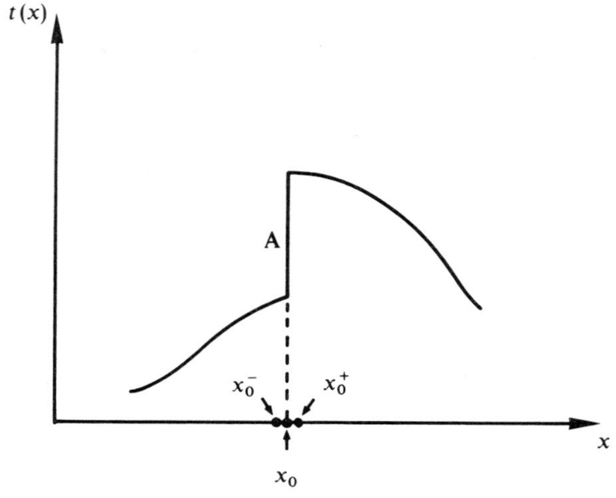

Fig. 1.3. A piecewise continuous function.

An integration by parts yields

$$\langle t', \phi\rangle = A\,\phi(x_0) + \int_{-\infty}^{x_0^-} \frac{dt}{dx}\phi\,dx + \int_{x_0^+}^{\infty} \frac{dt}{dx}\phi\,dx$$
$$= \int_{-\infty}^{\infty} \left(A\,\delta(x - x_0) + \left\{\frac{dt}{dx}\right\}\right)\phi\,dx. \tag{1.61}$$

The notation $\{dt/dx\}$, used frequently in the sequel, represents a function which is equal to the usual derivative anywhere but at x_0, where it remains undefined. The generating function of t' is therefore

$$\frac{dt}{dx} = A\,\delta(x - x_0) + \left\{\frac{dt}{dx}\right\}. \tag{1.62}$$

The derivative of the Heaviside step function, in particular, is

$$\frac{dY}{dx} = \delta(x). \tag{1.63}$$

The definition (1.58), applied to the Naperian logarithm, yields (Dirac, 1958, Schwartz, 1965, Lützen, 1982)

$$\frac{d}{dx}\log_e|x| = \mathrm{PV}\frac{1}{x}, \tag{1.64}$$

$$\frac{d}{dx}\log_e x = \frac{1}{x} - j\pi\delta(x) \quad \text{(on one branch).} \tag{1.65}$$

As an illustration of these concepts, consider the example of a particle of mass m which moves under the influence of a continuous force f, and experiences a sudden momentum increase $m\mathbf{v}_0$ (a kick) at $t = 0$. In the spirit of (1.62) we write the equation of motion of the particle as

$$m\frac{d\mathbf{v}}{dt} = \{f\} + m\mathbf{v}_0\,\delta(t). \tag{1.66}$$

The right-hand member is the generalized force.

1.9 Properties of the derivative

We will now list some of the important properties of the derivative in the distributional sense:

(1) A distribution has derivatives of all orders. Further, the ordering of differentiation in a partial derivative may always be permuted.

(2) A series of distributions which converges in the sense discussed in Section 1.2 may be differentiated term by term. This holds, for example, for

the Fourier series (Dirac 1958)

$$\sum_{k=-\infty}^{\infty} \mathrm{e}^{\mathrm{j}2\pi kx} = \sum_{k=-\infty}^{\infty} \delta(x-k). \tag{1.67}$$

The left-hand member is divergent in the classical sense but, in the sense of distributions, it yields periodic 'sharp' spectral lines at $x = 0, \pm 1, \pm 2, \ldots$. Differentiation yields

$$\mathrm{j}2\pi \sum_{k=-\infty}^{\infty} k\,\mathrm{e}^{\mathrm{j}2\pi kx} = \sum_{k=-\infty}^{\infty} \delta'(x-k). \tag{1.68}$$

(3) The usual differentiation formulas are valid, for instance:

$$\frac{\mathrm{d}}{\mathrm{d}x}(\alpha t) = \frac{\mathrm{d}\alpha}{\mathrm{d}x} t + \alpha \frac{\mathrm{d}t}{\mathrm{d}x}. \tag{1.69}$$

Also, $\mathrm{d}t/\mathrm{d}x = \mathrm{d}s/\mathrm{d}x$ implies that t and s differ by a constant.

(4) The operations of differentiation and passing to the limit may always be interchanged. Specifically, if f_m converges to f as $m \to \infty$, then

$$\lim_{m\to\infty} \left\langle \frac{\mathrm{d}f_m}{\mathrm{d}x}, \phi \right\rangle = \left\langle \frac{\mathrm{d}f}{\mathrm{d}x}, \phi \right\rangle. \tag{1.70}$$

(5) The chain rule for differentiation remains valid. Thus,

$$\frac{\mathrm{d}}{\mathrm{d}x} t[f(x)] = t'[f(x)]\, f'(x). \tag{1.71}$$

Let us apply this formula to $t(x) = \mathrm{Y}(x)$ and $f(x) = x^2 - a^2$. From (1.54):

$$\begin{aligned}\frac{\mathrm{d}}{\mathrm{d}x}\mathrm{Y}(x^2-a^2) &= \delta(x^2-a^2)2x = \frac{x}{|a|}[\delta(x-a)+\delta(x+a)]\\ &= \delta(x-a)-\delta(x+a).\end{aligned} \tag{1.72}$$

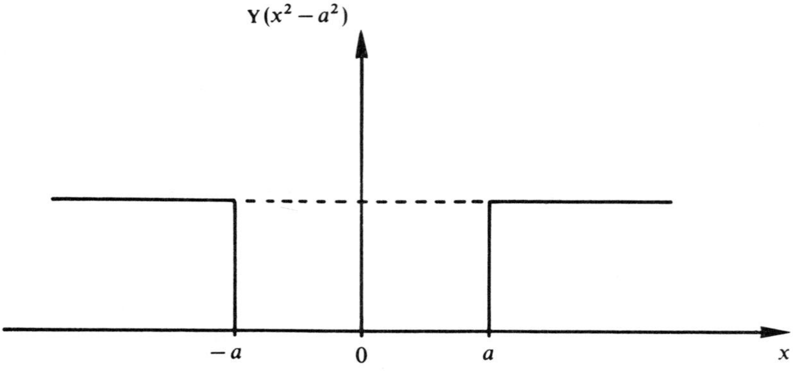

Fig. 1.4. An example of a piecewise continuous function.

This result could have been predicted from the graphical representation of $\mathrm{Y}(x^2 - a^2)$, given in Fig. 1.4.

1.10 Partial derivatives of scalar functions

In three dimensions the partial derivative $\partial t/\partial x_i$ is defined by the functional

$$\left\langle \frac{\partial t}{\partial x_i}, \phi \right\rangle \stackrel{\text{def}}{=} -\iiint t \frac{\partial \phi}{\partial x_i} \mathrm{d}V = \iiint \frac{\partial t}{\partial x_i} \phi \, \mathrm{d}V. \tag{1.73}$$

This definition can serve to express grad t in the sense of distributions. Applying (1.73) successively to $\partial t/\partial x$, $\partial t/\partial y$, and $\partial t/\partial z$ yields

$$\langle \operatorname{grad} t, \phi \rangle \stackrel{\text{def}}{=} -\iiint t \operatorname{grad} \phi \, \mathrm{d}V = \iiint \phi \operatorname{grad} t \, \mathrm{d}V. \tag{1.74}$$

Let us apply this formula to the three-dimensional Heaviside function Y_s, equal to one in V_1, and to zero in V_2 (Fig. 1.5). From (1.74) (Bouix 1964):

$$\begin{aligned}\iiint \phi \operatorname{grad} \mathrm{Y}_\mathrm{s} \, \mathrm{d}V &= -\iiint_{V_1} \mathrm{Y}_\mathrm{s} \operatorname{grad} \phi \, \mathrm{d}V \\ &= -\iiint_{V_1} \operatorname{grad}(\phi \, \mathrm{Y}_\mathrm{s}) \mathrm{d}V + \iiint_{V_1} \phi \operatorname{grad} \mathrm{Y}_\mathrm{s} \mathrm{d}V \\ &= \iint_S \phi \, \mathrm{Y}_\mathrm{s} \boldsymbol{u}_{\mathrm{n}1} \, \mathrm{d}S = \iint_S \phi \, \boldsymbol{u}_{\mathrm{n}1} \, \mathrm{d}S.\end{aligned} \tag{1.75}$$

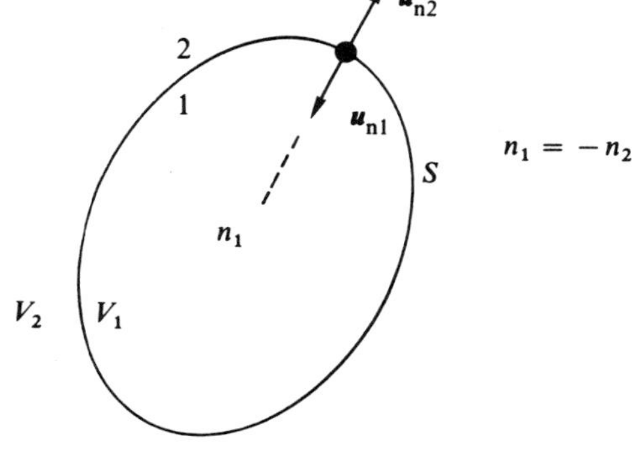

Fig. 1.5. A surface across which discontinuities may occur

This manipulation shows that, in the $\delta(n)$ formalism,

$$\operatorname{grad} Y_s = \delta(n_1)\boldsymbol{u}_{n1} \tag{1.76}$$

Applied to a test-function triple $\boldsymbol{\phi} = (\phi_x, \phi_y, \phi_z)$, (1.76) leads to the relationship

$$\langle \operatorname{grad} Y_s, \boldsymbol{\phi} \rangle = \iiint \boldsymbol{\phi} \cdot \operatorname{grad} Y_s \, \mathrm{d}V = \iint \boldsymbol{\phi} \cdot \boldsymbol{u}_{n1} \, \mathrm{d}S. \tag{1.77}$$

The extension of the concept 'derivative' to higher orders is immediate. In one-dimensional space, for example,

$$\left\langle \frac{\mathrm{d}^m t}{\mathrm{d}x^m}, \phi \right\rangle \stackrel{\text{def}}{=} (-1)^m \left\langle t, \frac{\mathrm{d}^m \phi}{\mathrm{d}x^m} \right\rangle. \tag{1.78}$$

Let us apply this formula to the second derivative of $|x|$. The steps are elementary:

$$\begin{aligned} \left\langle \frac{\mathrm{d}^2 |x|}{\mathrm{d}x^2}, \phi \right\rangle &= -\int_{-\infty}^{0^-} x \frac{\mathrm{d}^2\phi}{\mathrm{d}x^2} \mathrm{d}x + \int_{0^+}^{\infty} x \frac{\mathrm{d}^2\phi}{\mathrm{d}x^2} \mathrm{d}x \\ &= -\int_{-\infty}^{0^-} \left[\frac{\mathrm{d}}{\mathrm{d}x}\left(x \frac{\mathrm{d}\phi}{\mathrm{d}x} \right) - \frac{\mathrm{d}\phi}{\mathrm{d}x} \right] \mathrm{d}x + \int_{0^+}^{\infty} \left[\frac{\mathrm{d}}{\mathrm{d}x}\left(x \frac{\mathrm{d}\phi}{\mathrm{d}x} \right) - \frac{\mathrm{d}\phi}{\mathrm{d}x} \right] \mathrm{d}x \\ &= \phi(0^-) + \phi(0^+) = 2\,\phi(0). \end{aligned} \tag{1.79}$$

The generating function of the second derivative is therefore

$$\frac{\mathrm{d}^2 |x|}{\mathrm{d}x^2} = 2\delta(x). \tag{1.80}$$

Higher derivatives in n dimensions are defined along analogous lines. A linear differential operator in n dimensions is typically a summation of the form

$$\mathscr{L} = \sum_{\boldsymbol{p}} A_{\boldsymbol{p}} \left(\frac{\partial}{\partial x_1} \right)^{p_1} \left(\frac{\partial}{\partial x_2} \right)^{p_2} \cdots \left(\frac{\partial}{\partial x_n} \right)^{p_n}, \tag{1.81}$$

where $\boldsymbol{p} = (p_1, \ldots, p_n)$. The adjoint of $\mathscr{L}$ is

$$\mathscr{L}^\dagger = \sum_{\boldsymbol{p}} A_{\boldsymbol{p}} (-1)^{p_1 + \cdots + p_n} \left(\frac{\partial}{\partial x_1} \right)^{p_1} \left(\frac{\partial}{\partial x_2} \right)^{p_2} \cdots \left(\frac{\partial}{\partial x_n} \right)^{p_n}, \tag{1.82}$$

and the meaning of $\mathscr{L}t$ follows from

$$\langle \mathscr{L}t, \phi \rangle \stackrel{\text{def}}{=} \langle t, \mathscr{L}^\dagger \phi \rangle. \tag{1.83}$$

Applied to the Laplacian, this gives

$$\langle \nabla^2 t, \phi \rangle \stackrel{\text{def}}{=} \langle t, \nabla^2 \phi \rangle \tag{1.84}$$

In particular (Schwartz 1965; Bass 1971; Petit 1987):

$$\begin{aligned} \nabla^2 \log_e 1/|\boldsymbol{r} - \boldsymbol{r}'| &= -2\pi\delta(\boldsymbol{r} - \boldsymbol{r}') \quad \text{in 2 dimensions,} \\ \nabla^2 (1/|\boldsymbol{r} - \boldsymbol{r}'|) &= -4\pi\delta(\boldsymbol{r} - \boldsymbol{r}') \quad \text{in 3 dimensions.} \end{aligned} \tag{1.85}$$

The 'weak' definition of the derivative given above allows recasting a differential equation such as

$$\nabla^2 f = g \tag{1.86}$$

in the form

$$\langle \nabla^2 f, \phi \rangle = \langle g, \phi \rangle = \langle f, \nabla^2 \phi \rangle. \tag{1.87}$$

This formulation transfers the operator ∇^2 from the unknown f to the test function ϕ. It further avoids the difficulties which arise with a classical differential equation such as

$$\frac{\partial}{\partial x}\left(\frac{\partial \phi}{\partial y}\right) = 0. \tag{1.88}$$

This equation is satisfied by every function of x alone, whereas

$$\frac{\partial}{\partial y}\left(\frac{\partial \phi}{\partial x}\right) = 0. \tag{1.89}$$

need not have a sense for such a function. One method to avoid this difficulty is to follow Schwartz' example, and supplement 'usual' functions with new objects: the distributions. These always allow differentiation, and in particular the exchange of the *order* of differentiation.

1.11 Derivatives of $\delta(x)$

According to the general definition (1.58) $\delta'(x)$ is the generating function of

$$\langle \delta', \phi \rangle = -\int_{-\infty}^{\infty} \delta(x) \frac{d\phi}{dx} dx = -\phi'(0). \tag{1.90}$$

One can visualize $\delta'(x)$ by considering the derivatives of the functions shown in Fig. 1.1, which represent the δ-function in the limit $n \to \infty$. A typical graph of the first derivative is sketched in Fig. 1.6a. In the case of the 'rectangular pulse' (Fig. 1.1a), the graph reduces to two delta functions: a positive one supported at $x = 1/n$, and a negative one at $x = -1/n$. Such a 'doublet' can be physically realized by two point charges separated by a short distance 2ε (Fig. 1.6b). If ε approaches zero while $2\varepsilon q$ keeps a fixed value p_e, the volume charge density of the doublet takes the form

$$\rho = q\delta[x - (x_0 + \varepsilon)] - q\delta[x - (x_0 - \varepsilon)]. \tag{1.91}$$

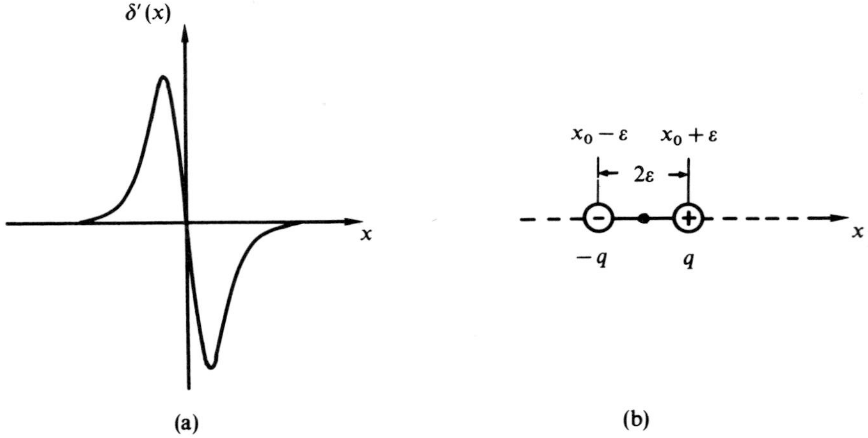

Fig. 1.6. Illustrations relating to the derivative of the delta-function.

Following Dirac's example, we treat the δ-function as a usual function and write

$$\rho = -q\{\delta[(x - x_0) + \varepsilon] - \delta[(x - x_0) - \varepsilon]\} = -q2\varepsilon\,\delta'(x - x_0)$$
$$= -p_e\,\delta'(x - x_0). \tag{1.92}$$

This is the volume density of an x-directed dipole. Schwartz (1965) remarked that the mathematical distributions constitute a correct description of the distributions encountered in physics (monopoles, dipoles, quadrupoles, etc.). This point of view is further discussed in Chapter 2.

Higher derivatives of $\delta(x)$ are defined by the relationship

$$\langle \delta^{(m)}, \phi \rangle \overset{\text{def}}{=} (-1)^m \phi^{(m)}(0). \tag{1.93}$$

The first derivative has the following useful property:

$$\alpha(x)\delta'(x) = \alpha(0)\delta'(x) - \alpha'(0)\delta(x), \tag{1.94}$$

where $\alpha(x)$ is infinitely differentiable. As an example:

$$x\,\delta'(x) = -\delta(x), \qquad x^2\,\delta'(x) = 0. \tag{1.95}$$

Other properties of interest are:

$$x\,\delta^{(m)}(x) = -m\,\delta^{(m-1)}(x), \qquad x^n\,\delta^{(m)}(x) = 0 \quad \text{when } n > m + 1, \tag{1.96}$$

$$\frac{d}{dx}\delta[g(x)] = \delta'[g(x)]\,g'(x).$$

If g is infinitely differentiable for $x < x_0$ and $x > x_0$, and if g and all its derivatives have left-hand and right-hand limits at $x = x_0$, then

$$g' = \{g'\} + \sigma_0 \delta(x - x_0), \qquad g'' = \{g''\} + \sigma_0 \delta'(x - x_0) + \sigma_1 \delta(x - x_0),$$
$$g^{(m)} = \{g^{(m)}\} + \sigma_0 \delta^{(m-1)}(x - x_0) + \cdots + \sigma_{m-1} \delta(x - x_0). \tag{1.97}$$

Here σ_k denotes the difference between the right-hand and the left-hand limit of the kth derivative, while $\{g^k\}$ denotes the distribution generated by a function equal to the usual kth derivative for $x \neq x_0$, but not defined at $x = x_0$. Equation (1.80) may be obtained directly from the value of g'' given above.

1.12 Partial derivatives of δ-functions

The first partial derivatives of $\delta(\boldsymbol{r} - \boldsymbol{r}_0)$ can be defined directly from (1.73) and (1.74). Thus,

$$\left\langle \frac{\partial \delta}{\partial x_i}, \phi \right\rangle = \iiint \frac{\partial \delta(\boldsymbol{r} - \boldsymbol{r}_0)}{\partial x_i} \phi \, \mathrm{d}V \overset{\text{def}}{=} -\iiint \delta(\boldsymbol{r} - \boldsymbol{r}_0) \frac{\partial \phi}{\partial x_i} \mathrm{d}V$$
$$= -\left[\frac{\partial \phi}{\partial x_i}\right]_{\boldsymbol{r}_0}, \tag{1.98}$$
$$\langle \operatorname{grad} \delta, \phi \rangle = \iiint \operatorname{grad} \delta(\boldsymbol{r} - \boldsymbol{r}_0) \mathrm{d}V \overset{\text{def}}{=} -\iiint \delta(\boldsymbol{r} - \boldsymbol{r}_0) \operatorname{grad} \phi \, \mathrm{d}V$$
$$= -[\operatorname{grad} \phi]_{\boldsymbol{r}_0}.$$

These relationships can serve to express the volume density of a concentrated dipole $\boldsymbol{p}_e$ as

$$\rho = -\boldsymbol{p}_e \cdot \operatorname{grad} \delta_{r_0}. \tag{1.99}$$

The value (1.92), derived for an x-oriented dipole, is a particular case of this formula. Derivatives of δ_s may also be defined in accordance with (1.73). In a direction $\boldsymbol{a}$, for example,

$$\left\langle \frac{\partial \delta_s}{\partial a}, \phi \right\rangle \overset{\text{def}}{=} -\iint_S \frac{\partial \phi}{\partial a} \mathrm{d}S = \iiint \phi \frac{\partial \delta_s}{\partial a} \mathrm{d}V. \tag{1.100}$$

It follows that

$$\langle \operatorname{grad} \delta_s, \phi \rangle \overset{\text{def}}{=} -\iint_S \operatorname{grad} \phi \, \mathrm{d}S = \iiint \phi \operatorname{grad} \delta_s \, \mathrm{d}V. \tag{1.101}$$

When $\boldsymbol{a}$ coincides with the normal to S, and τ is a function of the surface coordinates v_1 and v_2 only,

$$\left\langle \tau \frac{\partial \delta_s}{\partial n}, \phi \right\rangle \overset{\text{def}}{=} -\iint_S \tau \frac{\partial \phi}{\partial n} \mathrm{d}S = \iiint \tau \frac{\partial \delta_s}{\partial n} \phi \, \mathrm{d}V. \tag{1.102}$$

Since the potential generated by a dipole layer of density τ is

$$\phi(\boldsymbol{r}_0) = \frac{1}{4\pi\varepsilon_0}\iiint \frac{\rho(\boldsymbol{r})}{|\boldsymbol{r}_0 - \boldsymbol{r}|}\,\mathrm{d}V = \frac{1}{4\pi\varepsilon_0}\iint_S \tau(\boldsymbol{r})\frac{\partial}{\partial n}\frac{1}{|\boldsymbol{r}_0 - \boldsymbol{r}|}\,\mathrm{d}S, \tag{1.103}$$

it becomes clear that the volume density of a double layer can be represented as

$$\rho = -\,\tau(v_1, v_2)\frac{\partial \delta_s}{\partial n}. \tag{1.104}$$

In this formula, n is counted positive in the direction of the dipoles (Fig. 1.7a). In Appendix A, we show that the generalized function $\partial\delta_s/\partial n$ is not equivalent to $\delta'(n)$ when the surface is curved.

The distributional representation of a double layer of surface *currents* can be obtained by similar steps. Using the sign convention on Fig. 1.7b, we write

$$\boldsymbol{j} = -\,\boldsymbol{c}_s(v_1, v_2)\frac{\partial \delta_s}{\partial n}. \tag{1.105}$$

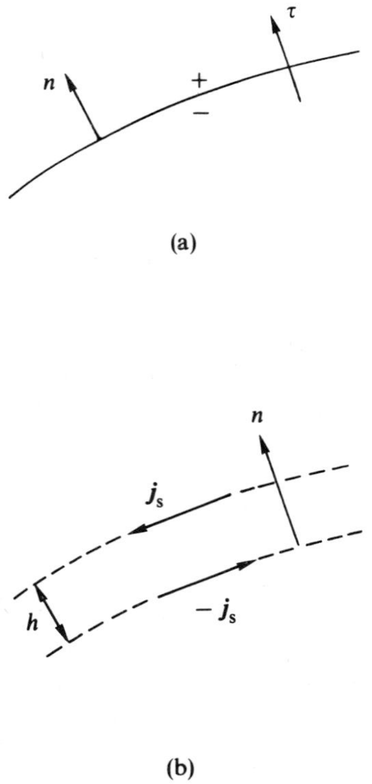

Fig. 1.7. Double layers on a surface.

In this equation, we have assumed that $c_s = j_s h$ approaches a well-defined (nonzero) limit when the distance h between the two layers approaches zero. The detailed nature of the limit process yielding (1.105) is discussed in Appendix A.

Analogous considerations hold for the distribution δ_c defined in (1.41). The gradient is now

$$\langle \operatorname{grad} \delta_c, \phi \rangle \stackrel{\text{def}}{=} -\int_C \operatorname{grad} \phi \, dC = \iiint \phi \operatorname{grad} \delta_c \, dV. \tag{1.106}$$

1.13 Piecewise continuous scalar functions

The derivatives of a one-dimensional piecewise continuous function have been discussed in Sections 1.8 and 1.11. We extend these concepts to three-dimensions by considering a function t which is continuous and has continuous derivatives in regions 1 and 2 (Fig. 1.5). Both t and its derivatives are assumed to approach well-defined limits on sides 1 and 2 of S. The gradient of such a function is

$$\operatorname{grad} t = \{\operatorname{grad} t\} + (t_2 - t_1)\delta(n_2)\boldsymbol{u}_{n2}. \tag{1.107}$$

This relationship is obtained by applying to V_1 and V_2 the formula for the integral of the gradient over a volume. In harmony with previous notation, the term between brackets represents the value of the gradient anywhere but on S.

Since the effect of $\delta(n)$ is to reduce the volume integral to a surface integral, (1.107) may be rewritten more elegantly as

$$\operatorname{grad} t = \{\operatorname{grad} t\} + (\boldsymbol{u}_{n1} t_1 + \boldsymbol{u}_{n2} t_2)\delta_s. \tag{1.108}$$

In particular,

$$\operatorname{grad} Y_s = \delta_s \boldsymbol{u}_{n1}, \tag{1.109}$$

from which (1.76) immediately follows. By similar arguments, combined with an application of Green's theorem to the regions V_1 and V_2, the distributional form of the Laplacian of t is found to be (Schwartz 1965)

$$\nabla^2 t = \{\nabla^2 t\} + \left(\frac{\partial t}{\partial n_1} + \frac{\partial t}{\partial n_2}\right)\delta_s + t_1 \frac{\partial \delta_s}{\partial n_1} + t_2 \frac{\partial \delta_s}{\partial n_2}. \tag{1.110}$$

In Electrostatics, t is the potential ϕ, in which case, (1.110) implies that a discontinuity of ϕ may be represented by a double layer of charge, and a discontinuity of $\partial\phi/\partial n$ (i.e. of the normal component of $\boldsymbol{e}$) by a single layer. Conversely, the boundary conditions of ϕ on S can be derived directly from (1.110). The proof is elementary. Assume that S carries a charge density ρ_s (a single layer) and a dipole density τ (a dipole layer). The corresponding volume density is, from (1.37) and (1.104),

$$\rho = \rho_s \delta_s - \tau \frac{\partial \delta_s}{\partial n_2}. \tag{1.111}$$

In the philosophy of distribution theory, Poisson's equation

$$\nabla^2 \phi = -\rho/\varepsilon_0 \tag{1.112}$$

is valid throughout space. Comparing (1.110), (1.111), and (1.112) shows that

$$\left(-\varepsilon_0 \frac{\partial \phi_1}{\partial n_1}\right) + \left(-\varepsilon_0 \frac{\partial \phi_2}{\partial n_2}\right) = \rho_s \quad \text{and} \quad \phi_2 - \phi_1 = \frac{\tau}{\varepsilon_0} \quad \text{on } S. \tag{1.113}$$

These are the classical boundary conditions.

1.14 Vector operators

Let $\boldsymbol{t}$ be a vector distribution, i.e. a triple of scalar distributions t_x, t_y, t_z. The operator $\operatorname{div} \boldsymbol{t}$ is defined, in classical vector analysis, by the expression

$$\operatorname{div} \boldsymbol{t} = \frac{\partial t_x}{\partial x} + \frac{\partial t_y}{\partial y} + \frac{\partial t_z}{\partial z}. \tag{1.114}$$

The distributional definition of $\operatorname{div} \boldsymbol{t}$ follows by applying (1.73) to the three derivatives shown above. More specifically:

$$\langle \operatorname{div} \boldsymbol{t}, \phi \rangle \overset{\text{def}}{=} -\iiint \boldsymbol{t} \cdot \operatorname{grad} \phi \, \mathrm{d}V = \iiint \phi \operatorname{div} \boldsymbol{t} \, \mathrm{d}V. \tag{1.115}$$

Such a definition gives a well-defined meaning to the equation

$$\operatorname{div} \boldsymbol{d} = \rho; \tag{1.116}$$

according to (1.115) it is

$$-\iiint \boldsymbol{d} \cdot \operatorname{grad} \phi \, \mathrm{d}V = \iiint \rho \phi \, \mathrm{d}V. \tag{1.117}$$

This relationship, which must hold for all test functions ϕ, remains valid when $\boldsymbol{d}$ does not possess everywhere the derivatives shown in (1.114). In consequence, Maxwell's equation

$$\operatorname{div} \boldsymbol{b} = 0 \tag{1.118}$$

now is interpreted as requiring

$$\iiint \boldsymbol{b} \cdot \operatorname{grad} \phi \, \mathrm{d}V = 0 \tag{1.119}$$

to hold for all ϕ.

As mentioned previously, integral formulations such as (1.116) and (1.118) make physical sense because macroscopic experimental evidence is obtained on an 'average' basis, rather than at a point. In addition, switching the differential operator to the test function has the advantage of broadening the class of admissible solutions to those which do not have a divergence in the classical sense of the word. This holds, for example, for the fields which exist at the leading front of a pulsed disturbance.

The distributional definition of curl $\boldsymbol{t}$ follows analogously from the classical value

$$\operatorname{curl} \boldsymbol{t} = \left(\frac{\partial t_z}{\partial y} - \frac{\partial t_y}{\partial z}, \frac{\partial t_x}{\partial z} - \frac{\partial t_z}{\partial x}, \frac{\partial t_y}{\partial x} - \frac{\partial t_x}{\partial y}\right). \tag{1.120}$$

The corresponding functionals are

$$\begin{aligned}\langle \operatorname{curl} \boldsymbol{t}, \phi\rangle &\stackrel{\text{def}}{=} \iiint \boldsymbol{t} \times \operatorname{grad} \phi \, \mathrm{d}V = \iiint \phi \operatorname{curl} \boldsymbol{t} \, \mathrm{d}V,\\ \langle \operatorname{curl} \boldsymbol{t}, \boldsymbol{\phi}\rangle &\stackrel{\text{def}}{=} \iiint \boldsymbol{t} \cdot \operatorname{curl} \boldsymbol{\phi} \, \mathrm{d}V = \iiint \boldsymbol{\phi} \cdot \operatorname{curl} \boldsymbol{t} \, \mathrm{d}V.\end{aligned} \tag{1.121}$$

A relationship such as

$$\operatorname{curl} \boldsymbol{h} = \boldsymbol{j} \tag{1.122}$$

now means, in a distributional sense, that

$$\iiint \phi \boldsymbol{j} \, \mathrm{d}V = \iiint \boldsymbol{h} \times \operatorname{grad} \phi \, \mathrm{d}V, \qquad \iiint \boldsymbol{\phi} \cdot \boldsymbol{j} \, \mathrm{d}V = \iiint \boldsymbol{h} \cdot \operatorname{curl} \boldsymbol{\phi} \, \mathrm{d}V. \tag{1.123}$$

An irrotational vector is therefore characterized by the properties

$$\iiint \boldsymbol{t} \times \operatorname{grad} \phi \, \mathrm{d}V = 0, \qquad \iiint \boldsymbol{t} \cdot \operatorname{curl} \boldsymbol{\phi} \, \mathrm{d}V = 0. \tag{1.124}$$

Extension to vector operators involving higher derivatives than the first proceeds in an analogous fashion. For example:

$$\begin{aligned}\langle \operatorname{curl} \operatorname{curl} \boldsymbol{t}, \phi\rangle &\stackrel{\text{def}}{=} -\iiint \boldsymbol{t} \, \nabla^2 \phi \, \mathrm{d}V + \iiint \boldsymbol{t} \cdot \operatorname{grad} \operatorname{grad} \phi \, \mathrm{d}V\\ &= \iiint \phi \operatorname{curl} \operatorname{curl} \boldsymbol{t} \, \mathrm{d}V,\\ \langle \operatorname{curl} \operatorname{curl} \boldsymbol{t}, \boldsymbol{\phi}\rangle &\stackrel{\text{def}}{=} \iiint \boldsymbol{t} \cdot \operatorname{curl} \operatorname{curl} \boldsymbol{\phi} \, \mathrm{d}V\\ &= \iiint \boldsymbol{\phi} \cdot \operatorname{curl} \operatorname{curl} \boldsymbol{t} \, \mathrm{d}V.\end{aligned} \tag{1.125}$$

The meaning of the symbol grad grad is discussed in Appendix A. Similarly the operator grad div is defined by the functionals

$$\langle \operatorname{grad}\operatorname{div} \boldsymbol{t}, \phi\rangle \overset{\text{def}}{=} \iiint \boldsymbol{t}\cdot \operatorname{grad}\operatorname{grad}\phi \,\mathrm{d}V$$
$$= \iiint \phi \operatorname{grad}\operatorname{div}\boldsymbol{t}\,\mathrm{d}V, \tag{1.126}$$
$$\langle \operatorname{grad}\operatorname{div} \boldsymbol{t}, \boldsymbol{\phi}\rangle \overset{\text{def}}{=} \iiint \boldsymbol{t}\cdot \operatorname{grad}\operatorname{div}\boldsymbol{\phi} \,\mathrm{d}V$$
$$= \iiint \boldsymbol{\phi}\cdot \operatorname{grad}\operatorname{div}\boldsymbol{t}\,\mathrm{d}V.$$

Combining (1.125) and (1.126) yields, for the vector Laplacian,

$$\langle \nabla^2 \boldsymbol{t}, \phi\rangle \overset{\text{def}}{=} \iiint \boldsymbol{t}\, \nabla^2\phi\,\mathrm{d}V = \iiint \phi\, \nabla^2\boldsymbol{t}\,\mathrm{d}V,$$
$$\langle \nabla^2 \boldsymbol{t}, \boldsymbol{\phi}\rangle \overset{\text{def}}{=} \iiint \boldsymbol{t}\cdot \nabla^2\boldsymbol{\phi}\,\mathrm{d}V = \iiint \boldsymbol{\phi}\cdot \nabla^2\boldsymbol{t}\,\mathrm{d}V. \tag{1.127}$$

1.15 Piecewise continuous vector functions

Let $\boldsymbol{t}$ be a continuous vector function which suffers jumps across a surface S. The distributional formula for its *divergence* is (Fig. 1.5)

$$\operatorname{div}\boldsymbol{t} = \{\operatorname{div}\boldsymbol{t}\} + (\boldsymbol{u}_{n1}\cdot\boldsymbol{t}_1 + \boldsymbol{u}_{n2}\cdot\boldsymbol{t}_2)\delta_s. \tag{1.128}$$

This equation is obtained by applying the divergence theorem to volumes V_1 and V_2. When used to interpret Maxwell's equation (1.116) in the sense of distributions, expression (1.128) implies that d_n suffers a jump of ρ_s across S. It also allows writing the equation of continuity of charge in the form (Idemen 1973)

$$\{\operatorname{div}\boldsymbol{j}\} - (\boldsymbol{u}_n\cdot\boldsymbol{j})\delta_s + \left\{\frac{\partial\rho}{\partial t}\right\} + \frac{\partial\rho_s}{\partial t}\delta_s = 0. \tag{1.129}$$

In consequence (Fig. 1.8):

$$\operatorname{div}\boldsymbol{j} + \frac{\partial\rho}{\partial t} = 0 \quad \text{in } V, \qquad \frac{\partial\rho_s}{\partial t} = \boldsymbol{u}_n\cdot\boldsymbol{j} \quad \text{on } S. \tag{1.130}$$

If the surface S carries surface currents of density $\boldsymbol{j}_s$, not flowing beyond a curve C, the equation of conservation of charge becomes (Foissac 1975)

$$\operatorname{div}\boldsymbol{j} + \frac{\partial\rho}{\partial t} = \{\operatorname{div}_s\boldsymbol{j}_s\}\delta_s - \boldsymbol{u}_m\cdot\boldsymbol{j}_s\delta_c + \left\{\frac{\partial\rho_s}{\partial t}\right\}\delta_s + \frac{\partial\rho_c}{\partial t}\delta_c = 0, \tag{1.131}$$

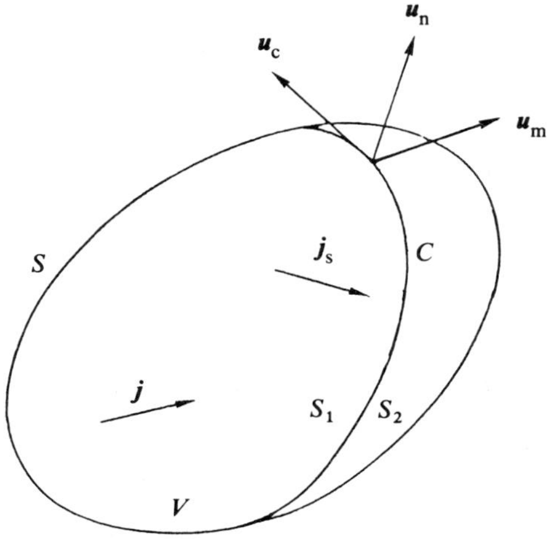

Fig. 1.8. Volume and surface electric currents.

where the brackets indicate the value at all points of S, contour C excluded. Unit vector $\boldsymbol{u}_m$ lies in the tangent plane, and is perpendicular to C. The symbol div_s denotes the surface divergence. In the (v_1, v_2) coordinates defined in Section 1.5, this operator is

$$\text{div}_s \, \boldsymbol{j}_s = \frac{1}{h_1 h_2}\left(\frac{\partial}{\partial v_1}(h_2 j_{s1}) + \frac{\partial}{\partial v_2}(h_1 j_{s2})\right). \tag{1.132}$$

Equating to zero the terms in δ_s and δ_c in (1.131) yields

$$\text{div}_s \, \boldsymbol{j}_s + \frac{\partial \rho_s}{\partial t} = 0 \quad \text{on } S, \qquad \frac{\partial \rho_c}{\partial t} = \boldsymbol{u}_m \cdot \boldsymbol{j}_s \quad \text{on } C. \tag{1.133}$$

These results are relevant for the 'physical optics' approximation, where S_1 is the illuminated part of a conductor, and S_2 the shadow region. They also apply to an open surface bounded by a rim (e.g. a parabolic reflector).

The distributional value of $\text{curl}\, \boldsymbol{t}$ is similarly obtained by applying the integral theorem

$$\iiint_V \text{curl}\, \boldsymbol{a} \, dV = \iint_S (\boldsymbol{u}_n \times \boldsymbol{a}) \, dS \tag{1.134}$$

to volumes V_1 and V_2 (Fig. 1.5). The result is

$$\text{curl}\, \boldsymbol{t} = \{\text{curl}\, \boldsymbol{t}\} + (\boldsymbol{u}_{n1} \times \boldsymbol{t}_1 + \boldsymbol{u}_{n2} \times \boldsymbol{t}_2)\delta_s . \tag{1.135}$$

Applied to the magnetostatic equation (1.122), this formula gives, when the surface S carries a tangential current $\boldsymbol{j}_s$,

$$\{\text{curl}\,\boldsymbol{h}\} + (\boldsymbol{u}_{n1} \times \boldsymbol{h}_1 + \boldsymbol{u}_{n2} \times \boldsymbol{h}_2)\delta_s = \boldsymbol{j}_s\delta_s. \tag{1.136}$$

Such a relationship implies that

$$(\boldsymbol{h}_2)_{\text{tang}} - (\boldsymbol{h}_1)_{\text{tang}} = \boldsymbol{j}_s \times \boldsymbol{u}_{n2}, \tag{1.137}$$

which is the traditional boundary condition.

Operators of a higher order acting on $\boldsymbol{t}$ may be defined by analogous methods. For example (Gagnon 1970):

$$\begin{aligned}\text{curl}\,\text{curl}\,\boldsymbol{t} = {} & \{\text{curl}\,\text{curl}\,\boldsymbol{t}\} + \{\boldsymbol{u}_{n1} \times \text{curl}\,\boldsymbol{t}_1 + \boldsymbol{u}_{n2} \times \text{curl}\,\boldsymbol{t}_2)\delta_s \\ & + \text{curl}\,[(\boldsymbol{u}_{n1} \times \boldsymbol{t}_1 + \boldsymbol{u}_{n2} \times \boldsymbol{t}_2)\delta_s],\end{aligned} \tag{1.138}$$

$$\begin{aligned}\text{grad}\,\text{div}\,\boldsymbol{t} = {} & \{\text{grad}\,\text{div}\,\boldsymbol{t}\} + (\boldsymbol{u}_{n1}\,\text{div}\,\boldsymbol{t}_1 + \boldsymbol{u}_{n2}\text{div}\,\boldsymbol{t}_2)\delta_s \\ & + \text{grad}\,[(\boldsymbol{u}_{n1}\cdot\boldsymbol{t}_1 + \boldsymbol{u}_{n2}\cdot\boldsymbol{t}_2)\delta_s].\end{aligned} \tag{1.139}$$

Equation (1.138) is obtained by two successive applications of (1.135), whereas (1.139) follows from a combination of (1.108) and (1.128). We delay until Chapter 2 a discussion of the meaning of a term such as $\text{curl}\,(\boldsymbol{a}\delta_s)$.

References

ARSAC, J. (1961). *Transformation de Fourier et théorie des distributions.* Dunod, Paris.

BASS, J. (1971). *Cours de Mathématiques, Tome III.* Masson, Paris.

BOUIX, M. (1964). *Les fonctions généralisées ou distributions.* Masson, Paris.

CONSTANTINESCU, F. (1974). *Distributionen und ihre Anwendung in der Physik.* Teubner, Stuttgart.

CRISTESCU, R., and MARINESCU, G. (1973). *Applications of the theory of distributions* (trans. S. Taleman). Wiley, London.

DE JAGER, E.M. (1969). *Applications of distributions in mathematical physics.* Math. Centrum, Amsterdam.

DE JAGER, E.M. (1970). Theory of distributions. Ch. 2 in *Mathematics applied to physics.* Springer, Berlin.

DIRAC, P.A.M. (1958). *The principles of quantum mechanics,* 4th edn. Oxford University Press, Oxford.

FOISSAC, Y. (1975). L'application de l'algèbre extérieure et de la théorie des distributions à l'étude du rayonnement électromagnétique. *Comptes Rendus de l'Académie des Sciences de Paris,* **281 B**(13), 13–6.

FRIEDLANDER, F.G. (1982). *Introduction to the theory of distributions.* Cambridge University Press, Cambridge.

FRIEDMAN, B. (1969). *Lectures on applications-oriented mathematics.* Holden Day, San Francisco.

GAGNON, R.J. (1970). Distribution theory of vector fields, *American Journal of Physics,* **38**, 879–91.

Gel'fand, I.M., and Shilov, G.E. (1964). *Generalized functions.* Academic Press, New York.

Halperin, I. (1952). *Introduction to the theory of distributions.* University of Toronto Press.

Hoskins, R.H. (1979). *Generalized functions.* Ellis Horwood, Chichester.

Idemen, M. (1973). The Maxwell's equations in the sense of distributions. *IEEE Transactions on Antennas and Propagation*, **21**, 736–8.

Jones, D.S. (1982). *The theory of generalised functions*, 2nd edn. Cambridge University Press, Cambridge.

Kecs, W., and Teodorescu, P.P. (1974). *Applications of the theory of distributions in mechanics.* Abacus Press, Tunbridge Wells.

Korevaar, J. (1968). *Mathematical Methods*, Vol. 1. Academic Press, New York.

Lighthill, M.J. (1959). *An introduction to Fourier analysis and generalised functions.* Cambridge University Press.

Lützen, J. (1982). *The prehistory of the theory of distributions.* Springer, New York.

Marchand, J.P. (1962). *Distributions: an outline.* North Holland, Amsterdam.

Petit, R. (1987). *L'outil mathématique*, 2nd edn. Masson, Paris.

Preuss, W., Bleyer, A., and Preuss, H. (1985). *Distributionen und Operatoren: ihre Anwendung in Naturwissenschaft und Technik.* Springer, Wien.

Richtmyer, R.D. (1978). *Principles of advanced mathematical physics*, Vol. 1. Springer, New York.

Schwartz, L. (1948). Généralisation de la notion de fonction et de dérivation. Théorie des distributions. *Annales des Telecommunications*, **3**, 135–40.

Schwartz, L. (1950). *Théorie des distributions.* Hermann et Cie, Paris.

Schwartz, L. (1965). *Méthodes mathématiques pour les sciences physiques*, Hermann et Cie, Paris. An English translation has been published by Addison-Wesley in 1966.

Skinner, R., and Weil, J.A. (1989). An introduction to generalized functions and their application to static electromagnetic point dipoles, including hyperfine interactions. *American Journal of Physics*, **57**, 777–91.

Van Bladel, J. (1985). *Electromagnetic Fields.* Appendix 6. Hemisphere Publ. Co., Washington. Reprinted, with corrections, from a text published in 1964 by McGraw-Hill, New York.

van der Pol, B., and Bremmer, H. (1950). *Operational calculus based on the two-sided Laplace integral*, pp. 62–66. Cambridge University Press.

Vladimirov, V. (1979). *Distributions en physique mathématique.* Editions Mir, Moscou.

Zemanian, A.H. (1987). *Distribution theory and transform analysis.* Dover Publications, New York. First published in 1965 by McGraw-Hill, New York.

2

CONCENTRATED SOURCES

A *point* charge at $\boldsymbol{r}_0$ may be represented by a volume density $\rho = q\delta(\boldsymbol{r} - \boldsymbol{r}_0)$. In the present Chapter, we extend this distributional representation to *clouds* of charge and *volumes* of current, both strongly concentrated around $\boldsymbol{r}_0$. Our analysis leads to the multipole expansion of sources and fields, and to a representation of the various multipole terms by combinations of δ-functions and derivatives. The 'spherical harmonics' aspects of multipole theory, discussed at length in numerous textbooks, are left aside (see e.g. Stratton 1941; Van Bladel 1985). We shall investigate the multipole expansion for magnetic as well as electric currents. We shall also pay much attention to the equivalence between these sources, and in particular to the formulas connecting $\boldsymbol{J}$ to its equivalent $\boldsymbol{K}$, and conversely.

2.1 Scalar sources

Integrals of the form

$$I_1 = \iiint_V \rho(\boldsymbol{r}')\phi(\boldsymbol{r}')\,\mathrm{d}V' \tag{2.1}$$

are often encountered in mathematical physics. The factor $\rho(\boldsymbol{r})$ typically represents a scalar density, and $\phi(\boldsymbol{r})$ a function which, together with its derivatives, is continuous in V (Fig. 2.1). A well-known example is the volume potential

$$I_1(\boldsymbol{r}) = \frac{1}{4\pi\varepsilon_0}\iiint_V \rho(\boldsymbol{r}')\,\frac{1}{|\boldsymbol{r}-\boldsymbol{r}'|}\,\mathrm{d}V' \quad (\boldsymbol{r} \text{ outside } V). \tag{2.2}$$

In electrostatics, $\rho(\boldsymbol{r})$ is the electric charge density.

The integral I_1 in (2.1) may be conveniently expanded when $\phi(\boldsymbol{r})$ varies little over V. In this limit it becomes appropriate to represent ϕ by its power-series expansion

$$\begin{aligned}\phi(\boldsymbol{r}) &= \phi_o + \sum_{i=1}^{3}\left(\frac{\partial\phi}{\partial x_i}\right)_o x_i + \tfrac{1}{2}\sum_{i,j=1}^{3}\left(\frac{\partial^2\phi}{\partial x_i\partial x_j}\right)_o x_i x_j + \cdots \\ &= \phi_o + \boldsymbol{r}\cdot\mathrm{grad}_o\,\phi + \tfrac{1}{2}\boldsymbol{r}\cdot(\mathrm{grad}\,\mathrm{grad}_o\,\phi)\cdot\boldsymbol{r} + \cdots.\end{aligned} \tag{2.3}$$

The subscript o refers to the value of ϕ (and derivatives) at an arbitrarily chosen origin O in V. From the definition of the gradient of a vector given in

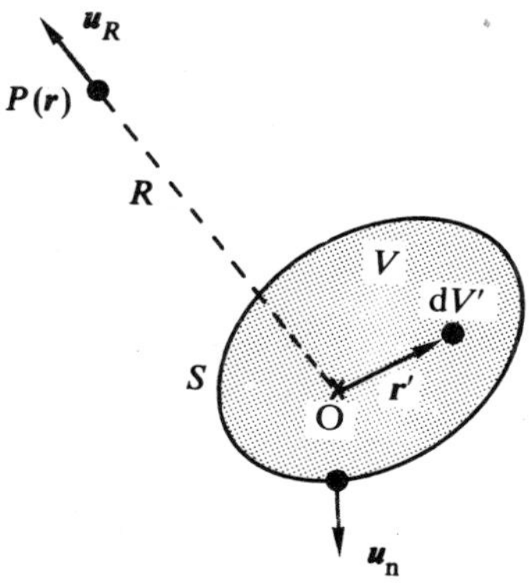

Fig. 2.1. A volume containing sources.

(B. 22), we write

$$\operatorname{grad}\operatorname{grad}\phi = \sum_{i,j=1}^{3} \frac{\partial^2\phi}{\partial x_i \partial x_j} \boldsymbol{u}_i\boldsymbol{u}_j\,. \tag{2.4}$$

Inserting expansion (2.3) into (2.1) yields

$$I_1 = q\phi_o + \boldsymbol{p}_{e1}\cdot\operatorname{grad}_o\phi + \tfrac{1}{2}\mathbf{q}_e:\operatorname{grad}\operatorname{grad}_o\phi + \cdots\,. \tag{2.5}$$

The 'colon' double product is defined in (B.14). The various moments of ρ are given by

$$q = \iiint_V \rho\,\mathrm{d}V, \qquad \boldsymbol{p}_{e1} = \iiint_V \rho\boldsymbol{r}\,\mathrm{d}V, \qquad \mathbf{q}_e = \iiint_V \rho\boldsymbol{r}\boldsymbol{r}\,\mathrm{d}V. \tag{2.6}$$

In Electrostatics, these moments are respectively the total charge, the electric dipole moment, and the electric quadrupole moment. The reason for the notation $\boldsymbol{p}_{e1}$, instead of the more familiar $\boldsymbol{p}_e$, is explained in Sections 2.5 and 2.7. The source distribution ρ must be understood in the sense of distributions, i.e. it may incorporate surface and line sources. When these are present, the volume integrals in (2.6) must be supplemented by surface and line integrals.

A point charge at the origin is represented distributionally by $\rho(\boldsymbol{r}) = q\delta(\boldsymbol{r})$. More generally, the density of a 'concentrated' cloud of charge may be written in the form (Namias 1977; Van Bladel 1977; Kocher 1978).

$$\rho(\boldsymbol{r}) = q\delta(\boldsymbol{r}) - \boldsymbol{p}_{e1}\cdot\operatorname{grad}\delta(\boldsymbol{r}) + \tfrac{1}{2}\mathbf{q}_e:\operatorname{grad}\operatorname{grad}\delta(\boldsymbol{r}) + \cdots\,. \tag{2.7}$$

The distributional definition of $\operatorname{grad}\delta(\boldsymbol{r})$ is given in (1.98). The dyadic $\operatorname{grad}\operatorname{grad}\delta$ may similarly be defined as the collection of second derivatives shown in (2.4) or, equivalently, by the functional

$$\langle \text{grad grad } \delta, \phi \rangle \stackrel{\text{def}}{=} \text{grad grad}_o \phi = \iiint \phi \, \text{grad grad } \delta \, dV. \tag{2.8}$$

When the charge cloud is concentrated around point $\boldsymbol{r}_o$ instead of around the origin:

$$\begin{aligned} \rho(\boldsymbol{r}) &= q\delta(\boldsymbol{r} - \boldsymbol{r}_o) - \boldsymbol{p}_{e1} \cdot \text{grad } \delta(\boldsymbol{r} - \boldsymbol{r}_o) \\ &\quad + \tfrac{1}{2}\mathbf{q}_e : \text{grad grad } \delta(\boldsymbol{r} - \boldsymbol{r}_o) + \cdots. \end{aligned} \tag{2.9}$$

Inserting this expression into the integral I_1 given in (2.1) leads immediately to expansion (2.5). The method is time-saving and elegant.

2.2 Application to the electrostatic potential

The electrostatic potential (2.2) at a point P is of the form (2.1), where ϕ is given by

$$\phi(\boldsymbol{r}') = \frac{1}{4\pi\varepsilon_0} \frac{1}{|\boldsymbol{r} - \boldsymbol{r}'|}. \tag{2.10}$$

When field point P is far away from the charges, $\phi(\boldsymbol{r})$ varies little over V, hence the formalism of the previous section may be applied. In the limit $R = |\boldsymbol{r} - \boldsymbol{r}'| \to 0$ (Fig. 2.1):

$$\begin{aligned} \phi_o &= \frac{1}{4\pi\varepsilon_0 R}, \qquad \text{grad}_o \phi = \frac{\boldsymbol{u}_R}{4\pi\varepsilon_0 R^2}, \\ \text{grad grad}_o \phi &= \frac{1}{4\pi\varepsilon_0 R^3}(3\boldsymbol{u}_R \boldsymbol{u}_R - \mathbf{I}). \end{aligned} \tag{2.11}$$

The last equation is obtained from (B.23), applied to a vector which has only the radial component $a_R = 1/4\pi\varepsilon_0 R^2$. Insertion of (2.11) into (2.5) gives, taking (B.18) and (B.19) into account, the potential $I_1(\boldsymbol{r})$ in P as

$$I_1(\boldsymbol{r}) = \frac{q}{4\pi\varepsilon_0 R} + \frac{\boldsymbol{p}_{e1} \cdot \boldsymbol{u}_R}{4\pi\varepsilon_0 R^2} + \frac{1}{8\pi\varepsilon_0 R^3} \boldsymbol{u}_R \cdot (3\mathbf{q}_e - \mathbf{I} \, \text{tr} \, \mathbf{q}_e) \cdot \boldsymbol{u}_R + \cdots. \tag{2.12}$$

The symbol tr, which denotes the trace (of a dyadic), is defined in (B.18). The classical multipole expansion (2.12) clearly shows the hierarchy of the various terms of the series. At large distance R, the term in q dominates. If the total charge is zero, it is the dipole term in $\boldsymbol{p}_{e1}$ which predominates. More generally, the first nonzero term determines the law according to which the potential decreases at large distances. These considerations are of major importance in physical chemistry, where the clouds of charge are formed by atoms and molecules (Gray *et al.* 1984).

The ranking obtained in the particular case of the potential also holds, mutatis mutandis, for the general form (2.5). Indeed, let L be a typical length for the variation of ϕ, and D the maximum dimension of the charged volume. With that notation, and calling $\phi_{\max}$ the maximum value of ϕ in V:

$$\mathrm{grad}_o\,\phi \text{ is of the order of } \phi_{\max}/L;$$

$$\mathrm{grad}\,\mathrm{grad}_o\,\phi \text{ is of the order of } \phi_{\max}/L^2.$$

We observe that q, $\mathbf{p}_{e1}$, and $\mathbf{q}_e$ are respectively of the orders of $\rho_{\max}D^3$, $\rho_{\max}D^4$, and $\rho_{\max}D^5$. Simple arithmetic now shows that the successive terms in (2.5) are in the ratios $1{:}D/L{:}D^2/L^2$. The first nonzero term therefore becomes dominant when D/L approaches zero, i.e. when the charge distribution becomes progressively more concentrated in the vicinity of the origin 0 (Fig. 2.1).

To illustrate the use of an expansion such as (2.12), consider the square pattern of point charges shown in Fig. 2.2. Evaluation of the moments gives

$$q = 0, \qquad \mathbf{p}_{e1} = 0, \quad \mathbf{q}_e = qd^2(\mathbf{u}_x\mathbf{u}_y + \mathbf{u}_y\mathbf{u}_x), \qquad \mathrm{tr}\,\mathbf{q}_e = 0. \qquad (2.13)$$

The dominant term in the multipole expansion is therefore the electric quadrupole term. The associated large-distance potential in the (x, y) plane is:

$$\text{Potential in } P = (3qd^2/8\pi\varepsilon_0 R^3)\sin 2\theta. \qquad (2.14)$$

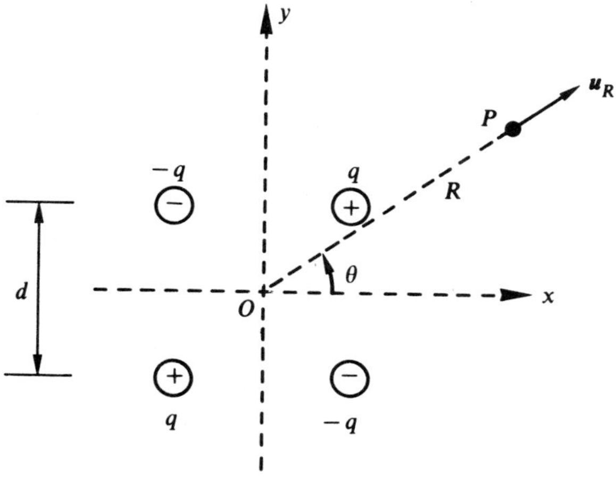

Fig. 2.2. Point charges forming a square pattern.

2.3 Multipole expansion of an integral: the first term

A typical 'moment' integral for the current density is

$$\boldsymbol{I}_2 = \iiint_V \phi(\boldsymbol{r})\boldsymbol{J}(\boldsymbol{r})\,\mathrm{d}V. \tag{2.15}$$

We shall assume that the current is time-harmonic, and that $\boldsymbol{J}$ is its complex representation. The use of capital letters for such representations is general throughout this text.

The integral $\boldsymbol{I}_2$ can be transformed by applying the power-series expansion (2.3) to the three components of $\boldsymbol{J}$. This gives

$$\begin{aligned}\boldsymbol{I}_2 = \phi_{\mathrm{o}} \iiint_V \boldsymbol{J}\,\mathrm{d}V + \sum_{i=1}^{3}\left(\frac{\partial\phi}{\partial x}\right)_{\mathrm{o}} \iiint_V x_i \boldsymbol{J}\,\mathrm{d}V \\ + \tfrac{1}{2}\sum_{i,j=1}^{3}\left(\frac{\partial^2\phi}{\partial x_i \partial x_j}\right)_{\mathrm{o}} \iiint_V x_i \boldsymbol{J} x_j\,\mathrm{d}V + \cdots .\end{aligned} \tag{2.16}$$

We evaluate the first integral by writing its x component as

$$\iiint_V (\boldsymbol{J}\cdot\boldsymbol{u}_x)\mathrm{d}V = \iiint_V (\boldsymbol{J}\cdot\operatorname{grad} x)\,\mathrm{d}V. \tag{2.17}$$

This expression is a particular case of the more general relationship

$$\iiint_V \boldsymbol{J}\cdot\operatorname{grad}\Theta\,\mathrm{d}V = \iiint_V \operatorname{div}(\Theta\boldsymbol{J})\,\mathrm{d}V - \iiint_V \Theta \operatorname{div}\boldsymbol{J}\,\mathrm{d}V. \tag{2.18}$$

The current density $\boldsymbol{J}$ satisfies the equations of conservation of charge, viz. (Fig. 2.1)

$$\operatorname{div}\boldsymbol{J} = -\mathrm{j}\omega P = -\mathrm{j}kcP \quad \text{in } V, \qquad \boldsymbol{u}_{\mathrm{n}}\cdot\boldsymbol{J} = \mathrm{j}\omega P_{\mathrm{s}} = \mathrm{j}kcP_{\mathrm{s}} \quad \text{on } S. \tag{2.19}$$

In these equations we have used the convention $\mathrm{e}^{\mathrm{j}\omega t}$ for the time factor. Inserting (2.19) into (2.18) yields

$$\iiint_V \boldsymbol{J}\cdot\operatorname{grad}\Theta\,\mathrm{d}V = \mathrm{j}kc\left(\iint_S \Theta P_{\mathrm{s}}\,\mathrm{d}S + \iiint_V \Theta P\,\mathrm{d}V\right). \tag{2.20}$$

Applied successively to $\Theta = x$, y, and z, this relationship leads to

$$\iiint_V \boldsymbol{J}\,\mathrm{d}V = \mathrm{j}\omega\boldsymbol{P}_{\mathrm{e}1} = \mathrm{j}kc\boldsymbol{P}_{\mathrm{e}1}, \tag{2.21}$$

where $\boldsymbol{P}_{\mathrm{e}1}$ is the electric dipole moment defined in (2.6), i.e.

$$\boldsymbol{P}_{\mathrm{e}1} = \iiint_V P\boldsymbol{r}\,\mathrm{d}V + \iint_S P_{\mathrm{s}}\boldsymbol{r}\,\mathrm{d}S. \tag{2.22}$$

2.4 Multipole expansion of an integral: the second term

The second term in (2.16) can be written as

$$\sum_{i=1}^{3} \frac{\partial \phi_o}{\partial x_i} \iiint_V (\boldsymbol{u}_i \cdot \boldsymbol{r}) \boldsymbol{J} \, dV = \text{grad}_o \phi \cdot \iiint_V \boldsymbol{r}\boldsymbol{J} \, dV. \tag{2.23}$$

It is useful to split the dyadic $\boldsymbol{rJ}$ into its symmetric and antisymmetric parts. Thus,

$$\boldsymbol{rJ} = \tfrac{1}{2}(\boldsymbol{rJ} + \boldsymbol{Jr}) + \tfrac{1}{2}(\boldsymbol{rJ} - \boldsymbol{Jr}). \tag{2.24}$$

Consider first the antisymmetric part. A vector dotted in such a dyadic gives rise to a cross-product. In the present case,

$$\tfrac{1}{2}\text{grad}_o \phi \cdot (\boldsymbol{rJ} - \boldsymbol{Jr}) = -\tfrac{1}{2}\text{grad}_o \phi \times (\boldsymbol{r} \times \boldsymbol{J}). \tag{2.25}$$

Integration over V now introduces the magnetic dipole moment

$$\boldsymbol{P}_m = \tfrac{1}{2} \iiint_V \boldsymbol{r} \times \boldsymbol{J} \, dV, \tag{2.26}$$

in terms of which the integral can be written as

$$\tfrac{1}{2}\text{grad}_o \phi \cdot \iiint_V (\boldsymbol{rJ} - \boldsymbol{Jr}) \, dV = -\,\text{grad}_o \phi \times \boldsymbol{P}_m. \tag{2.27}$$

We next evaluate the contribution of the symmetric part. Its (i, j) component is

$$\begin{aligned} \tfrac{1}{2}\boldsymbol{u}_i \cdot (\boldsymbol{rJ} + \boldsymbol{Jr}) \cdot \boldsymbol{u}_j &= \tfrac{1}{2}[x_i(\boldsymbol{u}_j \cdot \boldsymbol{J}) + x_j(\boldsymbol{u}_i \cdot \boldsymbol{J})] \\ &= \tfrac{1}{2}(x_i \,\text{grad}\, x_j + x_j \,\text{grad}\, x_i) \cdot \boldsymbol{J} \\ &= \tfrac{1}{2}\boldsymbol{J} \cdot \text{grad}(x_i x_j). \end{aligned} \tag{2.28}$$

When this expression is inserted into the integral in (2.23), an integral of type (2.20) is obtained, with $\Theta = x_i x_j$. Simple algebra now shows that

$$\tfrac{1}{2}\text{grad}\, \phi_o \cdot \iiint_V (\boldsymbol{rJ} + \boldsymbol{Jr}) \, dV = \tfrac{1}{2}j\omega \,\text{grad}\, \phi_o \cdot \mathbf{Q}_e, \tag{2.29}$$

where $\mathbf{Q}_e$ is the (symmetric) electric quadrupole dyadic defined in (2.6), viz.

$$\mathbf{Q}_e = \iiint_V P\boldsymbol{rr} \, dV + \iint_S P_s \boldsymbol{rr} \, dS = \frac{1}{j\omega} \iiint_V (\boldsymbol{rJ} + \boldsymbol{Jr}) \, dV. \tag{2.30}$$

The term in $\mathbf{Q}_e$ in the expansion is clearly of the order of D/L with respect to that in $\boldsymbol{P}_{e1}$.

2.5 Multipole expansion of an integral: the third term

The third term in (2.16) is the summation

$$\boldsymbol{I}_3 = \tfrac{1}{2} \sum_{i,\,j=1}^{3} \left(\frac{\partial^2 \phi}{\partial x_i \partial x_j}\right)_{\mathrm{o}} \iiint_V x_i \boldsymbol{J} x_j \,\mathrm{d}V. \tag{2.31}$$

We shall endeavour to identify, in this expression, the dominant term in the small parameter D/L. On the basis of (2.20), the kth component of the integral can be written as

$$\begin{aligned}(\boldsymbol{I}_3)_k &= \iiint_V x_i x_j \operatorname{grad} x_k \cdot \boldsymbol{J} \,\mathrm{d}V \\ &= \iint_S \mathrm{j}\omega P_s x_i x_j x_k \,\mathrm{d}S + \iiint_V \mathrm{j}\omega P x_i x_j x_k \,\mathrm{d}V \\ &\quad - \iiint_V x_k \boldsymbol{J} \cdot \operatorname{grad} x_i x_j \,\mathrm{d}V.\end{aligned} \tag{2.32}$$

The first two integrals are of the order of D^2/L^2 with respect to the term in $\boldsymbol{P}_{e1}$ in (2.16). We shall therefore neglect them. Applying (2.32) to the three space components of the third integral yields

$$\begin{aligned}\boldsymbol{I}_3 &= -\iiint_V \boldsymbol{r}[x_i(\boldsymbol{u}_j \cdot \boldsymbol{J}) + x_j(\boldsymbol{u}_i \cdot \boldsymbol{J})] \,\mathrm{d}V \\ &= -\iiint_V x_i[x_j \boldsymbol{J} + \boldsymbol{u}_j \times (\boldsymbol{r} \times \boldsymbol{J})] \,\mathrm{d}V \\ &\quad - \iiint_V x_j[x_i \boldsymbol{J} + \boldsymbol{u}_i \times (\boldsymbol{r} \times \boldsymbol{J})] \,\mathrm{d}V.\end{aligned} \tag{2.33}$$

In the second member, we detect the presence of two more integrals of the $\boldsymbol{I}_3$ type. Collecting together terms in $\boldsymbol{I}_3$ now gives

$$\boldsymbol{I}_3 = -\tfrac{1}{3} \iiint_V [(x_i \boldsymbol{u}_j + x_j \boldsymbol{u}_i) \times (\boldsymbol{r} \times \boldsymbol{J})] \,\mathrm{d}V. \tag{2.34}$$

After some elementary algebra, we obtain

$$\begin{aligned}\boldsymbol{I}_3 = \iiint_V x_i \boldsymbol{J} x_j \,\mathrm{d}V &= -\tfrac{1}{2} \boldsymbol{u}_i \times (\boldsymbol{u}_j \cdot \mathsf{Q}_{\mathrm{m}}) - \tfrac{1}{2} \boldsymbol{u}_j \times (\boldsymbol{u}_i \cdot \mathsf{Q}_{\mathrm{m}}) \\ &\quad - \boldsymbol{u}_i \times (\boldsymbol{u}_j \times \boldsymbol{P}_{e2}) - \boldsymbol{u}_j \times (\boldsymbol{u}_i \times \boldsymbol{P}_{e2}),\end{aligned} \tag{2.35}$$

where

$$\boldsymbol{P}_{e2} = \tfrac{1}{6} \iiint_V \boldsymbol{r} \times \boldsymbol{J} \times \boldsymbol{r} \,\mathrm{d}V, \quad \mathsf{Q}_{\mathrm{m}} = \tfrac{1}{3} \iiint_V [(\boldsymbol{r} \times \boldsymbol{J})\boldsymbol{r} + \boldsymbol{r}(\boldsymbol{r} \times \boldsymbol{J})] \,\mathrm{d}V. \tag{2.36}$$

The reason for the notation $\boldsymbol{P}_{e2}$, which suggests a term belonging to the electric dipole family, is clarified in Section 2.7. The quantity $\mathbf{Q}_m$ is the magnetic quadrupole dyadic, a symmetric traceless dyadic, which can be written in several equivalent forms (Van Bladel 1977; Gray 1979, 1980). The terms in $\boldsymbol{P}_{e2}$ and $\mathbf{Q}_m$ are clearly of the order of D/L with respect to the term in $\boldsymbol{P}_m$, and must therefore be retained in our expansion. Working out double cross products such as $\boldsymbol{u}_i \times (\boldsymbol{u}_j \times \boldsymbol{P}_{e2})$ shows that the contribution from $\boldsymbol{P}_{e2}$ is of the form

$$(\nabla_o^2 \phi)\boldsymbol{P}_{e2} - \boldsymbol{P}_{e2}\cdot \operatorname{grad}\operatorname{grad}_o \phi = \boldsymbol{P}_{e2}\cdot(\mathbf{I}\nabla_o^2\phi - \operatorname{grad}\operatorname{grad}_o\phi). \tag{2.37}$$

After some elementary algebra involving the term in $\mathbf{Q}_m$, we consolidate the results of Sections (2.3) to (2.5) into a single formula

$$\iiint_V \phi \boldsymbol{J}\,\mathrm{d}V = \mathrm{j}\omega\phi_o \boldsymbol{P}_{e1} - \operatorname{grad}_o\phi \times \boldsymbol{P}_m + \tfrac{1}{2}\mathrm{j}\omega \operatorname{grad}_o\phi\cdot\mathbf{Q}_e + (\mathbf{I}\nabla_o^2\phi - \operatorname{grad}\operatorname{grad}_o\phi)\cdot\boldsymbol{P}_{e2} - \tfrac{1}{2}\operatorname{curl}(\operatorname{grad}_o\phi\cdot\mathbf{Q}_m) + \cdots. \tag{2.38}$$

2.6 Application to the magnetostatic vector potential

The loop of direct current shown in Fig. 2.3 generates a vector potential

$$\boldsymbol{a}(\boldsymbol{r}) = \frac{\mu_0}{4\pi}\iiint_V \frac{\boldsymbol{j}(\boldsymbol{r}')}{|\boldsymbol{r}-\boldsymbol{r}'|}\,\mathrm{d}V'. \tag{2.39}$$

This integral is of the general form (2.15), with $\phi = \mu_0/4\pi|\boldsymbol{r}-\boldsymbol{r}'|$. Expansion (2.38) is therefore applicable but, because $\boldsymbol{j}$ is time-independent, ω must be set equal to zero. The dominant part is the term in $\boldsymbol{p}_m$. At large distances R, using (2.11), $\boldsymbol{a}$ is asymptotically given by

$$\boldsymbol{a} \underset{R}{\sim} \frac{\mu_0}{4\pi R^2}\boldsymbol{u}_R \times \boldsymbol{p}_m = \frac{\mu_0}{4\pi}\frac{\boldsymbol{p}_m \times \boldsymbol{r}}{R^3} = \frac{\mu_0}{4\pi}\operatorname{grad}\left(\frac{1}{R}\right) \times \boldsymbol{p}_m. \tag{2.40}$$

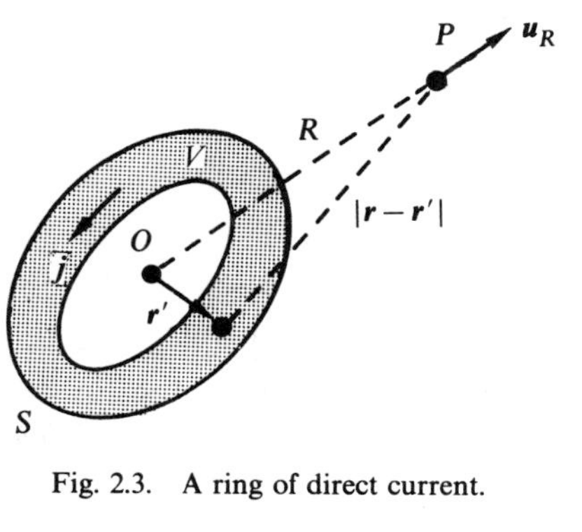

Fig. 2.3. A ring of direct current.

This expression may be obtained directly from the delta-function representation of $\boldsymbol{j}$, which is (Van Bladel 1977; Kocher 1978)

$$\boldsymbol{j} = \operatorname{curl}[\delta(\boldsymbol{r} - \boldsymbol{r}_o)\boldsymbol{p}_m] = \operatorname{grad}\delta(\boldsymbol{r} - \boldsymbol{r}_o) \times \boldsymbol{p}_m. \tag{2.41}$$

Applying (1.121) yields

$$\begin{aligned}\iiint_V \phi \boldsymbol{j}\,dV &= \iiint_V \phi \operatorname{curl}[\delta(\boldsymbol{r} - \boldsymbol{r}_o)\boldsymbol{p}_m]\,dV \\ &= \iiint_V \delta(\boldsymbol{r} - \boldsymbol{r}_o)\boldsymbol{p}_m \times \operatorname{grad}\phi\,dV = \boldsymbol{p}_m \times \operatorname{grad}_o \phi.\end{aligned} \tag{2.42}$$

This is recognized as the contribution from $\boldsymbol{p}_m$ in (2.38).

Relationship (2.41) is the most convenient mathematical representation for the current of a small loop (a 'frill'). It is of a more general nature than the often-used representation for a circular current i, viz. (Fig. 2.4)

$$\boldsymbol{j} = i\delta(r - a)\delta(z)\boldsymbol{u}_\varphi. \tag{2.43}$$

The multipole moments of this current are

$$\begin{aligned}\boldsymbol{p}_m &= \tfrac{1}{2}\iiint_V \boldsymbol{r} \times \boldsymbol{j}\,dV = \tfrac{1}{2}\iint a\boldsymbol{u}_r \times i\delta(r - a)\delta(z)\boldsymbol{u}_\varphi 2\pi r\,dr\,dz \\ &= \pi a^2 i\boldsymbol{u}_z, \\ \boldsymbol{p}_{e2} &= 0, \qquad \mathbf{q}_m = 0.\end{aligned} \tag{2.44}$$

The value of $\boldsymbol{p}_m$ is independent of the choice of the origin. This property, which holds for all direct currents, may be proved by shifting the origin by $\boldsymbol{r}_o$, and noticing that such a shift introduces an additional term in $\boldsymbol{p}_m$, viz.

$$\tfrac{1}{2}\iiint \boldsymbol{r}_0 \times \boldsymbol{j}\,dV = \tfrac{1}{2}\boldsymbol{r}_0 \times \iiint \boldsymbol{j}\,dV. \tag{2.45}$$

This term is zero because of (2.21).

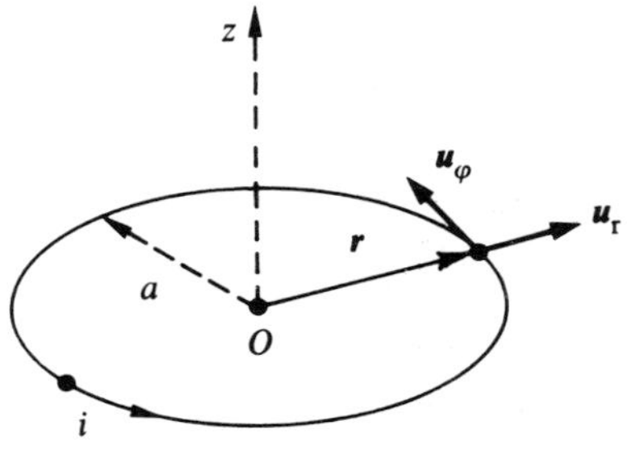

Fig. 2.4. A circular current loop.

Once the expansion of the vector potential $\boldsymbol{a}$ is known, a corresponding expansion for the magnetic induction $\boldsymbol{b}$ follows easily by evaluating the curl of $\boldsymbol{a}$. Outside the source, $\boldsymbol{b}$ may also be derived from a *scalar* potential ψ, a property which leads to an alternate expansion of $\boldsymbol{b}$, based on the multipole expansion of ψ. The details follow the pattern discussed in Section 2.2 (Gray 1978, 1979).

2.7 Radiation fields

The vector potential of the time-harmonic current $\boldsymbol{J}$, evaluated in the far field and in a direction of unit vector $\boldsymbol{u}$, is given by

$$\lim_{R\to\infty} \boldsymbol{A}(R,u) = \lim_{R\to\infty} \frac{\mu_0}{4\pi}\iiint \boldsymbol{J}(r')\frac{\mathrm{e}^{-\mathrm{j}k|r-r'|}}{|\boldsymbol{r}-\boldsymbol{r}'|}\,\mathrm{d}V'$$
$$= \frac{\mu_0}{4\pi}\frac{\mathrm{e}^{-\mathrm{j}kR}}{R}\iiint_V \boldsymbol{J}(\boldsymbol{r}')\mathrm{e}^{\mathrm{j}k\boldsymbol{u}\cdot\boldsymbol{r}'}\,\mathrm{d}V'. \tag{2.46}$$

The scale of variation of the exponents is λ. A concentrated source, by its very definition, must have dimensions small with respect to all relevant lengths. It follows that $k\boldsymbol{u}\cdot\boldsymbol{r}$ is a small quantity. The exponential may therefore be expanded in a series in $\mathrm{j}k\boldsymbol{u}\cdot\boldsymbol{r}$. This procedure yields the same results as a direct application of (2.38). By both methods:

$$\boldsymbol{A} = \frac{\mu_0}{4\pi}\frac{\mathrm{e}^{-\mathrm{j}kR}}{R}\Big[\mathrm{j}kc\boldsymbol{P}_{\mathrm{e}1} - \mathrm{j}k\boldsymbol{u}\times\boldsymbol{P}_{\mathrm{m}} - \tfrac{1}{2}k^2c\boldsymbol{u}\cdot\mathbf{Q}_{\mathrm{e}}$$
$$- k^2\boldsymbol{u}\times\boldsymbol{P}_{\mathrm{e}2}\times\boldsymbol{u} + \tfrac{1}{2}k^2\boldsymbol{u}\times(\boldsymbol{u}\cdot\mathbf{Q}_{\mathrm{m}}) + \cdots\Big]. \tag{2.47}$$

The corresponding electric and magnetic fields are

$$\boldsymbol{E} = \frac{R_{\mathrm{c}}}{4\pi}\frac{\mathrm{e}^{-\mathrm{j}kR}}{R}\{-k^2c\boldsymbol{u}\times(\boldsymbol{u}\times\boldsymbol{P}_{\mathrm{e}1}) - k^2\boldsymbol{u}\times\boldsymbol{P}_{\mathrm{m}}$$
$$-\tfrac{1}{2}\mathrm{j}k^3c\boldsymbol{u}\times[\boldsymbol{u}\times(\boldsymbol{u}\cdot\mathbf{Q}_{\mathrm{e}})] - \mathrm{j}k^3\boldsymbol{u}\times(\boldsymbol{u}\times\boldsymbol{P}_{\mathrm{e}2}) - \tfrac{1}{2}\mathrm{j}k^3\boldsymbol{u}\times(\boldsymbol{u}\cdot\mathbf{Q}_{\mathrm{m}}) + \cdots\} \tag{2.48}$$
$$\boldsymbol{H} = \frac{1}{4\pi}\frac{\mathrm{e}^{-\mathrm{j}kR}}{R}\{k^2c\boldsymbol{u}\times\boldsymbol{P}_{\mathrm{e}1} - k^2\boldsymbol{u}\times(\boldsymbol{u}\times\boldsymbol{P}_{\mathrm{m}}) + \tfrac{1}{2}\mathrm{j}k^3c\boldsymbol{u}\times(\boldsymbol{u}\cdot\mathbf{Q}_{\mathrm{e}})$$
$$+\mathrm{j}k^3\boldsymbol{u}\times\boldsymbol{P}_{\mathrm{e}2} - \tfrac{1}{2}\mathrm{j}k^3\boldsymbol{u}\times[\boldsymbol{u}\times(\boldsymbol{u}\cdot\mathbf{Q}_{\mathrm{m}})] + \cdots\}.$$

Here $R_{\mathrm{c}} = \sqrt{(\mu_0/\varepsilon_0)}$ is the characteristic impedance of free space. We observe that the contributions of $\boldsymbol{P}_{\mathrm{e}1}$ and $\boldsymbol{P}_{\mathrm{e}2}$ are of the same form, which justifies the notation $\boldsymbol{P}_{\mathrm{e}}$ used for both terms. We also notice the duality between the contributions of $(\boldsymbol{P}_{\mathrm{e}1}, \mathbf{Q}_{\mathrm{e}})$ on the one side, and $(\boldsymbol{P}_{\mathrm{m}}, \mathbf{Q}_{\mathrm{m}})$ on the other.

2.8 Hierarchy of terms in the multipole expansion

When the various multipole moments in (2.48) are frequency-independent, the terms in k^2 clearly become dominant at low frequencies. Those in k^3 take over if $\boldsymbol{P}_{e1}$ and $\boldsymbol{P}_m$ happen to vanish. These fairly trivial statements lose their validity when the current distribution (together with its associated moments) varies with frequency. A meaningful example is that of a dielectric resonator, in which the terms in $\boldsymbol{P}_{e2}$ may turn out to be of the same order of magnitude as those in $\boldsymbol{P}_{e1}$. As an illustration, consider the *confined* modes of a resonator with rotational symmetry (Fig. 2.5a). In the limit $\varepsilon_r = N^2 \to \infty$ the fields of these modes are completely contained within the dielectric volume, i.e. they do not leak outside V (Van Bladel 1975, 1988a). The modal magnetic field, expanded in a series in $1/N$, is of the form

$$\boldsymbol{H} = \left(\beta_0(r, z) + \frac{1}{N^2}\beta_2(r, z) + \cdots\right)\boldsymbol{u}_\varphi = \boldsymbol{H}_0 + \frac{1}{N^2}\boldsymbol{H}_2 + \cdots. \tag{2.49}$$

The term in β_0 is the already mentioned limit $\varepsilon_r \to \infty$. The currents $\boldsymbol{J}$ are the polarization currents, obtainable from Maxwell's equation

$$\operatorname{curl}\boldsymbol{H} = \mathrm{j}\omega\varepsilon\boldsymbol{E} = \mathrm{j}\omega\varepsilon_0\boldsymbol{E} + \boldsymbol{J}. \tag{2.50}$$

This equation yields

$$\begin{aligned}\boldsymbol{J} &= \left(1 - \frac{1}{N^2}\right)\operatorname{curl}\boldsymbol{H} \\ &= \operatorname{curl}(\beta_0\boldsymbol{u}_\varphi) + \frac{1}{N^2}[-\operatorname{curl}(\beta_0\boldsymbol{u}_\varphi) + \operatorname{curl}(\beta_2\boldsymbol{u}_\varphi)] + \cdots.\end{aligned} \tag{2.51}$$

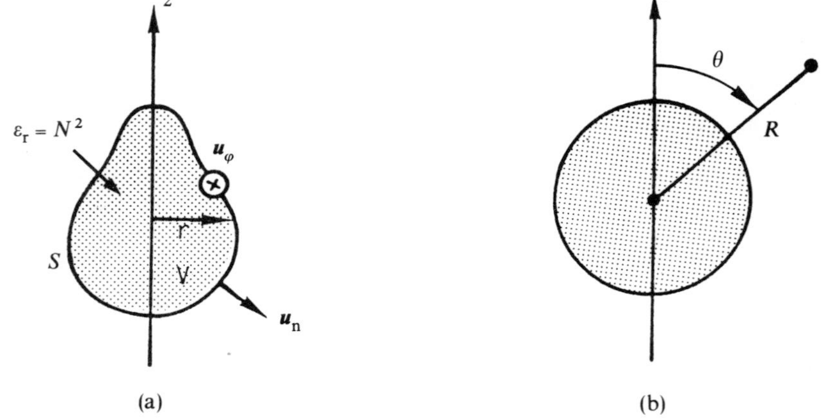

Fig. 2.5. Dielectric resonators endowed with symmetry of revolution.

Resonances occur for discrete values of k_d, the wavenumber in the dielectric. This parameter is equal to kN; hence, when ε_r grows, the resonant frequency decreases proportionally to $1/N$.

It is easy to check that the special symmetry of the mode currents results in $\boldsymbol{P}_m = 0$ and $\mathbf{Q}_m = 0$, and that $\mathbf{Q}_e$ is of the order of $1/N$. An evaluation of $\boldsymbol{P}_{e1}$ according to (2.22) shows that the terms in $\beta_0 \boldsymbol{u}_\varphi$ do not contribute to the integral, and consequently that $\boldsymbol{P}_{e1}$ stems from the term in $\beta_2 \boldsymbol{u}_\varphi$. The resulting moment turns out to be (Van Bladel 1975)

$$\boldsymbol{P}_{e1} = \varepsilon_0 \iint_S z E_n \, dS \, \boldsymbol{u}_z, \tag{2.52}$$

where E_n is the normal electric field just outside the resonator. This field is of the order of $1/N$. The *second* dipole moment is

$$\boldsymbol{P}_{e2} = -\pi \iint_S \beta_0 r^2 \, dS \, \boldsymbol{u}_z. \tag{2.53}$$

It does not depend on N, i.e. on the frequency. On the other hand, $\boldsymbol{P}_{e1}$ decreases proportionally to the frequency. The terms in $\boldsymbol{P}_{e1}$ and $\boldsymbol{P}_{e2}$ in (2.45) are therefore *both* of the order of $1/N^3$, while the term in $\mathbf{Q}_e$ is of the order of $1/N^4$.

As an illustration, consider the confined modes of a spherical resonator (Fig. 2.5b). The function β_0 is here (Van Bladel 1975, 1988a)

$$\beta_0 = B \sin\theta \left(\frac{\sin k_d R}{R^2} - k_d \frac{\cos k_d R}{R} \right). \tag{2.54}$$

The eigenvalues $k_d a$ follow from the boundary condition

$$\sin k_d a - k_d a \cos k_d a = 0. \tag{2.55}$$

In (2.54) B is an arbitrary constant, of dimension A (m^{-1}). Detailed calculations show that the fields outside the resonator are those of a dipole moment

$$\boldsymbol{P}_e = j(4\pi/Nc) k_d a^2 (\sin k_d a) B \boldsymbol{u}_z. \tag{2.56}$$

The separate contributions of $\boldsymbol{P}_{e1}$ and $\boldsymbol{P}_{e2}$ to the radiation field can be evaluated from (2.52) and (2.53). These formulas give

$$\begin{aligned} \boldsymbol{P}_{e1} &= j(8\pi/3Nc) k_d a^2 (\sin k_d a) B \boldsymbol{u}_z = \tfrac{2}{3} \boldsymbol{P}_e, \\ j(k/c)\boldsymbol{P}_{e2} &= j(4\pi/3Nc) k_d a^2 (\sin k_d a) B \boldsymbol{u}_z = \tfrac{1}{3} \boldsymbol{P}_e. \end{aligned} \tag{2.57}$$

The sum of these two expressions reproduces, as expected, the value of $\boldsymbol{P}_e$ given in (2.56). We observe that the two contributions are comparable, hence that neglecting $\boldsymbol{P}_{e2}$ would lead to unacceptable results. In the present case, the radiated power would be too low by a factor of $\frac{4}{9}$, and the quality factor Q too high by a factor of $\frac{9}{4}$.

2.9 Multipole expansion of the current

In (2.38), we have obtained the multipole expansion of an integral of the type $\iiint \phi \boldsymbol{J} \,\mathrm{d}V$. This expansion can also be derived, quite elegantly, from the representation

$$\boldsymbol{J}(\boldsymbol{r}) = \mathrm{j}\omega\delta(\boldsymbol{r}-\boldsymbol{r}_{\mathrm{o}})\boldsymbol{P}_{\mathrm{e}1} + \operatorname{curl}[\delta(\boldsymbol{r}-\boldsymbol{r}_{\mathrm{o}})\boldsymbol{P}_{\mathrm{m}}] - \tfrac{1}{2}\mathrm{j}\omega \operatorname{grad}\delta(\boldsymbol{r}-\boldsymbol{r}_{\mathrm{o}})\cdot \mathbf{Q}_{\mathrm{e}}$$
$$- \operatorname{curl}\operatorname{curl}[\delta(\boldsymbol{r}-\boldsymbol{r}_{\mathrm{o}})\boldsymbol{P}_{\mathrm{e}2}] - \tfrac{1}{2}\operatorname{curl}[\operatorname{grad}\delta(\boldsymbol{r}-\boldsymbol{r}_{\mathrm{o}})\cdot \mathbf{Q}_{\mathrm{m}}] + \cdots. \quad (2.58)$$

The same 'distributional' series may also serve to obtain the multipole expansion of another frequently encountered integral, namely $\iiint \boldsymbol{\phi}\cdot\boldsymbol{J}\,\mathrm{d}V$. The formula is

$$\iiint_V \boldsymbol{\phi}\cdot\boldsymbol{J}\,\mathrm{d}V = \mathrm{j}\omega\boldsymbol{\phi}_{\mathrm{o}}\cdot\boldsymbol{P}_{\mathrm{e}1} + (\operatorname{curl}_{\mathrm{o}}\boldsymbol{\phi})\cdot\boldsymbol{P}_{\mathrm{m}} + \tfrac{1}{2}\mathrm{j}\omega(\operatorname{grad}_{\mathrm{o}}\boldsymbol{\phi}):\mathbf{Q}_{\mathrm{e}}$$
$$- (\operatorname{curl}\operatorname{curl}_{\mathrm{o}}\boldsymbol{\phi})\cdot\boldsymbol{P}_{\mathrm{e}2} + \tfrac{1}{2}(\operatorname{grad}\operatorname{curl}_{\mathrm{o}}\boldsymbol{\phi}):\mathbf{Q}_{\mathrm{m}}. \quad (2.59)$$

Let us check, for example, how the term in $\boldsymbol{P}_{\mathrm{e}2}$ in (2.59) follows from the corresponding term in (2.58). Applying (1.125) yields immediately

$$\iiint_V \boldsymbol{\phi}\cdot\operatorname{curl}\operatorname{curl}[\delta(\boldsymbol{r}-\boldsymbol{r}_{\mathrm{o}})\boldsymbol{P}_{\mathrm{e}2}]\,\mathrm{d}V = \iiint_V \delta(\boldsymbol{r}-\boldsymbol{r}_{\mathrm{o}})\boldsymbol{P}_{\mathrm{e}2}\cdot\operatorname{curl}\operatorname{curl}\boldsymbol{\phi}\,\mathrm{d}V$$
$$= \boldsymbol{P}_{\mathrm{e}2}\cdot\operatorname{curl}\operatorname{curl}_{\mathrm{o}}\boldsymbol{\phi}. \quad (2.60)$$

This simple distributional approach is also appropriate for deriving the term in $\boldsymbol{P}_{\mathrm{e}2}$ in the expansion (2.38) for $\iiint \phi\boldsymbol{J}\,\mathrm{d}V$. From (1.125):

$$\iiint_V \phi \operatorname{curl}\operatorname{curl}[\delta(\boldsymbol{r}-\boldsymbol{r}_{\mathrm{o}})\boldsymbol{P}_{\mathrm{e}2}]\,\mathrm{d}V$$
$$= \iiint_V \delta(\boldsymbol{r}-\boldsymbol{r}_{\mathrm{o}})\boldsymbol{P}_{\mathrm{e}2}\cdot\operatorname{grad}\operatorname{grad}\phi\,\mathrm{d}V - \iiint \delta(\boldsymbol{r}-\boldsymbol{r}_{\mathrm{o}})\boldsymbol{P}_{\mathrm{e}2}\,\nabla^2\phi\,\mathrm{d}V$$
$$= \boldsymbol{P}_{\mathrm{e}2}\cdot(\operatorname{grad}\operatorname{grad}_{\mathrm{o}}\phi - \nabla_{\mathrm{o}}^2\phi). \quad (2.61)$$

Integrals of the form (2.59) are frequently encountered in the technical literature, e.g. in the evaluation of the fields in a resonator. The modal expansions of the fields, given in Appendix C, contain integrals which are precisely of the type (2.59). Let, for example, the source be a short linear antenna of length l, carrying a time-harmonic current I (Fig. 2.6). From (2.21) and (2.58), the current density is

$$\boldsymbol{J} = Il\delta(\boldsymbol{r}-\boldsymbol{r}_1)\boldsymbol{u}_{\mathrm{a}}$$
$$= \mathrm{j}\omega P_{\mathrm{e}1}\delta(\boldsymbol{r}-\boldsymbol{r}_1)\boldsymbol{u}_{\mathrm{a}} = \mathrm{j}\omega\boldsymbol{P}_{\mathrm{e}1}\delta(\boldsymbol{r}-\boldsymbol{r}_1), \quad (2.62)$$

where $\boldsymbol{u}_{\mathrm{a}}$ is a unit vector in the direction of the antenna. The current density $\boldsymbol{J}$ is linearly polarized, since its time-dependent form $\boldsymbol{j}$ vibrates along the

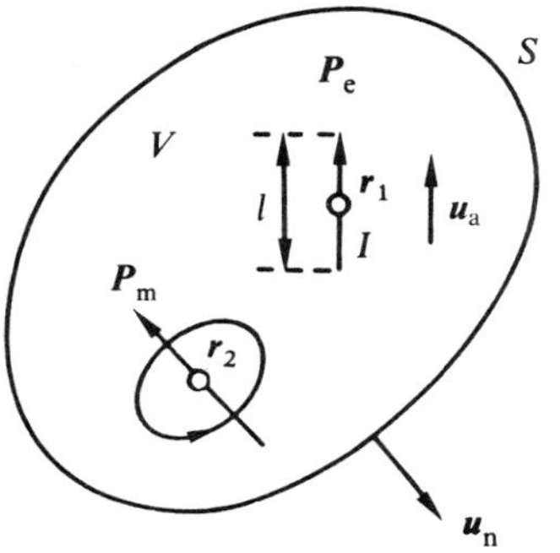

Fig. 2.6. A short linear antenna and frill of electric current.

antenna axis. Inserting (2.61) into the coupling integrals gives

$$\iiint_V \boldsymbol{J}\cdot\boldsymbol{f}_m\,dV = Il\boldsymbol{u}_a\cdot\boldsymbol{f}_m(\boldsymbol{r}_1), \qquad \iiint_V \boldsymbol{J}\cdot\boldsymbol{e}_m\,dV = Il\boldsymbol{u}_a\cdot\boldsymbol{e}_m(\boldsymbol{r}_1). \tag{2.63}$$

These formulas show that an 'm' mode is not excited when the antenna is perpendicular to the electric field of the mode.

In an analogous fashion, a frill located in $\boldsymbol{r}_2$ couples to the m mode by way of the integrals

$$\iiint_V \boldsymbol{J}\cdot\boldsymbol{f}_m\,dV = \iiint_V \operatorname{curl}[\delta(\boldsymbol{r}-\boldsymbol{r}_2)\boldsymbol{P}_m]\cdot\boldsymbol{f}_m\,dV = \boldsymbol{P}_m\cdot(\operatorname{curl}\boldsymbol{f}_m)_2 = 0,$$

$$\iiint_V \boldsymbol{J}\cdot\boldsymbol{e}_m\,dV = \boldsymbol{P}_m\cdot(\operatorname{curl}\boldsymbol{e}_m)_2 = k_m\boldsymbol{P}_m\cdot(\boldsymbol{h}_m)_2. \tag{2.64}$$

The first equation shows that the frill excites only the $\boldsymbol{e}_m$ and $\boldsymbol{h}_m$ modes.

2.10 More examples of concentrated sources

We have already given, in (2.41) and (2.62), the δ-function representation of the currents of a frill and a short linear antenna. In this Section, we perform the same task for three additional concentrated sources. The first one consists of two collinear short antennas, fed in phase opposition (Fig. 2.7a). With respect to the origin 0 of the figure, $\boldsymbol{P}_{e1}$, $\boldsymbol{P}_{e2}$, $\boldsymbol{P}_m$, and $\mathbf{Q}_m$ vanish. The first nonzero multipole term is, from (2.30),

$$j\omega\mathbf{Q}_e = 2Ild\boldsymbol{u}_z\boldsymbol{u}_z = 2Idl\boldsymbol{u}_z. \tag{2.65}$$

According to (2.58), the corresponding current density is

$$\boldsymbol{J} = -Id\frac{\partial\delta(\boldsymbol{r}-\boldsymbol{r}_o)}{\partial z}\,l\boldsymbol{u}_z. \tag{2.66}$$

This value can also be obtained by the simple operation

$$\begin{aligned} \boldsymbol{J} &= Il\delta(x)\delta(y)\delta(z - \tfrac{1}{2}d)\boldsymbol{u}_z - Il\delta(x)\delta(y)\delta(z + \tfrac{1}{2}d)\boldsymbol{u}_z \\ &= -Il\delta(x)\delta(y)d\,\frac{\partial\delta(z)}{\partial z}\,\boldsymbol{u}_z. \end{aligned} \tag{2.67}$$

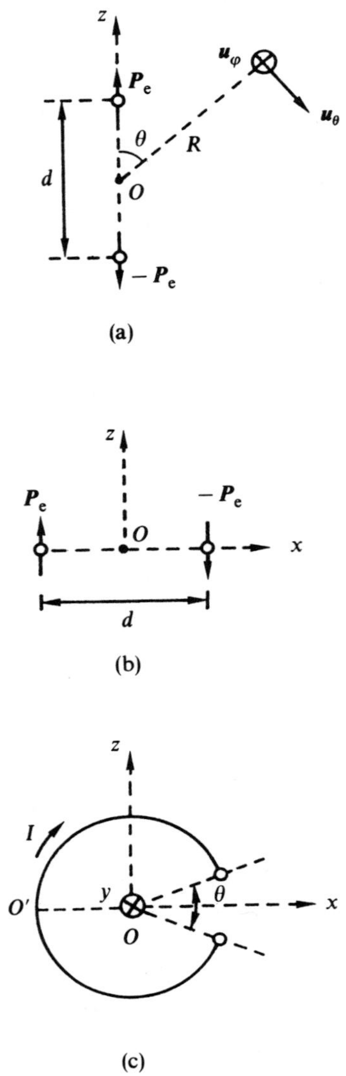

Fig. 2.7. Typical concentrated sources.

The current (2.66) generates a vector potential

$$\boldsymbol{A} = \frac{\mathrm{j}k\mu_0}{4\pi}(Ild\cos\theta)\frac{\mathrm{e}^{-\mathrm{j}kR}}{R}\boldsymbol{u}_z \tag{2.68}$$

with resulting radiation fields

$$\boldsymbol{E} = -(k^2\sin\theta\cos\theta)Ild\boldsymbol{u}_\theta, \qquad \boldsymbol{H} = -(k^2\sin\theta\cos\theta)Ild\boldsymbol{u}_\varphi. \tag{2.69}$$

The power radiation density is proportional to $\sin^2 2\theta$, which means that shadow zones exist in both the z direction and in the equatorial plane. The maximum gain, equal to $\frac{15}{8}$, is obtained in the cones of opening angles $\theta = 45°$ and $\theta = 135°$.

Another interesting radiating system, shown in Fig. 2.7b, consists of two antiparallel short antennas. For this system (a 'doublet'), the multipoles $\boldsymbol{P}_{e1}$, $\boldsymbol{P}_{e2}$, and $\mathbf{Q}_m$ vanish, but the magnetic dipole $\boldsymbol{P}_m$ survives. Intuitively, this is because the two antiparallel arrows may be interpreted as forming the embryo of a current loop. The nonzero moments are

$$\boldsymbol{P}_m = -\tfrac{1}{2}dIl\boldsymbol{u}_y, \qquad \mathrm{j}\omega\mathbf{Q}_e = -Ild(\boldsymbol{u}_x\boldsymbol{u}_z + \boldsymbol{u}_z\boldsymbol{u}_x). \tag{2.70}$$

The current density representing the radiating system is now

$$\boldsymbol{J} = Ild\left(\frac{\partial\delta(x)}{\partial x}\right)_0 \delta(y)\delta(z)\boldsymbol{u}_z. \tag{2.71}$$

The third example concerns the incomplete circular loop shown in Fig. 2.7c. The 'loop' part is expected to generate a *magnetic* dipole moment, while the 'linear' component of the structure should give an *electric* dipole moment. Detailed calculations confirm these expectations. Indeed,

$$\mathrm{j}\omega\boldsymbol{P}_{e1} = (2Ia\sin\tfrac{1}{2}\theta)\boldsymbol{u}_z, \qquad \boldsymbol{P}_m = Ia^2(\pi - \tfrac{1}{2}\theta)\boldsymbol{u}_y. \tag{2.72}$$

We observe that the electric dipole moment vanishes when the loop is closed (i.e. when $\theta = 0$). The magnetic dipole moment, on the other hand, decreases progressively (and vanishes ultimately) when the loop opens up to become, in the limit, a short dipole antenna.

The value of $\boldsymbol{P}_m$ in (2.72) is taken with respect to a given origin 0. When the latter is shifted to 0′, the moment becomes

$$\boldsymbol{P}'_m = Ia^2(\pi - \tfrac{1}{2}\theta + \sin\tfrac{1}{2}\theta)\boldsymbol{u}_y. \tag{2.73}$$

The value of $\boldsymbol{P}_m$ therefore depends on the choice of the origin. Such behaviour is in harmony with the definition (2.26) of $\boldsymbol{P}_m$, which implies that shifting the origin by a distance $\boldsymbol{a}$ introduces an additional moment

$$\Delta\boldsymbol{P}_m = \tfrac{1}{2}\boldsymbol{a} \times \iiint \boldsymbol{J}\,\mathrm{d}V = \tfrac{1}{2}\mathrm{j}\omega\boldsymbol{a} \times \boldsymbol{P}_{e1}. \tag{2.74}$$

This shift may be exploited to optimize the truncated expansion which obtains by keeping only the first two terms in (2.47). A solution based on complex values of $\boldsymbol{a}$ has recently been proposed (Lindell *et al.* 1986; Lindell 1987; Lindell *et al.* 1987).

2.11 Application to waveguides

In many problems of mathematical physics, the moment integrals are not volume integrals, as in (2.59), but *surface* integrals of the type

$$I_4 = \iint_S \boldsymbol{\phi}(x, y) \cdot \boldsymbol{J}(x, y, z)\, \mathrm{d}x\, \mathrm{d}y. \tag{2.75}$$

These integrals are encountered, in particular, in waveguide theory, where S is the cross-section of the guide (Fig. 2.8). The fields in a waveguide are classically expanded in a set of normal modes, and the differential equations for the expansion coefficients, given in Appendix D, contain integrals of the type I_4. We shall determine the value of I_4 when the current source $\boldsymbol{J}$ shrinks to a very small volume. To this effect, we insert (2.58) into I_4 and obtain, for the first term of the expansion,

$$\begin{aligned} \iint_S \boldsymbol{J} \cdot \boldsymbol{\phi}\, \mathrm{d}x\, \mathrm{d}y &= \mathrm{j}\omega \boldsymbol{P}_{\mathrm{e}1} \cdot \iint_S \boldsymbol{\phi} \delta(x - x_{\mathrm{o}}) \delta(y - y_{\mathrm{o}}) \delta(z - z_{\mathrm{o}}) \mathrm{d}x\, \mathrm{d}y \\ &= \mathrm{j}\omega \boldsymbol{P}_{\mathrm{e}1} \cdot \boldsymbol{\phi}_{\mathrm{o}} \delta(z - z_{\mathrm{o}}). \end{aligned} \tag{2.76}$$

The term in $\boldsymbol{P}_{\mathrm{m}}$ gives

$$\iint_S \boldsymbol{\phi} \cdot \mathrm{curl}\,[\delta(\boldsymbol{r} - \boldsymbol{r}_{\mathrm{o}}) \boldsymbol{P}_{\mathrm{m}}]\, \mathrm{d}x\, \mathrm{d}y = \iint_S \boldsymbol{\phi} \cdot [\mathrm{grad}\, \delta(\boldsymbol{r} - \boldsymbol{r}_{\mathrm{o}}) \times \boldsymbol{P}_{\mathrm{m}}]\, \mathrm{d}x\, \mathrm{d}y. \tag{2.77}$$

But the gradient may be written as

$$\mathrm{grad}\, \delta = \delta'(x)\delta(y)\delta(z) \boldsymbol{u}_x + \delta(x)\delta'(y)\delta(z) \boldsymbol{u}_y + \delta(x)\delta(y)\delta'(z) \boldsymbol{u}_z. \tag{2.78}$$

Repeated use of the sifting property of $\delta'(x_i)$ leads to

$$\iint_S \boldsymbol{J} \cdot \boldsymbol{\phi}\, \mathrm{d}x\, \mathrm{d}y = \boldsymbol{P}_{\mathrm{m}} \cdot (\mathrm{curl}_{\mathrm{t}}\, \boldsymbol{\phi}_{\mathrm{o}}) \delta(z - z_{\mathrm{o}}) + \boldsymbol{P}_{\mathrm{m}} \cdot (\boldsymbol{\phi}_{\mathrm{o}} \times \boldsymbol{u}_z) \delta'(z - z_{\mathrm{o}}). \tag{2.79}$$

The subscript o means evaluation at $\boldsymbol{r}_{\mathrm{o}}$, and the subscript t indicates that only derivatives with respect to x and y are involved. Thus,

$$\begin{aligned} \mathrm{grad}_{\mathrm{t}}\, \boldsymbol{\phi} &= \boldsymbol{u}_x \frac{\partial \boldsymbol{\phi}}{\partial x} + \boldsymbol{u}_y \frac{\partial \boldsymbol{\phi}}{\partial y} = \mathrm{grad}\, \boldsymbol{\phi} - \boldsymbol{u}_z \frac{\partial \boldsymbol{\phi}}{\partial z}, \\ \mathrm{curl}_{\mathrm{t}}\, \boldsymbol{\phi} &= \boldsymbol{u}_x \times \frac{\partial \boldsymbol{\phi}}{\partial x} + \boldsymbol{u}_y \times \frac{\partial \boldsymbol{\phi}}{\partial y} = \mathrm{curl}\, \boldsymbol{\phi} - \boldsymbol{u}_z \times \frac{\partial \boldsymbol{\phi}}{\partial z}. \end{aligned} \tag{2.80}$$

Analogous calculations for the other multipole moments lead to the final result (Van Bladel 1977)

$$\begin{aligned}\iint_S \boldsymbol{J}\cdot\boldsymbol{\phi}\,\mathrm{d}x\,\mathrm{d}y &= \mathrm{j}\omega\boldsymbol{P}_{e1}\cdot\boldsymbol{\phi}_o\delta(z-z_o) + \boldsymbol{P}_m\cdot(\mathrm{curl}_t\,\boldsymbol{\phi}_o)\delta(z-z_o)\\ &+ \boldsymbol{P}_m\cdot(\boldsymbol{\phi}_o\times\boldsymbol{u}_z)\delta'(z-z_o)\\ &+ \tfrac{1}{2}\mathrm{j}\omega(\mathbf{Q}_e:\mathrm{grad}_t\,\boldsymbol{\phi}_o)\delta(z-z_o) - \tfrac{1}{2}\mathrm{j}\omega(\boldsymbol{u}_z\cdot\mathbf{Q}_e)\cdot\boldsymbol{\phi}_o\delta'(z-z_o)\\ &+ \boldsymbol{P}_{e2}\cdot[\nabla_t^2\boldsymbol{\phi}_{oz} + \boldsymbol{u}_z\times\mathrm{grad\,div}(\boldsymbol{\phi}_o\times\boldsymbol{u}_z)]\delta(z-z_o) \qquad (2.81)\\ &+ \boldsymbol{P}_{e2}\cdot[\mathrm{grad}_t\,\phi_{oz} + \boldsymbol{u}_z\,\mathrm{div}_t\,\boldsymbol{\phi}_o]\delta'(z-z_o) + \boldsymbol{P}_{e2}\cdot\boldsymbol{\phi}_{ot}\delta''(z-z_o)\\ &+ \tfrac{1}{2}\mathbf{Q}_m:\mathrm{grad}_t\,\mathrm{curl}_t\,\boldsymbol{\phi}_o\delta(z-z_o) - \tfrac{1}{2}(\boldsymbol{u}_z\cdot\mathbf{Q}_m)\cdot(\mathrm{curl}_t\,\boldsymbol{\phi}_o)\delta'(z-z_o)\\ &- \tfrac{1}{2}[\mathrm{grad}_t\,\boldsymbol{\phi}_o:(\mathbf{Q}_m\times\boldsymbol{u}_z)]\delta'(z-z_o)\\ &- \tfrac{1}{2}(\boldsymbol{u}_z\cdot\mathbf{Q}_m)\cdot(\boldsymbol{\phi}_o\times\boldsymbol{u}_z)\delta''(z-z_o) + \cdots .\end{aligned}$$

As an elementary application of (2.81), let us evaluate the fields in an infinite waveguide excited by a magnetic dipole (Fig. 2.8). The lowest mode is always a TE mode, characterized by a pattern function ψ_1. For this mode the only nonzero second member in (D.4) is

$$-\iint_S \boldsymbol{J}\cdot\boldsymbol{\Phi}\,\mathrm{d}S = \boldsymbol{P}_m\cdot\mathrm{grad}\,\psi_1(x_o, y_o)\delta'(z-z_o) - v_1^2\boldsymbol{P}_m\cdot\psi_1(x_o, y_o)\boldsymbol{u}_z\delta(z-z_o), \qquad (2.82)$$

where $\boldsymbol{\Phi} = \mathrm{grad}\,\psi_1 \times \boldsymbol{u}_z$. Solution of (D.4) gives, when the mode is propagated,

$$\boldsymbol{E} = \tfrac{1}{2}\mathrm{j}\omega\mu_0[\mp \boldsymbol{P}_m\cdot\mathrm{grad}_o\,\psi_1 + \mathrm{j}(v_1^2/\gamma_1)\boldsymbol{P}_m\cdot(\psi_1\boldsymbol{u}_z)](\mathrm{grad}\,\psi_1\times\boldsymbol{u}_z)\mathrm{e}^{-\mathrm{j}\gamma_1|z-z_o|}. \qquad (2.83)$$

In this expression,

$$\gamma_1 = \sqrt{(k^2 - v_1^2)}. \qquad (2.84)$$

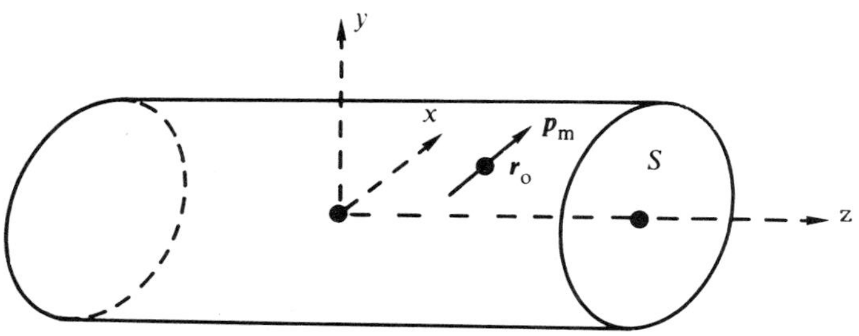

Fig. 2.8. A waveguide excited by a magnetic dipole.

The upper and lower signs in (2.83) correspond to positive and negative z, respectively.

2.12 Equivalence of electric and magnetic currents

The fields generated by electric currents in vacuum satisfy Maxwell's equations

$$\operatorname{curl} \boldsymbol{E}_a = -j\omega\mu_0 \boldsymbol{H}_a, \qquad \operatorname{curl} \boldsymbol{H}_a = j\omega\varepsilon_0 \boldsymbol{E}_a + \boldsymbol{J}. \tag{2.85}$$

The fields generated by magnetic currents, on the other hand, satisfy

$$\operatorname{curl} \boldsymbol{E}_b = -j\omega\mu_0 \boldsymbol{H}_b - \boldsymbol{K}, \qquad \operatorname{curl} \boldsymbol{H}_b = j\omega\varepsilon_0 \boldsymbol{E}_b. \tag{2.86}$$

To find the $\boldsymbol{K}$ equivalent to $\boldsymbol{J}$, we subtract (2.86) from (2.85), and obtain (Mayes 1958)

$$\operatorname{curl}(\boldsymbol{E}_a - \boldsymbol{E}_b) = -j\omega\mu_0(\boldsymbol{H}_a - \boldsymbol{H}_b) + \boldsymbol{K}, \qquad \operatorname{curl}(\boldsymbol{H}_a - \boldsymbol{H}_b) = j\omega\varepsilon_0(\boldsymbol{E}_a - \boldsymbol{E}_b) + \boldsymbol{J}. \tag{2.87}$$

We now eliminate $\boldsymbol{H}_a - \boldsymbol{H}_b$ by taking the curl of the first equation, and subsequently substituting $\operatorname{curl}(\boldsymbol{H}_a - \boldsymbol{H}_b)$ from the second. There results an equation for $\boldsymbol{E}_a - \boldsymbol{E}_b$ alone, viz.

$$-\operatorname{curl}\operatorname{curl}(\boldsymbol{E}_a - \boldsymbol{E}_b) + \omega^2\varepsilon_0\mu_0(\boldsymbol{E}_a - \boldsymbol{E}_b) = -\operatorname{curl}\boldsymbol{K} + j\omega\mu_0\boldsymbol{J}. \tag{2.88}$$

Similarly

$$-\operatorname{curl}\operatorname{curl}(\boldsymbol{H}_a - \boldsymbol{H}_b) + \omega^2\varepsilon_0\mu_0(\boldsymbol{H}_a - \boldsymbol{H}_b) = -\operatorname{curl}\boldsymbol{J} - j\omega\varepsilon_0\boldsymbol{K}. \tag{2.89}$$

In deriving an equation such as (2.89), we should remember that $\boldsymbol{J}$ is discontinuous on S (Fig. 2.9a). In the sense of distributions, therefore,

$$\operatorname{curl}\boldsymbol{J} = \{\operatorname{curl}\boldsymbol{J}\} - (\boldsymbol{u}_n \times \boldsymbol{J})\delta_s. \tag{2.90}$$

If $\boldsymbol{J}$ and $\boldsymbol{K}$ are to create the same magnetic field, $\boldsymbol{H}_a - \boldsymbol{H}_b$ must vanish everywhere. This requires the second member in (2.89) to vanish, which leads to the equivalence relationships

$$\{\boldsymbol{K}\} = -(1/j\omega\varepsilon_0)\{\operatorname{curl}\boldsymbol{J}\}, \qquad \boldsymbol{K}_s = (1/j\omega\varepsilon_0)\boldsymbol{u}_n \times \boldsymbol{J}. \tag{2.91}$$

The electric current $\boldsymbol{J}$ may therefore be replaced by a volume magnetic current $\boldsymbol{K}$, augmented by a surface magnetic current $\boldsymbol{K}_s$. The need for the latter is confirmed by the example of a homogeneous current density $\boldsymbol{J}$ filling a spherical volume (Fig. 2.9b). Because $\boldsymbol{J}$ is uniform, $\boldsymbol{K}$ vanishes, and there would be no magnetic equivalent if we would restrict our attention to volume currents. In the present case, the equivalence comes from the surface current

$$\boldsymbol{K}_s = (1/j\omega\varepsilon_0)\boldsymbol{u}_R \times \boldsymbol{J} = -(J/j\omega\varepsilon_0)(\sin\theta)\boldsymbol{u}_\varphi. \tag{2.92}$$

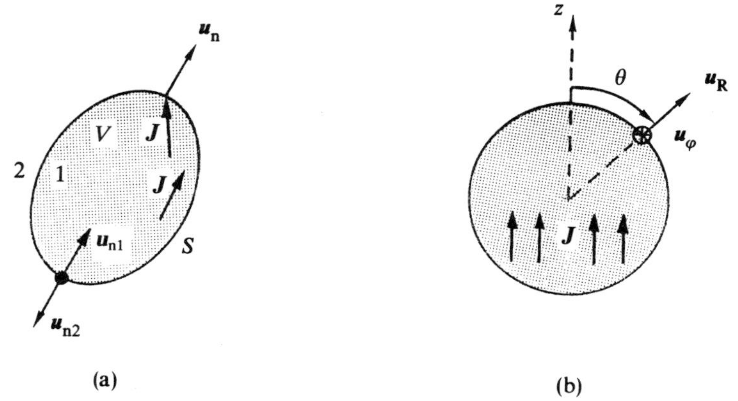

Fig. 2.9. Illustration of the equivalence between electric and magnetic currents.

The equivalence of $\boldsymbol{J}$ and $(\boldsymbol{K}, \boldsymbol{K}_s)$ holds only for the $\boldsymbol{H}$ field. Indeed, from (2.87),

$$\boldsymbol{E}_a - \boldsymbol{E}_b = \boldsymbol{E}_J - \boldsymbol{E}_K = -\boldsymbol{J}/j\omega\varepsilon_0. \tag{2.93}$$

This equation shows that the electric fields coincide outside V, but that they differ inside the sources. These properties are confirmed, in the case of the sphere of Fig. 2.9(b), by a detailed evaluation of the fields inside and outside the volume (Van Bladel 1988b).

It should be underlined that the discontinuity implied by (2.93) respects the boundary conditions satisfied by $\boldsymbol{E}$. For the *normal* component, these are

$$\begin{aligned} \boldsymbol{u}_{n1}\cdot\boldsymbol{E}_{a1} + \boldsymbol{u}_{n2}\cdot\boldsymbol{E}_{a2} &= P_s/\varepsilon_0 = -(1/j\omega\varepsilon_0)\boldsymbol{u}_n\cdot\boldsymbol{J}, \\ \boldsymbol{u}_{n1}\cdot\boldsymbol{E}_{b1} + \boldsymbol{u}_{n2}\cdot\boldsymbol{E}_{b2} &= 0. \end{aligned} \tag{2.94}$$

The first equation results from the equation of conservation of charge. The second is a consequence of the absence of electric charges in the 'b' situation. The difference between the two yields the normal component of (2.93). The boundary conditions for the *tangential* components of $\boldsymbol{E}$, on the other hand, are

$$\boldsymbol{u}_{n1}\times\boldsymbol{E}_{a1} + \boldsymbol{u}_{n2}\times\boldsymbol{E}_{a2} = 0, \qquad \boldsymbol{u}_{n1}\times\boldsymbol{E}_{b1} + \boldsymbol{u}_{n2}\times\boldsymbol{E}_{b2} = -\boldsymbol{K}_s. \tag{2.95}$$

From (2.91), the difference between the two is the tangential component of (2.93), rotated over 90°.

It is clear, again from (2.91), that electric current distributions $\boldsymbol{J}$ having identical curl and tangential components may be replaced by the *same* magnetic currents, although the original currents $\boldsymbol{J}_1$ and $\boldsymbol{J}_2$ differ. This perhaps surprising statement becomes more acceptable upon noticing that the difference $\boldsymbol{J} = \boldsymbol{J}_1 - \boldsymbol{J}_2$ must satisfy

$$\operatorname{curl} \boldsymbol{J} = 0 \quad \text{in } V, \qquad \boldsymbol{u}_n \times \boldsymbol{J} = 0 \quad \text{on } S. \tag{2.96}$$

The associated magnetic field is the solution of

$$-\operatorname{curl}\operatorname{curl} \boldsymbol{H} + \omega^2 \varepsilon_0 \mu_0 \boldsymbol{H} = -\operatorname{curl} \boldsymbol{J}. \tag{2.97}$$

The second member, interpreted in the sense of distributions, vanishes, hence $\boldsymbol{H}$ vanishes everywhere. We conclude that $\boldsymbol{H}_1 = \boldsymbol{H}_2$ everywhere and, from

$$\operatorname{curl} \boldsymbol{H} = \mathrm{j}\omega\varepsilon_0 \boldsymbol{E} + \boldsymbol{J}, \tag{2.98}$$

that

$$\boldsymbol{E} = \boldsymbol{E}_1 - \boldsymbol{E}_2 = -(1/\mathrm{j}\omega\varepsilon_0)\boldsymbol{J}. \tag{2.99}$$

The property that a current distribution such as $\boldsymbol{J}$ in (2.96) does not create any field outside S is by no means exceptional. The currents in a cavity with perfectly conducting walls are a case in point, because the volume currents in the cavity, and the induced currents in the wall, form an ensemble that does not radiate. The same property obtains for currents of the general form (Devaney *et al.* 1973; Lindell 1988)

$$\boldsymbol{J} = \operatorname{curl}\operatorname{curl} \boldsymbol{F}(\boldsymbol{r}) - k^2 \boldsymbol{F}(\boldsymbol{r}), \tag{2.100}$$

where $\boldsymbol{F}$ is any vector field that is continuous, has continuous partial derivatives up to the third order, and vanishes at all points outside S.

We have derived, in (2.90), formulas for the magnetic currents equivalent to a $\boldsymbol{J}$. Analogous steps show that the electric currents

$$\{\boldsymbol{J}\} = (1/\mathrm{j}\omega\mu_0)\{\operatorname{curl} \boldsymbol{K}\}, \qquad \boldsymbol{J}_s = -(1/\mathrm{j}\omega\mu_0)\boldsymbol{u}_n \times \boldsymbol{K}, \tag{2.101}$$

are equivalent to the magnetic current $\boldsymbol{K}$. The electric sources create the same fields as the original $\boldsymbol{K}$, with the exception of the magnetic fields inside the sources, which differ by

$$\boldsymbol{H}_J - \boldsymbol{H}_K = (1/\mathrm{j}\omega\mu_0)\boldsymbol{K}. \tag{2.102}$$

The equivalence property, discussed above for sources in vacuum, is also applicable to sources embedded in material media. As an example, let us discuss the case of the 'chiral' materials. These media are characterized by either a left-handedness or a right-handedness in their microstructure. They have constitutive equations of the form

$$\boldsymbol{D} = \varepsilon(\boldsymbol{E} + \beta \operatorname{curl} \boldsymbol{E}), \qquad \boldsymbol{B} = \mu(\boldsymbol{H} + \beta \operatorname{curl} \boldsymbol{H}), \tag{2.103}$$

where β is the chirality parameter. In chiral materials, left- and right-circularly polarized waves propagate with different phase velocities. This property, observed at optical frequencies since the early 19th century, is of potential interest for the design of telecommunication devices at suboptical frequencies. A serious effort is being made to develop chiral materials (either artificial, or belonging to the polymer family) suitable for use at those frequencies.

In materials governed by (2.103), equations (2.88) and (2.89) are replaced by (Lakhtakia *et al.* 1989)

$$\begin{aligned}-\operatorname{curl}\operatorname{curl}(\mathbf{E}_a - \mathbf{E}_b) + 2\beta\gamma^2 \operatorname{curl}(\mathbf{E}_a - \mathbf{E}_b) + \gamma^2(\mathbf{E}_a - \mathbf{E}_b) \\ = (\gamma^2/k^2)[j\omega\mu(\mathbf{J} + \beta \operatorname{curl}\mathbf{J}) - \operatorname{curl}\mathbf{K}], \\ -\operatorname{curl}\operatorname{curl}(\mathbf{H}_a - \mathbf{H}_b) + 2\beta\gamma^2 \operatorname{curl}(\mathbf{H}_a - \mathbf{H}_b) + \gamma^2(\mathbf{H}_a - \mathbf{H}_b) \\ = -(\gamma^2/k^2)[j\omega\varepsilon(\mathbf{K} + \beta \operatorname{curl}\mathbf{K}) + \operatorname{curl}\mathbf{J}],\end{aligned} \tag{2.104}$$

where $k^2 = \omega^2\varepsilon\mu$ and $\gamma^2 = k^2(1 - \beta^2 k^2)^{-1}$. Identical electric fields at any point in space are therefore obtained when the sources are connected by

$$\mathbf{J} + \beta \operatorname{curl}\mathbf{J} = (1/j\omega\mu)\operatorname{curl}\mathbf{K}. \tag{2.105}$$

Conversely, identical magnetic fields obtain when

$$\mathbf{K} + \beta \operatorname{curl}\mathbf{K} = -(1/j\omega\varepsilon)\operatorname{curl}\mathbf{J}. \tag{2.106}$$

2.13 The curl of a surface field

In the previous Section, we investigated the magnetic equivalent of an electric current flowing in a volume. In many situations, however, the currents flow on a surface, for example at the boundary of a perfectly conducting scatterer immersed in an incident field. To find the magnetic equivalent of such a $\mathbf{J}_s$, the first step is to write the curl operator in the (v_1, v_2, n) coordinates defined in Section 1.5. Thus,

$$\begin{aligned}\operatorname{curl}\mathbf{A} = \frac{1}{h_2}\frac{\partial A_n}{\partial v_2}\mathbf{u}_1 - \frac{\partial A_2}{\partial n}\mathbf{u}_1 - \frac{1}{h_2}\frac{\partial h_2}{\partial n}A_2\mathbf{u}_1 \\ - \frac{1}{h_1}\frac{\partial A_n}{\partial v_1}\mathbf{u}_2 + \frac{\partial A_1}{\partial n}\mathbf{u}_2 + \frac{1}{h_1}\frac{\partial h_1}{\partial n}A_1\mathbf{u}_2 + \mathbf{u}_n \operatorname{div}_s(\mathbf{A} \times \mathbf{u}_n).\end{aligned} \tag{2.107}$$

The surface divergence has been defined in (1.132). Because the normal derivatives of $\mathbf{u}_1$, $\mathbf{u}_2$, and $\mathbf{u}_n$ vanish, the curl can be written more concisely as

$$\operatorname{curl}\mathbf{A} = \operatorname{grad}_s A_n \times \mathbf{u}_n + \mathbf{u}_n \times \frac{\partial \mathbf{A}_t}{\partial n} + \mathbf{\Gamma}\cdot(\mathbf{A}_t \times \mathbf{u}_n) + \mathbf{u}_n \operatorname{div}_s(\mathbf{A}_t \times \mathbf{u}_n). \tag{2.108}$$

The subscript t denotes the tangential component of $\mathbf{A}$ on S. The surface gradient operator is given by

$$\operatorname{grad}_s A = \frac{1}{h_1}\frac{\partial A}{\partial v_1}\mathbf{u}_1 + \frac{1}{h_2}\frac{\partial A}{\partial v_2}\mathbf{u}_2, \tag{2.109}$$

and $\boldsymbol{\Gamma}\cdot(\boldsymbol{A}_t \times \boldsymbol{u}_n)$ is the linear form

$$\boldsymbol{\Gamma}\cdot(\boldsymbol{A}_t \times \boldsymbol{u}_n) = -\frac{1}{h_2}\frac{\partial h_2}{\partial n} A_2 \boldsymbol{u}_1 + \frac{1}{h_1}\frac{\partial h_1}{\partial n} A_1 \boldsymbol{u}_2, \tag{2.110}$$

where

$$\begin{aligned}\boldsymbol{\Gamma}(v_1, v_2) &= -\frac{1}{h_2}\frac{\partial h_2}{\partial n}\boldsymbol{u}_1\boldsymbol{u}_1 - \frac{1}{h_1}\frac{\partial h_1}{\partial n}\boldsymbol{u}_2\boldsymbol{u}_2 \\ &= \frac{1}{R_2}\boldsymbol{u}_1\boldsymbol{u}_1 + \frac{1}{R_1}\boldsymbol{u}_2\boldsymbol{u}_2.\end{aligned} \tag{2.111}$$

This dyadic vanishes for a planar surface, and therefore represents a curvature effect.

After these preliminaries, we are ready to derive an expression for the curl of $\boldsymbol{J}_s$, where $\boldsymbol{J}_s$ is allowed to have both normal and tangential components. The basis for our derivation is the distributional definition of the curl, viz.

$$\iiint \boldsymbol{\phi}\cdot \operatorname{curl} \boldsymbol{J}\, \mathrm{d}V \stackrel{\text{def}}{=} \iiint \boldsymbol{J}\cdot \operatorname{curl} \boldsymbol{\phi}\, \mathrm{d}V, \tag{2.112}$$

where $\boldsymbol{\phi}$ is a test vector. To obtain $\operatorname{curl}\boldsymbol{J}$, we start from the right-hand side, and endeavour to put it in a form corresponding to the left-hand side, where $\boldsymbol{\phi}$ appears as a factor. The value of $\operatorname{curl}\boldsymbol{J}$ will then be found by inspection.

Because the current is concentrated in a very thin shell, the element of volume is equal to $\mathrm{d}V = \mathrm{d}S\,\mathrm{d}n$. The current density becomes a surface density (Fig. 2.10)

$$\boldsymbol{J}_s = \int_0^h \boldsymbol{J}\,\mathrm{d}n. \tag{2.113}$$

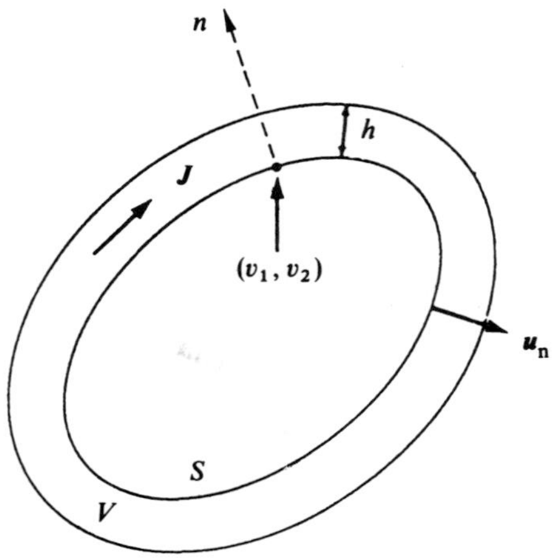

Fig. 2.10. A shell-like volume.

This gives, from (2.108),

$$\iiint \boldsymbol{\phi}\cdot\operatorname{curl}\boldsymbol{J}\,\mathrm{d}V = \iint_S \boldsymbol{J}_s\cdot\operatorname{curl}\boldsymbol{\phi}\,\mathrm{d}S \tag{2.114}$$

$$= \iint_S \boldsymbol{J}_s\cdot\left(\operatorname{grad}_s\phi_n \times \boldsymbol{u}_n + \frac{\partial}{\partial n}(\boldsymbol{u}_n \times \boldsymbol{\phi}) + \boldsymbol{\Gamma}\cdot(\boldsymbol{\phi}\times\boldsymbol{u}_n) + \boldsymbol{u}_n \operatorname{div}_s(\boldsymbol{\phi}\times\boldsymbol{u}_n)\right)\mathrm{d}S.$$

We shall transform the various terms in the second member successively. For the first term, we rely on a relationship valid on a closed surface S, viz. (Van Bladel 1985).

$$\iint_S \boldsymbol{A}_t\cdot\operatorname{grad}_s B\,\mathrm{d}S = -\iint_S B\operatorname{div}_s\boldsymbol{A}_t\,\mathrm{d}S. \tag{2.115}$$

Applied to the first term, this equation gives

$$\iint_S \operatorname{grad}_s\phi_n\cdot(\boldsymbol{u}_n\times\boldsymbol{J}_s)\,\mathrm{d}S = -\iint_S \phi_n\operatorname{div}_s(\boldsymbol{u}_n\times\boldsymbol{J}_s)\,\mathrm{d}S \tag{2.116}$$

$$= \iint_S \boldsymbol{\phi}\cdot\boldsymbol{u}_n\operatorname{div}_s(\boldsymbol{J}_s\times\boldsymbol{u}_n)\,\mathrm{d}S.$$

Relationship (2.115) also allows us to transform the fourth term as follows:

$$\iint_S J_{sn}\operatorname{div}_s(\boldsymbol{\phi}\times\boldsymbol{u}_n)\,\mathrm{d}S = -\iint_S (\boldsymbol{\phi}\times\boldsymbol{u}_n)\cdot\operatorname{grad}_s J_{sn}\,\mathrm{d}S \tag{2.117}$$

$$= \iint_S \boldsymbol{\phi}\cdot(\operatorname{grad}_s J_{sn}\times\boldsymbol{u}_n)\,\mathrm{d}S.$$

Turning now to the second term, we write, from (1.102),

$$\iint_S \boldsymbol{J}_s\cdot\left(\boldsymbol{u}_n\times\frac{\partial\boldsymbol{\phi}}{\partial n}\right)\mathrm{d}S = \iint_S (\boldsymbol{J}_s\times\boldsymbol{u}_n)\cdot\frac{\partial\boldsymbol{\phi}}{\partial n}\,\mathrm{d}S \tag{2.118}$$

$$= \iiint \boldsymbol{\phi}\cdot(\boldsymbol{u}_n\times\boldsymbol{J}_s)\frac{\partial\delta_s}{\partial n}\,\mathrm{d}V.$$

Finally, the third term can be written as

$$\boldsymbol{J}_s\cdot[\boldsymbol{\Gamma}\times(\boldsymbol{\phi}\times\boldsymbol{u}_n)] = \boldsymbol{\phi}\cdot[\boldsymbol{\pi}\cdot(\boldsymbol{u}_n\times\boldsymbol{J}_s)], \tag{2.119}$$

where $\boldsymbol{\pi}$ is the dyadic

$$\boldsymbol{\pi} = \frac{1}{R_1}\boldsymbol{u}_1\boldsymbol{u}_1 + \frac{1}{R_2}\boldsymbol{u}_2\boldsymbol{u}_2. \tag{2.120}$$

We have now reached the point where the various terms can be joined to lead to the value of curl $\boldsymbol{J}$. In our derivation, $\boldsymbol{J}$ is not necessarily a current density.

We shall therefore write the formula in terms of a general surface field $\boldsymbol{A}_s$. The result is

$$\operatorname{curl}(\boldsymbol{A}_s\delta_s) = \\ [\operatorname{grad}_s A_{sn} \times \boldsymbol{u}_n + \boldsymbol{\pi}\cdot(\boldsymbol{u}_n \times \boldsymbol{A}_s) + \boldsymbol{u}_n \operatorname{div}_s(\boldsymbol{A}_s \times \boldsymbol{u}_n)]\delta_s + (\boldsymbol{u}_n \times \boldsymbol{A}_s)\frac{\partial \delta_s}{\partial n}. \tag{2.121}$$

This important formula can also be derived on the basis of the $\delta(n)$ formulation. The steps are elementary:

$$\begin{aligned} \operatorname{curl}[\delta(n)\boldsymbol{A}_s(v_1, v_2)] &= \delta(n)\operatorname{curl}\boldsymbol{A}_s(v_1, v_2) + \operatorname{grad}\delta(n) \times \boldsymbol{A}_s \\ &= \delta(n)[\operatorname{grad} A_{sn} \times \boldsymbol{u}_n + \boldsymbol{\Gamma}\cdot(\boldsymbol{A}_s \times \boldsymbol{u}_n) + \boldsymbol{u}_n \operatorname{div}_s(\boldsymbol{A}_s \times \boldsymbol{u}_n)] \\ &\quad + \delta'(n)(\boldsymbol{u}_n \times \boldsymbol{A}_s). \end{aligned} \tag{2.122}$$

The equivalence of the two expressions, (2.121) and (2.122) is an immediate consequence of (A.12) and (A.26).

2.14 The equivalence of electric and magnetic surface sources

The magnetic equivalent of a $\boldsymbol{J}_s$ can be obtained from the general equivalence equations (2.91) by replacing the curl by its expression (2.121). Thus,

$$\begin{aligned} \boldsymbol{K} = &-(1/j\omega\varepsilon_0)(\operatorname{grad}_s J_{sn} \times \boldsymbol{u}_n)\delta_s \\ &-(1/j\omega\varepsilon_0)\boldsymbol{u}_n \operatorname{div}_s(\boldsymbol{J}_{st} \times \boldsymbol{u}_n)\delta_s \\ &-(1/j\omega\varepsilon_0)\left[(\boldsymbol{u}_n \times \boldsymbol{J}_{st})\frac{\partial \delta_s}{\partial n} + \boldsymbol{\pi}\cdot(\boldsymbol{u}_n \times \boldsymbol{J}_{st})\delta_s\right]. \end{aligned} \tag{2.123}$$

Similarly, the $\boldsymbol{J}$ equivalent to a $\boldsymbol{K}_s$ is

$$\begin{aligned} \boldsymbol{J} = &(1/j\omega\mu_0)(\operatorname{grad}_s K_{sn} \times \boldsymbol{u}_n)\delta_s \\ &+ (1/j\omega\mu_0)\boldsymbol{u}_n \operatorname{div}_s(\boldsymbol{K}_{st} \times \boldsymbol{u}_n)\delta_s \\ &+ (1/j\omega\mu_0)\left[(\boldsymbol{u}_n \times \boldsymbol{K}_{st})\frac{\partial \delta_s}{\partial n} + \boldsymbol{\pi}\cdot(\boldsymbol{u}_n \times \boldsymbol{K}_{st})\delta_s\right]. \end{aligned} \tag{2.124}$$

It is an easy matter to identify the three partial equivalents in (2.124):

(a) the first line represents a surface electric current. It is the contribution of the normal component of $\boldsymbol{K}_s$.

(b) the last two lines are generated by the tangential component of $\boldsymbol{K}_s$. The second line represents a normal layer of electric currents, and the third one a double layer of tangential electric currents. From (A.30), the double layer is of the 'zero' type, and its 'strength' is

$$\boldsymbol{C}_s = (1/j\omega\mu_0)(\boldsymbol{K}_{st} \times \boldsymbol{u}_n). \tag{2.125}$$

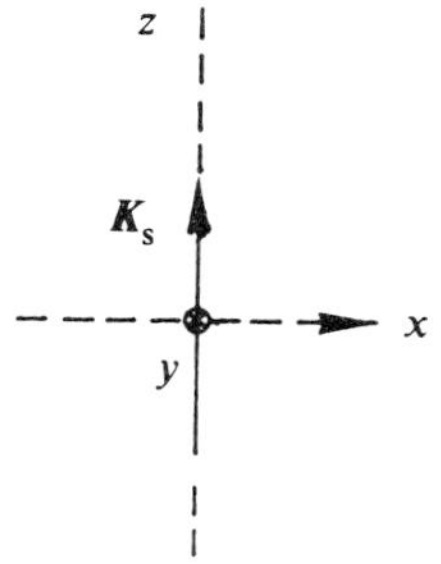

Fig. 2.11. Plane distribution of magnetic surface currents.

As an example, let us apply the equivalence formula to a plane distribution of magnetic surface currents

$$\boldsymbol{K}_s = K_s \boldsymbol{u}_z. \tag{2.126}$$

For a planar surface, $\boldsymbol{\pi}$ vanishes; hence the equivalent electric current becomes (Fig. 2.11)

$$\begin{aligned} \boldsymbol{J} &= (1/\mathrm{j}\omega\mu_0)(\boldsymbol{u}_x \times \boldsymbol{K}_s)\frac{\partial \delta_s}{\partial x} \\ &= -(1/\mathrm{j}\omega\mu_0)K_s \boldsymbol{u}_y \frac{\partial \delta_s}{\partial x}. \end{aligned} \tag{2.127}$$

This current is that of a double sheet (of which the type need not be specified because the three types coincide when the surface is plane).

It is evident, from (2.123) and (2.124), that a normal magnetic current $\boldsymbol{K}_{sn}$ can be replaced by an appropriate tangential electric current $\boldsymbol{J}_{st}$. But the converse does not hold, i.e. a $\boldsymbol{K}_{sn}$ alone does not suffice to replace an arbitrary $\boldsymbol{J}_{st}$.

2.15 Boundary conditions

The validity of an equivalence formula such as (2.124) can be verified by investigating the jump conditions satisfied by $\boldsymbol{K}_s$ and its electric equivalents. The discontinuities across S are the source of the electromagnetic fields, consequently the equality of the jumps is a condition 'sine qua non' for the equivalence to hold.

We shall consider the jumps only in the tangential components. Indeed, if the jumps in these components coincide for $\boldsymbol{K}_s$ and its electric equivalents, then the jumps in the normal components will coincide as well. This is because

$$j\omega\varepsilon_0 E_n = (\text{curl}\,\boldsymbol{H})_n = \text{div}_s(\boldsymbol{H} \times \boldsymbol{u}_n), \qquad j\omega\mu_0 H_n = -(\text{curl}\,\boldsymbol{E})_n = \text{div}_s(\boldsymbol{u}_n \times \boldsymbol{E}). \tag{2.128}$$

The jump conditions across the original layer of magnetic current are classical. For the tangential component $\boldsymbol{K}_{st}$:

$$\boldsymbol{E}_t^+ - \boldsymbol{E}_t^- = \boldsymbol{u}_n \times \boldsymbol{K}_{st}, \qquad \boldsymbol{H}_t^+ - \boldsymbol{H}_t^- = 0. \tag{2.129}$$

The + and − superscripts correspond respectively to the positive and negative sides of the normal. The normal component K_{sn} creates the discontinuity (Michalski 1986)

$$\boldsymbol{E}_t^+ - \boldsymbol{E}_t^- = 0, \qquad \boldsymbol{H}_t^+ - \boldsymbol{H}_t^- = -(1/j\omega\mu_0)\,\text{grad}_s K_{sn}. \tag{2.130}$$

The jump conditions corresponding to the electric equivalents must now be considered. For the first line of (2.124), which represents a surface electric current (Fig. 2.12),

$$\boldsymbol{E}_t^+ - \boldsymbol{E}_t^- = 0, \qquad \boldsymbol{H}_t^+ - \boldsymbol{H}_t^- = -(1/j\omega\mu_0)\,\text{grad}_s K_{sn}. \tag{2.131}$$

A normal electric surface current creates a discontinuity

$$\boldsymbol{E}_t^+ - \boldsymbol{E}_t^- = -(1/j\omega\varepsilon_0)\,\text{grad}_s J_{sn}, \qquad \boldsymbol{H}_t^+ - \boldsymbol{H}_t^- = 0. \tag{2.132}$$

Applied to the current on the second line of (2.124), this condition gives

$$\boldsymbol{E}_t^+ - \boldsymbol{E}_t^- = (1/k^2)\,\text{grad}_s\,\text{div}_s(\boldsymbol{K}_{st} \times \boldsymbol{u}_n), \qquad \boldsymbol{H}_t^+ - \boldsymbol{H}_t^- = 0. \tag{2.133}$$

Finally, the discontinuity relative to the third line of (2.124) is that of a double layer of the 'zero' type. The relevant formulas are (A.43) to (A.46), in which we set $\boldsymbol{J}_{st} = 0$. The strength of the double layer is

$$\boldsymbol{C}_s = (1/j\omega\mu_0)(\boldsymbol{K}_{st} \times \boldsymbol{u}_n); \tag{2.134}$$

hence

$$\boldsymbol{E}_t^+ - \boldsymbol{E}_t^- = -\boldsymbol{K}_{st} \times \boldsymbol{u}_n - (1/k^2)\,\text{grad}_s\,\text{div}_s(\boldsymbol{K}_{st} \times \boldsymbol{u}_n), \qquad \boldsymbol{H}_t^+ - \boldsymbol{H}_t^- = 0. \tag{2.135}$$

Adding (2.131), (2.133), and (2.135) together reproduces the original values (2.129) and (2.130). The equivalence is therefore confirmed.

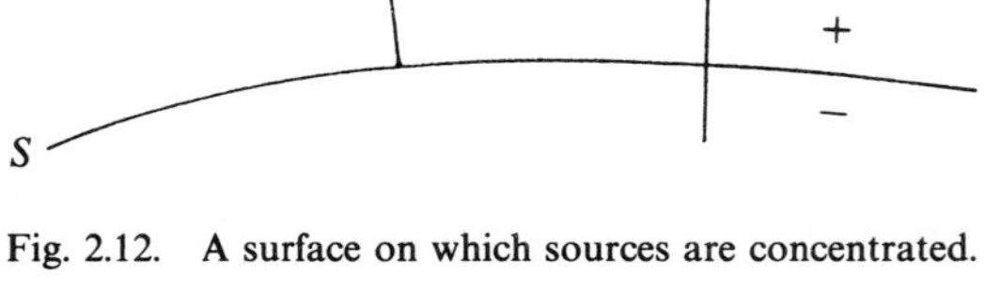

Fig. 2.12. A surface on which sources are concentrated.

2.16 Multipole expansion for magnetic currents

In vacuum, the far fields produced by a distribution of magnetic currents $\boldsymbol{K}$ can be derived from a vector potential $\boldsymbol{C}$. The formulas are

$$\boldsymbol{D} = -\operatorname{curl} \boldsymbol{C}, \qquad \boldsymbol{C} = \frac{\varepsilon_0}{4\pi} \frac{\mathrm{e}^{-jkR}}{R} \iiint_V \boldsymbol{K}(\boldsymbol{r}) \mathrm{e}^{jk\boldsymbol{u}\cdot\boldsymbol{r}} \, \mathrm{d}V. \tag{2.136}$$

The expression for $\boldsymbol{C}$ is of the same type as that obtained for the vector potential $\boldsymbol{A}$. From (2.47) and (2.48), therefore,

$$\boldsymbol{C} = \frac{\varepsilon_0}{4\pi} \frac{\mathrm{e}^{-jkR}}{R} [\mathrm{j}kc\boldsymbol{P}'_{\mathrm{m}1} + \mathrm{j}k\boldsymbol{u} \times \boldsymbol{P}'_{\mathrm{e}} - \tfrac{1}{2}k^2 c\boldsymbol{u} \cdot \mathbf{Q}'_{\mathrm{m}} \\ - k^2(\boldsymbol{u} \times \boldsymbol{P}'_{\mathrm{m}2} \times \boldsymbol{u}) - \tfrac{1}{2}k^2 \boldsymbol{u} \times (\boldsymbol{u} \cdot \mathbf{Q}'_{\mathrm{e}}) + \cdots], \tag{2.137}$$

where

$$\boldsymbol{P}'_{\mathrm{m}1} = \frac{1}{\mathrm{j}\omega} \iiint_V \boldsymbol{K} \, \mathrm{d}V, \quad \boldsymbol{P}'_{\mathrm{e}} = \tfrac{1}{2} \iiint_V \boldsymbol{K} \times \boldsymbol{r} \, \mathrm{d}V, \quad \boldsymbol{P}'_{\mathrm{m}2} = \tfrac{1}{6} \iiint_V \boldsymbol{r} \times \boldsymbol{K} \times \boldsymbol{r} \, \mathrm{d}V, \tag{2.138}$$

$$\mathbf{Q}'_{\mathrm{m}} = \frac{1}{\mathrm{j}\omega} \iiint_V (\boldsymbol{r}\boldsymbol{K} + \boldsymbol{K}\boldsymbol{r}) \, \mathrm{d}V, \qquad \mathbf{Q}'_{\mathrm{e}} = \tfrac{1}{3} \iiint_V [(\boldsymbol{K} \times \boldsymbol{r})\boldsymbol{r} + \boldsymbol{r}(\boldsymbol{K} \times \boldsymbol{r})] \, \mathrm{d}V.$$

The corresponding far fields are

$$\boldsymbol{E} = \frac{1}{4\pi} \frac{\mathrm{e}^{-jkR}}{R} \{-k^2 c\boldsymbol{u} \times \boldsymbol{P}'_{\mathrm{m}1} - k^2 \boldsymbol{u} \times (\boldsymbol{u} \times \boldsymbol{P}'_{\mathrm{e}}) - \tfrac{1}{2}\mathrm{j}k^3 c\boldsymbol{u} \times (\boldsymbol{u} \cdot \mathbf{Q}'_{\mathrm{m}}) \\ -\mathrm{j}k^3 \boldsymbol{u} \times \boldsymbol{P}'_{\mathrm{m}2} - \tfrac{1}{2}\mathrm{j}k^3 \boldsymbol{u} \times [\boldsymbol{u} \times (\boldsymbol{u} \cdot \mathbf{Q}'_{\mathrm{e}})] + \cdots \},$$

$$\boldsymbol{H} = \frac{1}{4\pi R_{\mathrm{c}}} \frac{\mathrm{e}^{-jkR}}{R} \{-k^2 c\boldsymbol{u} \times (\boldsymbol{u} \times \boldsymbol{P}'_{\mathrm{m}1}) + k^2 \boldsymbol{u} \times \boldsymbol{P}'_{\mathrm{e}} - \tfrac{1}{2}\mathrm{j}k^3 c\boldsymbol{u} \times [\boldsymbol{u} \times (\boldsymbol{u} \cdot \mathbf{Q}'_{\mathrm{m}})] \\ - \mathrm{j}k^3 \boldsymbol{u} \times (\boldsymbol{u} \times \boldsymbol{P}'_{\mathrm{m}2}) + \tfrac{1}{2}\mathrm{j}k^3 \boldsymbol{u} \times (\boldsymbol{u} \cdot \mathbf{Q}'_{\mathrm{e}}) + \cdots \} \tag{2.139}$$

As an interesting exercise, let us verify whether the radiation fields (2.48) of a given electric current $\boldsymbol{J}$ are the same as those of the equivalent magnetic currents (2.91). We limit this verification to the electric dipole term. From (2.48) and (2.139), the contributions of these dipoles will coincide, provided that

$$\boldsymbol{P}'_{\mathrm{e}} = (1/\varepsilon_0)\boldsymbol{P}_{\mathrm{e}1}. \tag{2.140}$$

To prove this equality, let us evaluate the x component of $\boldsymbol{P}'_{\mathrm{e}}$. From (2.91) and (2.138):

$$\begin{aligned}\boldsymbol{u}_x \cdot \boldsymbol{P}'_{\mathrm{e}} &= \tfrac{1}{2}\left(-\frac{1}{\mathrm{j}\omega\varepsilon_0}\iiint_V (\operatorname{curl}\boldsymbol{J}) \times \boldsymbol{r}\,\mathrm{d}V + \frac{1}{\mathrm{j}\omega\varepsilon_0}\iint_S (\boldsymbol{u}_{\mathrm{n}} \times \boldsymbol{J}) \times \boldsymbol{r}\,\mathrm{d}S\right)\cdot \boldsymbol{u}_x \\ &= \frac{1}{\mathrm{j}\omega\varepsilon_0}\left(\tfrac{1}{2}\iiint_V (\boldsymbol{u}_x \times \boldsymbol{r})\cdot \operatorname{curl}\boldsymbol{J}\,\mathrm{d}V - \tfrac{1}{2}\iint_S (\boldsymbol{u}_x \times \boldsymbol{r})\cdot(\boldsymbol{u}_{\mathrm{n}} \times \boldsymbol{J})\,\mathrm{d}S\right).\end{aligned} \tag{2.141}$$

The divergence theorem, applied to the first integral, yields

$$\begin{aligned}&\tfrac{1}{2}\iiint_V (\boldsymbol{u}_x \times \boldsymbol{r})\cdot \operatorname{curl}\boldsymbol{J}\,\mathrm{d}V \\ &= \tfrac{1}{2}\iiint_V \boldsymbol{J}\cdot \operatorname{curl}(\boldsymbol{u}_x \times \boldsymbol{r})\,\mathrm{d}V - \tfrac{1}{2}\iiint_V \operatorname{div}[(\boldsymbol{u}_x \times \boldsymbol{r}) \times \boldsymbol{J}]\,\mathrm{d}V \qquad (2.142) \\ &= \tfrac{1}{2}\iiint_V \boldsymbol{J}\cdot(2\boldsymbol{u}_x)\,\mathrm{d}V - \tfrac{1}{2}\iint_S \boldsymbol{u}_{\mathrm{n}}\cdot[(\boldsymbol{u}_x \times \boldsymbol{r}) \times \boldsymbol{J}]\,\mathrm{d}S.\end{aligned}$$

Inserting this value into (2.141) yields

$$\frac{1}{\mathrm{j}\omega\varepsilon_0}\iiint (\boldsymbol{u}_x \cdot \boldsymbol{J})\,\mathrm{d}V = \boldsymbol{u}_x \cdot \frac{\boldsymbol{P}_{\mathrm{e}1}}{\varepsilon_0}. \tag{2.143}$$

This is the x-component of the sought relationship (2.140). Analogous methods may be used to investigate the equivalence of the magnetic dipole terms. Detailed calculations, left to the reader, show that $\boldsymbol{P}'_{\mathrm{m}1}$ vanishes when the equivalent $\boldsymbol{K}$ are inserted, and that it is the term in $\boldsymbol{P}'_{\mathrm{m}2}$ which is responsible for the equivalence with $\boldsymbol{P}_{\mathrm{m}}$. The two moments are related by

$$\boldsymbol{P}_{\mathrm{m}} = (\mathrm{j}k/R_{\mathrm{c}})\boldsymbol{P}'_{\mathrm{m}2} = \mathrm{j}\omega\varepsilon_0 \boldsymbol{P}'_{\mathrm{m}2}. \tag{2.144}$$

2.17 Waveguide excitation by magnetic currents

In Section 2.11 we have evaluated the waveguide fields generated by a $\boldsymbol{P}_{\mathrm{m}}$ of electric origin. We now perform the same task for the magnetic dipole $\boldsymbol{P}'_{\mathrm{m}}$ associated with *magnetic* currents. Based on (2.138), these currents are

$$\boldsymbol{K} = \mathrm{j}\omega \boldsymbol{P}'_{\mathrm{m}}\delta(\boldsymbol{r} - \boldsymbol{r}_{\mathrm{o}}). \tag{2.145}$$

The equations satisfied by the modal fields are given in Appendix D. As in Section 2.11, we shall only investigate the excitation of the lowest mode, which belongs to the TE family. The relevant equations are (D.4), in which the only nonvanishing second members are

$$\iint_S \boldsymbol{K}\cdot\operatorname{grad}\psi_1\,\mathrm{d}S = \mathrm{j}\omega\boldsymbol{P}'_{\mathrm{m}}\cdot\operatorname{grad}\psi_1\,\delta(z-z_{\mathrm{o}}),$$
$$\iint_S (\boldsymbol{K}\cdot\boldsymbol{u}_z)\psi_1\,\mathrm{d}S = \mathrm{j}\omega(\boldsymbol{P}'_{\mathrm{m}}\cdot\boldsymbol{u}_z)\psi_1\,\delta(z-z_{\mathrm{o}}). \tag{2.146}$$

Solution of (D.4) gives, with the help of (D.5) to (D.13),

$$\boldsymbol{E} = \tfrac{1}{2}\mathrm{j}\omega[\mp\boldsymbol{P}'_{\mathrm{m}}\cdot\operatorname{grad}_{\mathrm{o}}\psi_1 + \mathrm{j}(\nu_1^2/\gamma_1)\boldsymbol{P}'_{\mathrm{m}}\cdot(\psi_1\boldsymbol{u}_z)_{\mathrm{o}}](\operatorname{grad}\psi_1\times\boldsymbol{u}_z)\mathrm{e}^{-\mathrm{j}\gamma_1|z-z_{\mathrm{o}}|}. \tag{2.147}$$

The corresponding magnetic field is $I_1(z)\operatorname{grad}\psi_1 + B_1(z)\psi_1\boldsymbol{u}_z$, where

$$I_1 = [-(\mathrm{j}\gamma_1/2\mu_0)\boldsymbol{P}'_{\mathrm{m}}\cdot\operatorname{grad}_{\mathrm{o}}\psi_1 \mp (\nu_1^2/2\mu_0)\boldsymbol{P}'_{\mathrm{m}}\cdot(\psi_1\boldsymbol{u}_z)_{\mathrm{o}}]\mathrm{e}^{-\mathrm{j}\gamma_1|z-z_{\mathrm{o}}|},$$
$$B_1 = [\pm(\nu_1^2/2\mu_0)\boldsymbol{P}'_{\mathrm{m}}\cdot\operatorname{grad}_{\mathrm{o}}\psi_1 - \tfrac{1}{2}\mathrm{j}(\nu_1^4/\gamma_1\mu_0)\boldsymbol{P}'_{\mathrm{m}}\cdot(\psi_1\boldsymbol{u}_z)_{\mathrm{o}} \tag{2.148}$$
$$- (\nu_1^2/\mu_0)\boldsymbol{P}'_{\mathrm{m}}\cdot\boldsymbol{u}_z\psi_1\,\delta(z-z_{\mathrm{o}})]\mathrm{e}^{-\mathrm{j}\gamma_1|z-z_{\mathrm{o}}|}.$$

It is interesting to compare these fields to those produced by the magnetic dipole moment

$$\boldsymbol{P}_{\mathrm{m}} = \tfrac{1}{2}\iiint \boldsymbol{r}\times\boldsymbol{J}\,\mathrm{d}V \tag{2.149}$$

generated by the electric currents equivalent to $\boldsymbol{K}$. A few elementary steps, similar to those used to prove (2.144), show that the two moments are related by

$$\boldsymbol{P}_{\mathrm{m}} = (1/\mu_0)\boldsymbol{P}'_{\mathrm{m}}. \tag{2.150}$$

A comparison of (2.83) and (2.147) immediately confirms that $\boldsymbol{P}'_{\mathrm{m}}$ and the equivalent $\boldsymbol{P}_{\mathrm{m}}$ produce the same electric field. The magnetic fields, however, do not coincide, which is in accordance with the conclusions reached in Section 2.12. The functions $I_1(z)$ and $B_1(z)$ relative to the field generated by $\boldsymbol{P}_{\mathrm{m}}$ are found to be

$$I_1 = [-\tfrac{1}{2}\mathrm{j}\gamma_1\boldsymbol{P}_{\mathrm{m}}\cdot\operatorname{grad}_{\mathrm{o}}\psi_1 \mp \tfrac{1}{2}\nu_1^2\boldsymbol{P}_{\mathrm{m}}\cdot(\psi_1\boldsymbol{u}_z)_{\mathrm{o}}$$
$$+ \boldsymbol{P}_{\mathrm{m}}\cdot\operatorname{grad}_{\mathrm{o}}\psi_1\,\delta(z-z_{\mathrm{o}})]\mathrm{e}^{-\mathrm{j}\gamma_1|z-z_{\mathrm{o}}|}, \tag{2.151}$$
$$B_1 = [\pm\tfrac{1}{2}\nu_1^2\boldsymbol{P}_{\mathrm{m}}\cdot\operatorname{grad}_{\mathrm{o}}\psi_1 - (\mathrm{j}\nu_1^4/2\gamma_1)\boldsymbol{P}_{\mathrm{m}}\cdot(\psi_1\boldsymbol{u}_z)_{\mathrm{o}}]\mathrm{e}^{-\mathrm{j}\gamma_1|z-z_{\mathrm{o}}|},$$

and these do not coincide with the corresponding functions in (2.148). The difference between the two is proportional to $\delta(z-z_{\mathrm{o}})$, a result which could have been expected from (2.102). The contributions of the other modes also differ by terms which contain $\delta(z-z_{\mathrm{o}})$. These contributions may not be neglected, because the other modes, although they are not propagated, exist at the level of the source in $\boldsymbol{r}_{\mathrm{o}}$. The sum of the various differences is

$$\boldsymbol{H}_J - \boldsymbol{H}_K = (1/\mu_{\mathrm{o}})\boldsymbol{P}'_{\mathrm{m}}\delta(\boldsymbol{r}-\boldsymbol{r}_{\mathrm{o}}) = \boldsymbol{P}_{\mathrm{m}}\delta(\boldsymbol{r}-\boldsymbol{r}_{\mathrm{o}}). \tag{2.152}$$

This result is in accordance with (2.102).

References

DEVANEY, A.J., and WOLF, E. (1973). Radiating and nonradiating classical current distributions and the fields they create. *Physical Review*, **D8**, 1044–7.

GRAY, C.G. (1978). Simplified derivation of the magnetostatic multipole expansion using the scalar potential. *American Journal of Physics*, **46**, 582–3.

GRAY, C.G. (1979). Magnetic multipole expansions using the scalar potential. *American Journal of Physics*, **47**, 457–9.

GRAY, C.G. (1980). Definition of the magnetic quadrupole moment. *American Journal of Physics*, **48**, 984–5.

GRAY, C.G., and GUBBINS, K. (1984). *Theory of molecular fluids*, Vol. 1: Fundamentals. Oxford University Press, Oxford.

KOCHER, C.A. (1978). Point-multipole expansions for charge and current distributions *American Journal of Physics*, **46**, 578–9.

LAKHTAKIA, A., VARADAN, V.K., and VARADAN, V.V. (1989). *Time-harmonic electromagnetic fields in chiral media*. Chapter 16. Springer, Berlin.

LINDELL, I.V., and NIKOSKINEN, K. (1986). Complex space multipole theory for scattering problems. *Proceedings of the URSI International Symposium of Electromagnetic Theory*, pp. 509–11. Akademiai Kado, Budapest.

LINDELL, I.V. (1987). Complex space multipole expansion theory with application to scattering from dielectric bodies. *IEEE Transactions on Antennas and Propagation*, **35**, 683–9.

LINDELL, I.V., and NIKOSKINEN, K. (1987). Complex space multipole theory for scattering and diffraction problems. *Radio Science*, **22**, 963–7.

LINDELL, I.V. (1988). TE-TM decomposition of electromagnetic sources. *IEEE Transactions on Antennas and Propagation*, **36**, 1382–8.

MAYES, P.E. (1958). The equivalence of electric and magnetic sources. *IRE Transactions on Antennas and Propagation*, **6**, 295–6.

MICHALSKI, K.A. (1986). Missing boundary conditions of electromagnetics. *Electronics Letters*, **22**, 921–2.

NAMIAS, V. (1977). Application of the Dirac delta function to electric charge and multipole distributions. *American Journal of Physics*, **45**, 624–30.

STRATTON, J.A. (1941). *Electromagnetic Theory*, pp. 431–8. McGraw-Hill, New York.

VAN BLADEL, J. (1975). On the resonances of a dielectric resonator of very high permittivity. *IEEE Transactions on Microwave Theory and Techniques*, **23**, 199–208.

VAN BLADEL, J. (1977). The multipole expansion revisited, *Archiv für Elektronik und Übertragungstechnik*, **31**, 407–11.

VAN BLADEL, J. (1985). *Electromagnetic Fields*. pp. 55–7, Hemisphere Publ. Co. Washington. Reprinted, with corrections, from a text published in 1964 by McGraw-Hill, New York.

VAN BLADEL, J. (1988*a*). On the hierarchy of terms in a multipole expansion. *Electronics Letters*, **24**, 492–3.

VAN BLADEL, J. (1988*b*). On the equivalence of electric and magnetic currents. *Archiv für Elektronik und Übertragungstechnik*, **42**, 314–5.

3
GREEN'S DYADICS

Infinitely concentrated sources create singular fields. A static point charge, for example, creates an electric field proportional to $1/R^2$, and this field becomes infinite at the source. In the present Chapter, we investigate the field singularities associated with other concentrated sources of charge, such as surface layers and layers of dipoles. We further extend the analysis to electric and magnetic sources of *current*. Central here is the problem of the Green's dyadic, a topic which has generated a flurry of publications since the early 1960s. The main characteristic of some of these dyadics is their strong singularity, of the order of $1/R^3$. There are several ways to take this singular behaviour into account. Our approach has been to take potentials and modal expansions as a basis, although other methods, e.g. those relying on spatial Fourier techniques, are equally useful. Because of space limitations, we shall focus our attention on the *basic* features of the theory, illustrated by the example of a time-harmonic source in vacuum. A more exhaustive presentation would necessarily include sources in material media, sources of arbitrary time-dependence, and even sources in motion (see e.g. Kastner 1987; De Zutter 1982; Faché *et al.* 1989; Weiglhofer 1989a; Lakhtakia *et al.* 1989).

The discussion of the properties of the Green's dyadic depends heavily on fundamental results obtained in potential theory. These are discussed at length in specialized text books, such as Sternberg *et al.* (1952), MacMillan (1958), Kellogg (1967), Günter (1967), Martensen (1968), Müller (1969), and Mikhlin (1970). We have summarized the most important aspects of the theory in Sections 3.1 to 3.9. The style is that of the applied mathematician, i.e. no attempt has been made to achieve advanced mathematical rigour.

3.1 Volume potential integrals

The problem of concern is the value of the volume integral

$$\phi(\mathbf{r}) = \frac{1}{4\pi\varepsilon_0} \iiint_V \frac{\rho(\mathbf{r}')}{|\mathbf{r} - \mathbf{r}'|} \, \mathrm{d}V' \tag{3.1}$$

when $\mathbf{r}$ is an interior point of the charged volume. In (3.1), ϕ is the potential, and $\rho(\mathbf{r})$ the volume charge density (Fig. 3.1). Integrals such as (3.1) are frequently encountered in mathematical physics. Their general form is

$$I = \iiint_V f(\mathbf{r}') \, \mathrm{d}V', \tag{3.2}$$

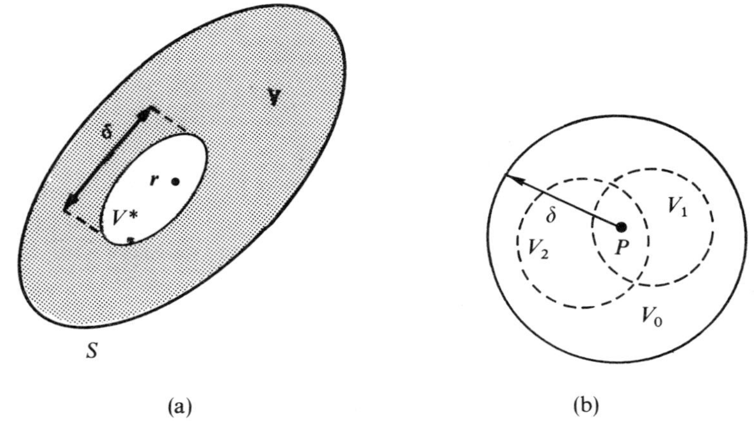

Fig. 3.1. (a) A volume V from which a small volume V^* has been removed. (b) details of small volumes surrounding P.

where $f(\mathbf{r}')$ is unbounded at $\mathbf{r}' = \mathbf{r}$. To give a meaning to I, we excise a small volume V^*, containing $\mathbf{r}$, from V. If the integral

$$I' = \iiint_{V-V^*} f(\mathbf{r}')\,\mathrm{d}V' \tag{3.3}$$

approaches a limit while the maximum chord δ of V^* approaches zero, we define this limit as the value of the (convergent) integral. To be unique, the limit must obviously be independent of the shape of V^*. Potential theory shows that the integral in (3.1) converges according to criterion (3.3) when $\rho(\mathbf{r}')$ is piecewise continuous. In fact, it converges for integrands containing the factor $|\mathbf{r} - \mathbf{r}'|^{-n}$, provided that $0 < n < 3$. We write

$$\phi(\mathbf{r}) = \lim_{\substack{V_\delta \to 0 \\ \delta \to 0}} \frac{1}{4\pi\varepsilon_0} \iiint_{V-V_\delta} \frac{\rho(\mathbf{r}')}{|\mathbf{r} - \mathbf{r}'|}\,\mathrm{d}V', \tag{3.4}$$

where V_δ is a small volume of maximum chord δ. The potential function thus defined turns out to be continuous in space. If the integral is convergent, then

$$\lim_{\delta \to 0} \frac{1}{4\pi\varepsilon_0} \iiint_{V_\delta} \frac{\rho(\mathbf{r}')}{|\mathbf{r} - \mathbf{r}'|}\,\mathrm{d}V' = 0. \tag{3.5}$$

This relationship means that the contribution from V_δ (the 'self cell') becomes negligibly small when δ approaches zero.

3.2 Single-layer potential

We shall assume, throughout the present chapter, that surfaces are 'well-behaved', a loose term which can be defined more accurately by means of concepts such as the 'regular surface' (Kellogg 1967) or the 'Lyapunov surface' (Mikhlin 1970). The most important examples of irregularities are edges and vertices: they are discussed extensively in Chapters 4 and 5. We shall also assume that the charge density ρ_s on the surface is piecewise continuous, leaving again to Chapters 4 and 5 the study of cases where ρ_s may become infinite. With these restrictions the potential at a point P_0 of S is defined as the improper, but convergent integral

$$\phi(\boldsymbol{r}) = \lim_{\delta \to 0} \frac{1}{4\pi\varepsilon_0} \iint_{S-\Sigma_\delta} \frac{\rho_s(\boldsymbol{r}')}{|\boldsymbol{r}-\boldsymbol{r}'|}\, \mathrm{d}S' . \tag{3.6}$$

Here, Σ_δ is a small area of arbitrary shape containing P_0, and of maximum chord δ. The convergence also holds when $|\boldsymbol{r}-\boldsymbol{r}'|$ is replaced by $|\boldsymbol{r}-\boldsymbol{r}'|^n$, but only if $0 < n < 2$.

The single-layer potential remains continuous throughout space. Its value (3.6) may be used to derive, for example, an integral equation for the charge density on a conductor raised to potential C. Thus,

$$\lim_{\delta \to 0} \frac{1}{4\pi\varepsilon_0} \iint_{S-\Sigma_\delta} \frac{\rho_s(\boldsymbol{r}')}{|\boldsymbol{r}-\boldsymbol{r}'|}\, \mathrm{d}S' = C \quad \text{for} \quad \boldsymbol{r} \text{ on } S. \tag{3.7}$$

In the numerical solution of this equation, δ must ideally be vanishingly small. In practice, however, S is divided into small elements S_i, one of which is taken to be Σ_δ. This element is *not* infinitely small, therefore its contribution to the integral might not be negligible. Fortunately, this contribution (the 'self-patch') is known for a few simple shapes of Σ_δ, associated with simple laws of variation of ρ_s. For a rectangle and a uniform ρ_s, for example (Fig. 3.2a) (Durand, 1964),

$$\phi(0) = \frac{\rho_s}{2\pi\varepsilon_0}\left(a \log_e \frac{b + \sqrt{(a^2+b^2)}}{a} + b \log_e \frac{a + \sqrt{(a^2+b^2)}}{b} \right), \tag{3.8}$$

$$\phi(Q) = \frac{\rho_s}{4\pi\varepsilon_0}\left(2a \log_e \frac{b + \sqrt{(4a^2+b^2)}}{2a} + b \log_e \frac{2a + \sqrt{(4a^2+b^2)}}{b} \right). \tag{3.9}$$

For a square, in particular,

$$\begin{aligned} \phi(0) &= (\rho_s a/\pi\varepsilon_0) \log_e (1 + \sqrt{2}) = 0.2805\, \rho_s a/\varepsilon_0, \\ \phi(Q) &= (\rho_s a/4\pi\varepsilon_0) \log_e (11 + 5\sqrt{5})/2 = 0.1915\, \rho_s a/\varepsilon_0 . \end{aligned} \tag{3.10}$$

On a circular disk carrying a uniform charge density (Fig. 3.2b),

$$\phi(r) = \rho_s a/\pi\varepsilon_0\, \mathrm{E}\,(r/a), \tag{3.11}$$

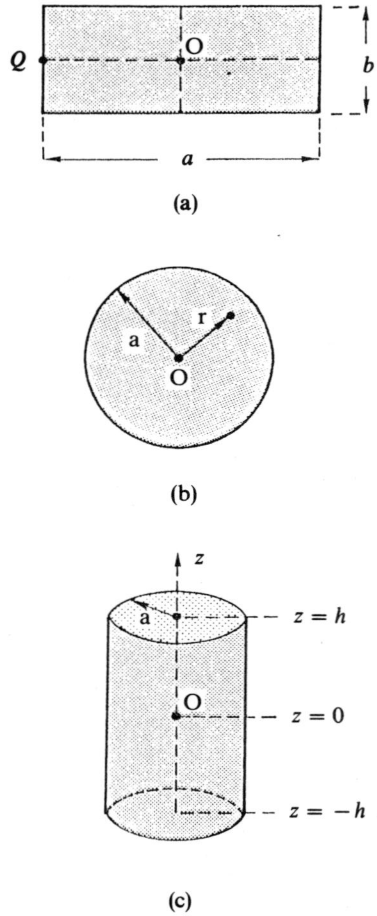

Fig. 3.2. Uniformly charged surfaces: (a) rectangular plate, (b) circular disk, (c) hollow metallic circular cylinder.

where

$$\mathrm{E}(x) = \int_0^{\frac{1}{2}\pi} \sqrt{(1 - x^2 \sin^2 \varphi)}\, \mathrm{d}\varphi. \tag{3.12}$$

At the centre of the disk,

$$\phi(0) = \rho_s a / 2\varepsilon_0. \tag{3.13}$$

At the central point 0 of a uniformly charged thin-walled cylinder (Fig. 3.2c),

$$\phi(0) = (\rho_s a/\varepsilon_0) \log_e \{h/a + \sqrt{[1 + (h/a)^2]}\}. \tag{3.14}$$

3.3 Double-layer potential

Let a surface S be covered with a layer of normal dipoles of density $\tau(\boldsymbol{r})$. Each element $\mathrm{d}S$ carries a dipole moment $\tau\,\boldsymbol{u}_{\mathrm{n}}\mathrm{d}S$, so that the potential at a point P located outside the double layer is given by (Fig. 3.3)

$$\begin{aligned}\phi(\boldsymbol{r}) &= \frac{1}{4\pi\varepsilon_0}\iint_S \frac{\tau(\boldsymbol{r}')\cos\theta'}{|\boldsymbol{r}-\boldsymbol{r}'|^2}\,\mathrm{d}S' \\ &= \frac{1}{4\pi\varepsilon_0}\iint_S \tau(\boldsymbol{r}')\frac{\partial}{\partial n'}\frac{1}{|\boldsymbol{r}-\boldsymbol{r}'|}\,\mathrm{d}S'.\end{aligned} \tag{3.15}$$

When $\tau(\boldsymbol{r})$ is piecewise continuous, the integral converges when $P = P_0$ is *on* S. The corresponding potential is sometimes called the *direct potential* $\phi_0(P_0)$. It is important to note that $\phi(\boldsymbol{r})$ does not approach the value $\phi_0(P_0)$ when P approaches the surface from points P_1 and P_2. The limits are, more precisely,

$$\begin{aligned}\lim_{P_1\to P_0}\ \phi(P_1) &= \tau(P_0)/2\varepsilon_0 + \phi_0(P_0), \\ \lim_{P_2\to P_0}\ \phi(P_2) &= -\tau(P_0)/2\varepsilon_0 + \phi_0(P_0),\end{aligned} \tag{3.16}$$

where

$$\phi_0(P_0) = \lim_{\delta\to 0}\frac{1}{4\pi\varepsilon_0}\iint_{S-\Sigma_\delta}\tau(\boldsymbol{r}')\frac{\partial}{\partial n'}\frac{1}{|\boldsymbol{r}-\boldsymbol{r}'|}\,\mathrm{d}S'. \tag{3.17}$$

Points P_1 and P_2 lie respectively on the positive and negative sides of S. In Fig. 3.3a the approach chosen follows the normal, but any other approach is equally valid. It is clear from (3.16) that $\phi(\boldsymbol{r})$ suffers a jump τ/ε_0 when passing through the layer from the negative to the positive side. This property is in harmony with the common conception of a double-layer representing a charged 'battery'.

In (3.16), the terms in $\tau/2\varepsilon_0$ stand for the contribution of the small element Σ_δ. These terms retain a nonzero value (i.e. they 'resist compression') when Σ_δ approaches zero. The reason becomes clear when (3.15) is rewritten in the equivalent form

$$\phi(\boldsymbol{r}) = \frac{1}{4\pi\varepsilon_0}\iint_S \tau(\boldsymbol{r}')\,\mathrm{d}\Omega'. \tag{3.18}$$

Here $\mathrm{d}\Omega'$ is the elementary solid angle subtended by $\mathrm{d}S'$ (Fig. 3.3b). This angle is positive when P is on the positive side of the dipoles, and negative in the opposite case. In the approach $P_1 \to P_0$, the solid angle subtended by Σ_δ approaches 2π, whereas for $P_2 \to P_0$ it approaches -2π. The (3.18) integrals then yield $\tau/2\varepsilon_0$ and $-\tau/2\varepsilon_0$ respectively, and this independently of the shape of Σ_δ. The argument may be extended to a point P_0 at the apex of a cone of opening solid angle Ω_0 (Fig. 3.3c). The limit solid angles are now $4\pi - \Omega_0$ for

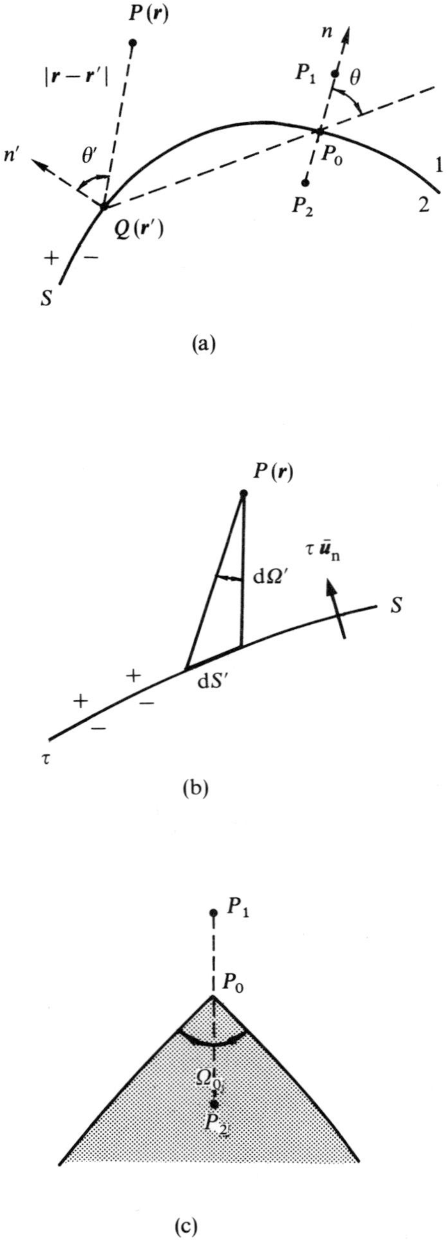

Fig. 3.3. (a) A surface S and relevant angles and distances. (b) A double layer and elementary solid angle. (c) A surface with a conical point P_0.

P_1, and $-\Omega_0$ for P_2. The first terms on the right-hand sides of (3.16) must now be written as

$$[\tau(P_0)/\varepsilon_0](1 - \Omega_0/4\pi) \quad \text{for } P_1, \qquad -[\tau(P_0)/\varepsilon_0](\Omega_0/4\pi) \quad \text{for } P_2. \tag{3.19}$$

The potential jump remains equal to τ/ε_0.

3.4 Extension to scalar radiation problems

The singular kernel encountered in scalar radiation problems is

$$G(\boldsymbol{r}|\boldsymbol{r}') = -\frac{1}{4\pi}\frac{\mathrm{e}^{-\mathrm{j}k|\boldsymbol{r}-\boldsymbol{r}'|}}{|\boldsymbol{r}-\boldsymbol{r}'|}. \tag{3.20}$$

At small distances $|\boldsymbol{r} - \boldsymbol{r}'|$,

$$G(\boldsymbol{r}|\boldsymbol{r}') = -\frac{1}{4\pi}\left(\frac{1}{|\boldsymbol{r}-\boldsymbol{r}'|} - \mathrm{j}k - \tfrac{1}{2}k^2|\boldsymbol{r}-\boldsymbol{r}'| + \cdots\right). \tag{3.21}$$

The singular behaviour of $G(\boldsymbol{r}|\boldsymbol{r}')$ is therefore identical to that of the kernel of potential theory, hence the results of the previous sections may be extended, mutatis mutandis, to the new kernel. To illustrate this remark by an application, consider a pressure wave P_i incident on an acoustically hard body (Fig. 3.4). The boundary condition is $\partial P/\partial n = 0$ on S, where P is the total pressure, sum of incident and scattered values. The scattered field $P_{\mathrm{sc}} = P - P_{\mathrm{i}}$ satisfies both Helmholtz' equation and the radiation condition at infinity. More precisely,

$$\nabla^2 P_{\mathrm{sc}} + k^2 P_{\mathrm{sc}} = 0 \quad (\boldsymbol{r} \text{ outside } S), \tag{3.22}$$

$$\lim_{R\to\infty} R\left(\frac{\partial P_{\mathrm{sc}}}{\partial R} + \mathrm{j}kP_{\mathrm{sc}}\right) = 0. \tag{3.23}$$

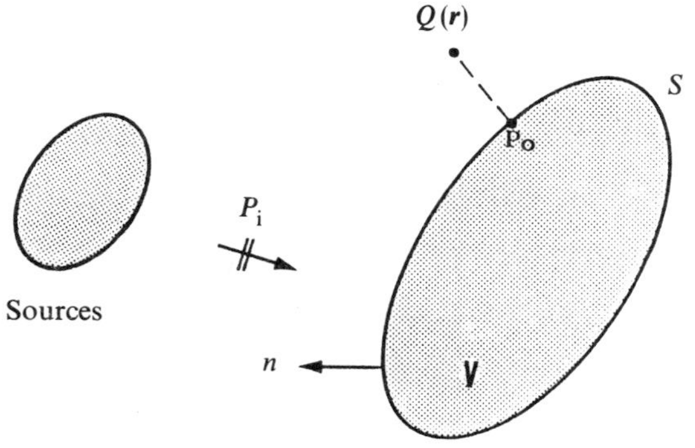

Fig. 3.4. The acoustic problem of a hard body immersed in an incident wave.

The Green's function $G(\boldsymbol{r}|\boldsymbol{r}')$, on the other hand, satisfies

$$\nabla^2 G + k^2 G = \delta(\boldsymbol{r} - \boldsymbol{r}'). \tag{3.24}$$

Multiplying (3.22) by G, (3.24) by P_{sc}, subtracting the results, and integrating outside S, yields

$$\begin{aligned}\iiint (G \nabla^2 P_{\text{sc}} - P_{\text{sc}} \nabla^2 G)\,\mathrm{d}V' &= -\iiint P_{\text{sc}}(\boldsymbol{r}')\delta(\boldsymbol{r} - \boldsymbol{r}')\,\mathrm{d}V' \\ &= -P_{\text{sc}}(\boldsymbol{r}) \quad (\boldsymbol{r} \text{ outside } S).\end{aligned} \tag{3.25}$$

We transform the first volume integral by means of Green's theorem. With the positive sense of the normal shown in Fig. 3.4, the theorem gives

$$P_{\text{sc}}(\boldsymbol{r}) = \iint_S \left(G \frac{\partial P_{\text{sc}}}{\partial n'} - P_{\text{sc}} \frac{\partial G}{\partial n'} \right) \mathrm{d}S'. \tag{3.26}$$

The integral over the outer boundary (a large sphere at infinity) does not appear here since it vanishes due to the radiation condition. A similar analysis may be applied to P_{i}, which satisfies (3.22) in V. The integration is now over V instead of over the exterior volume. The result is

$$0 = \iint_S \left(G \frac{\partial P_{\text{i}}}{\partial n'} - P_{\text{i}} \frac{\partial G}{\partial n'} \right) \mathrm{d}S'. \tag{3.27}$$

Adding (3.26) and (3.27) together gives, for $\boldsymbol{r}$ outside S,

$$P(\boldsymbol{r}) = P_{\text{i}}(\boldsymbol{r}) + \frac{1}{4\pi} \iint_S P(\boldsymbol{r}') \frac{\partial}{\partial n'} \frac{\mathrm{e}^{-\mathrm{j}k|\boldsymbol{r} - \boldsymbol{r}'|}}{|\boldsymbol{r} - \boldsymbol{r}'|} \mathrm{d}S'. \tag{3.28}$$

Let now Q approach a point P_0 on S. From (3.16), and for $\boldsymbol{r}$ on S,

$$\tfrac{1}{2}P(\boldsymbol{r}) - \frac{1}{4\pi} \lim_{\delta \to 0} \iint_{S - \Sigma_\delta} P(\boldsymbol{r}') \frac{\partial}{\partial n'} \frac{\mathrm{e}^{-\mathrm{j}k|\boldsymbol{r} - \boldsymbol{r}'|}}{|\boldsymbol{r} - \boldsymbol{r}'|} \mathrm{d}S' = P_{\text{i}}(\boldsymbol{r}). \tag{3.29}$$

This is a two-dimensional integral equation for the surface value of P. The reduction to two dimensions represents a significant simplification with respect to the original three-dimensional differential problem. Once (3.29) is solved (and several numerical methods are available for the purpose), the pressure outside S can be evaluated from (3.28).

3.5 Two-dimensional potential problems

The results of the preceding Sections can easily be extended to two-dimensional problems by means of the substitution replacing $1/4\pi|\boldsymbol{r} - \boldsymbol{r}'|$ by

$$\frac{1}{2\pi} \log_{\mathrm{e}} \frac{1}{|\boldsymbol{r} - \boldsymbol{r}'|} = \frac{1}{2\pi} \log_{\mathrm{e}} \frac{1}{\sqrt{[(x - x')^2 + (y - y')^2]}}. \tag{3.30}$$

The following relationships are often useful (Fig. 3.5a):

$$\frac{1}{2\pi}\frac{\partial}{\partial n}\log_e\frac{1}{|\boldsymbol{r}-\boldsymbol{r}'|} = -\frac{\cos\theta}{2\pi|\boldsymbol{r}-\boldsymbol{r}'|}, \tag{3.31}$$

$$\frac{1}{2\pi}\frac{\partial}{\partial n'}\log_e\frac{1}{|\boldsymbol{r}-\boldsymbol{r}'|} = -\frac{\cos\theta'}{2\pi|\boldsymbol{r}-\boldsymbol{r}'|}. \tag{3.32}$$

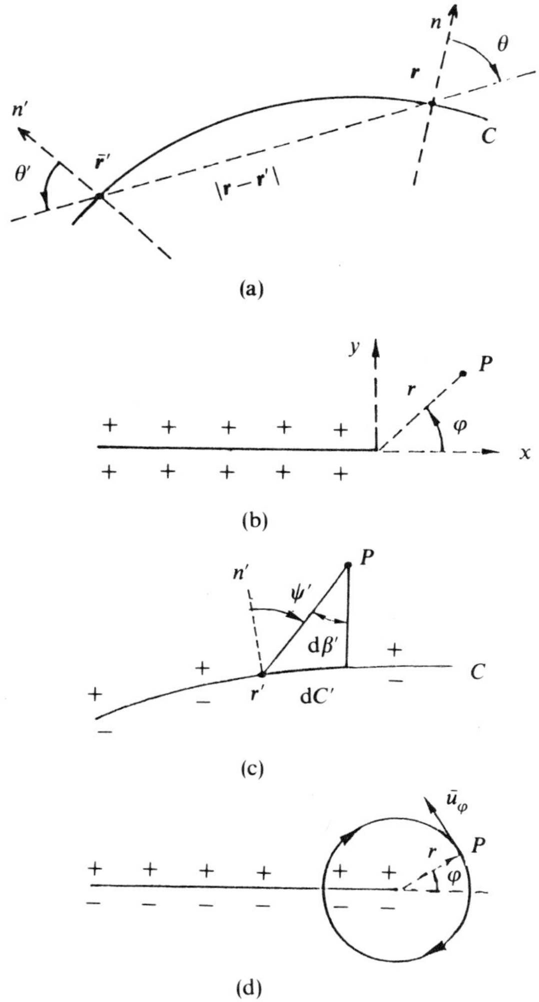

Fig. 3.5. Two-dimensional potentials: (a) curve C and relevant angles and distances, (b) half plane carrying a uniform ρ_s, (c) double layer on a curve, (d) half plane carrying a uniform τ.

The two kernels are obviously singular for $\mathbf{r} \to \mathbf{r}'$. Other types of singularity are also of interest. As a first example, take the field near a half plane carrying a uniform charge density ρ_s (Fig. 3.5b). Potential and field components near the edge are given by (Durand 1964)

$$\begin{gathered} \phi = (\rho_s/2\pi\varepsilon_0)(x \log_e r - y\varphi - x), \\ e_x = (\rho_s/2\pi\varepsilon_0) \log_e 1/r, \qquad e_y = (\rho_s/2\pi\varepsilon_0)\varphi. \end{gathered} \tag{3.33}$$

The singularities of ϕ and e_x are evident from the formulas. Another example is the potential of a dipole layer, given by (Fig. 3.5c)

$$\begin{aligned} \phi(\mathbf{r}) &= \frac{1}{2\pi\varepsilon_0} \int_C \tau(\mathbf{r}') \frac{\partial}{\partial n'} \log_e \frac{1}{|\mathbf{r} - \mathbf{r}'|} \, \mathrm{d}C' \\ &= \frac{1}{2\pi\varepsilon_0} \int_C \tau(\mathbf{r}') \frac{\cos \psi'}{|\mathbf{r} - \mathbf{r}'|} \, \mathrm{d}C' = \frac{1}{2\pi\varepsilon_0} \int_C \tau(\mathbf{r}') \, \mathrm{d}\beta'. \end{aligned} \tag{3.34}$$

Near the edge of a uniformly distributed layer on a half-plane, this expression yields (Fig. 3.5d)

$$\phi(P) = (\tau/2\pi\varepsilon_0)\varphi, \qquad \mathbf{e}(P) = -(\tau/2\pi\varepsilon_0 r)\mathbf{u}_\varphi. \tag{3.35}$$

The lines of force of $\mathbf{e}$ near the edge are clearly circular.

3.6 First derivatives of the volume potential

The electric field of a point charge is given by Coulomb's law. This law may be applied to elementary charges $\rho \, \mathrm{d}V$ to yield, for the field generated by a volume charge,

$$\mathbf{e}(\mathbf{r}) = \frac{1}{4\pi\varepsilon_0} \iiint_V \frac{\rho(\mathbf{r}')}{|\mathbf{r} - \mathbf{r}'|^3} (\mathbf{r} - \mathbf{r}') \, \mathrm{d}V'. \tag{3.36}$$

The integral in (3.36) is convergent and continuous throughout space. We shall exceptionally prove the convergence property in order to illustrate how such proofs are presented in potential theory (Kellogg 1967). We focus our attention on the x component of $\mathbf{e}$.

The necessary and sufficient condition for the existence of a limit is that, for any positive test number ε, there is a corresponding number $\delta > 0$ such that, if V_1 and V_2 are any two volumes contained in a sphere V_0 of radius δ about P (Fig. 3.1b),

$$\left| \iiint_{V-V_1} \frac{\rho(\mathbf{r}')(x - x')}{|\mathbf{r} - \mathbf{r}'|^3} \, \mathrm{d}V' - \iiint_{V-V_2} \frac{\rho(\mathbf{r}')(x - x')}{|\mathbf{r} - \mathbf{r}'|^3} \, \mathrm{d}V' \right| < \varepsilon. \tag{3.37}$$

Both volumes V_1 and V_2 must contain P. In (3.37) the part of V outside V_0, which is common to both integrals, may be excluded, and V replaced by V_0.

It now suffices to show that each integral in (3.37) can be made less in absolute value than $\frac{1}{2}\varepsilon$ by proper choice of δ. For the first integral in (3.37), for example, we use the inequality

$$\left|\iiint_{V_0-V_1} \frac{\rho(\mathbf{r}')(x-x')}{|\mathbf{r}-\mathbf{r}'|^3}\,\mathrm{d}V'\right| \leqslant A \iiint_{V_0-V_1} \frac{|x-x'|}{|\mathbf{r}-\mathbf{r}'|^3}\,\mathrm{d}V', \tag{3.38}$$

where A is an upper bound for $|\rho|$. The integral on the right hand side may be evaluated over V_0 by performing the integration in spherical coordinates centred on P. This yields $2\pi\delta$. An upper bound on the right-hand member of (3.38) is therefore $2\pi A\delta$, and this may be made less than $\frac{1}{2}\varepsilon$ by taking $\delta < \varepsilon/4\pi A$.

An important property of the electric field, given here without proof, is that $\boldsymbol{e}$ is minus the gradient of the potential. Thus,

$$\begin{aligned} \boldsymbol{e}(\mathbf{r}) &= -\operatorname{grad} \lim_{\delta\to 0} \frac{1}{4\pi\varepsilon_0} \iiint_{V-V_\delta} \frac{\rho(\mathbf{r}')}{|\mathbf{r}-\mathbf{r}'|}\,\mathrm{d}V' \\ &= \lim_{\delta\to 0} \frac{1}{4\pi\varepsilon_0} \iiint_{V-V_\delta} \frac{\rho(\mathbf{r}')}{|\mathbf{r}-\mathbf{r}'|^3}(\mathbf{r}-\mathbf{r}')\,\mathrm{d}V'. \end{aligned} \tag{3.39}$$

This property implies that the first derivatives of the volume potential are convergent. It also implies that differentiation and integration may be interchanged, i.e. that the derivative may be moved behind the integral sign. The right-hand side of (3.39) can indeed be written as

$$\iiint_V \frac{\rho(\mathbf{r}')(\mathbf{r}-\mathbf{r}')}{|\mathbf{r}-\mathbf{r}'|^3}\,\mathrm{d}V' = \iiint_V \rho(\mathbf{r}')\operatorname{grad}\left(-\frac{1}{|\mathbf{r}-\mathbf{r}'|}\right)\mathrm{d}V'. \tag{3.40}$$

The interchangeability of the two operators is remarkable. It may not, however, be extended blindly to derivatives higher than the first, a point discussed at length in Section 3.9.

3.7 First derivatives of the single- and double-layer potentials

The mere continuity of ρ_s is insufficient to ensure the existence of the limit of a *tangential* derivative of the single-layer potential when P_1 and P_2 approach P_0. The function must satisfy a Hölder condition in P_0, i.e. three positive constants c, A, and α must exist such that

$$|\rho_s(Q) - \rho_s(P_0)| < A|\overline{P_0Q}|^{\alpha}, \tag{3.41}$$

for all points Q on the surface for which $|\overline{P_0Q}| < c$. If this condition is satisfied (which normally is the case in practice, except at singular points) the tangential components of $\boldsymbol{e}$ exist, and are continuous across S. The *normal* component, however, suffers a jump equal to ρ_s/ε_0, a well-known boundary condition from electrostatics. More precisely (Fig. 3.3a),

$$\lim_{P_1 \to P_0} \frac{\partial}{\partial n}\left(\frac{1}{4\pi\varepsilon_0}\iint_S \frac{\rho_s(\boldsymbol{r}')}{|\boldsymbol{r}-\boldsymbol{r}'|}\,dS'\right) = -\frac{\rho_s(P_0)}{\varepsilon_0} + \lim_{\delta\to 0}\frac{1}{4\pi\varepsilon_0}\iint_{S-\Sigma_\delta} \rho_s(\boldsymbol{r}')\frac{\partial}{\partial n}\frac{1}{|\boldsymbol{r}-\boldsymbol{r}'|}\,dS', \tag{3.42}$$

$$\lim_{P_2 \to P_0} \frac{\partial}{\partial n}\left(\frac{1}{4\pi\varepsilon_0}\iint_S \frac{\rho_s(\boldsymbol{r}')}{|\boldsymbol{r}-\boldsymbol{r}'|}\,dS'\right) = \frac{\rho_s(P_0)}{\varepsilon_0} + \lim_{\delta\to 0}\frac{1}{4\pi\varepsilon_0}\iint_{S-\Sigma_\delta} \rho_s(\boldsymbol{r}')\frac{\partial}{\partial n}\frac{1}{|\boldsymbol{r}-\boldsymbol{r}'|}\,dS'. \tag{3.43}$$

The terms in $\rho_s(P_0)/\varepsilon_0$ represents the contribution from the small surface Σ_δ around P_0. In the integrals the kernel can be written as

$$\frac{\partial}{\partial n}\frac{1}{|\boldsymbol{r}-\boldsymbol{r}'|} = \boldsymbol{u}_n \cdot \text{grad}\,\frac{1}{|\boldsymbol{r}-\boldsymbol{r}'|} = -\frac{\cos\theta}{|\boldsymbol{r}-\boldsymbol{r}'|^2}. \tag{3.44}$$

Relationships (3.42) and (3.43) remain valid when $\rho_s(\boldsymbol{r})$ is merely continuous, but the approach $P_{1,2} \to P_0$ must now take place along the normal (Mikhlin 1970).

As an example of application of (3.43), let medium 2 in Fig. 3.3a be perfectly conducting, and consider the approach $P_2 \to P_0$. In the conductor the potential is constant, which implies that the left-hand member of (3.43) is zero. This condition leads to an integral equation for ρ_s, viz.

$$\rho_s(\boldsymbol{r}) + \frac{1}{4\pi}\lim_{\delta\to 0}\iint_{S-\Sigma_\delta} \rho_s(\boldsymbol{r}')\frac{\partial}{\partial n}\frac{1}{|\boldsymbol{r}-\boldsymbol{r}'|}\,dS' = 0 \quad (\boldsymbol{r} \text{ on } S). \tag{3.45}$$

This is Robin's integral equation, which determines ρ_s to within a multiplicative factor which can easily be found once the total charge is given (MacMillan 1958; Martensen 1968).

The normal derivative of the double-layer potential is continuous, i.e. it approaches identical limits when P_1 and P_2 approach P_0. Because the potential jump is equal to τ/ε_0, there must be a jump in the *tangential* derivatives of the potential when τ is variable along the surface. Thus, in the limit $P_{1,2} \to P_0$,

$$\begin{aligned} \boldsymbol{e}_{\text{tang}}(P_1) - \boldsymbol{e}_{\text{tang}}(P_2) &= (-\,\text{grad}_s\,\phi)_{P_1} - (-\,\text{grad}_s\,\phi)_{P_2} \\ &= -\,\text{grad}_s[\phi(P_1) - \phi(P_2)] \\ &= -\,(1/\varepsilon_0)\,\text{grad}_s\,\tau(P_0). \end{aligned} \tag{3.46}$$

3.8 First derivatives of the vector potential

The vector potential of time-independent currents is given by (2.39), rewritten here for convenience:

$$\boldsymbol{a}(\boldsymbol{r}) = \frac{\mu_0}{4\pi} \iiint_V \frac{\boldsymbol{j}(\boldsymbol{r}')}{|\boldsymbol{r} - \boldsymbol{r}'|} \, dV'. \tag{3.47}$$

The components of $\boldsymbol{a}$ have the form of scalar potentials, which means that their derivatives are governed by the rules discussed in Sections 3.6 and 3.7. Let us investigate how the magnetic field generated by a surface current $\boldsymbol{j}_s$ behaves in the vicinity of S. At an exterior point $\boldsymbol{r}$, the magnetic field $\boldsymbol{h}$ is given by

$$\boldsymbol{h}(\boldsymbol{r}) = \operatorname{curl} \boldsymbol{a} = \frac{1}{4\pi} \iint_S \boldsymbol{j}_s(\boldsymbol{r}') \times \operatorname{grad}' \frac{1}{|\boldsymbol{r} - \boldsymbol{r}'|} \, dS', \tag{3.48}$$

where grad′ implies differentiation with respect to the primed coordinates. The magnetic field approaches different limits, depending on whether the surface is approached from P_1 or P_2. More precisely (Fig. 3.3a),

$$\lim_{P_1 \to P_0} \boldsymbol{h}(P_1) = \tfrac{1}{2}\boldsymbol{j}_s \times \boldsymbol{u}_n(P_0) + \lim_{\delta \to 0} \frac{1}{4\pi} \iint_{S-\Sigma_\delta} \boldsymbol{j}_s(\boldsymbol{r}') \times \operatorname{grad}' \frac{1}{|\boldsymbol{r} - \boldsymbol{r}'|} \, dS, \tag{3.49}$$

$$\lim_{P_2 \to P_0} \boldsymbol{h}(P_2) = -\tfrac{1}{2}\boldsymbol{j}_s \times \boldsymbol{u}_n(P_0) + \lim_{\delta \to 0} \frac{1}{4\pi} \iint_{S-\Sigma_\delta} \boldsymbol{j}_s(\boldsymbol{r}') \times \operatorname{grad}' \frac{1}{|\boldsymbol{r} - \boldsymbol{r}'|} \, dS'. \tag{3.50}$$

Clearly, the normal component of $\boldsymbol{h}$ is continuous, but the tangential components experience a jump

$$\lim_{P_1, P_2 \to P_0} [\boldsymbol{h}(P_1) - \boldsymbol{h}(P_2)] = \boldsymbol{j}_s \times \boldsymbol{u}_n(P_0). \tag{3.51}$$

This is the well-known boundary condition for $\boldsymbol{h}_{\text{tang}}$.

As an application of (3.49), let us derive an integral equation for the current density on a perfect conductor immersed in a time-harmonic incident field (Fig. 3.6). We use capital letters for the complex representation of fields and sources. The scattered magnetic field $\boldsymbol{H}_s$ is given by (3.48), with one modification: the static kernel $|\boldsymbol{r} - \boldsymbol{r}'|^{-1}$ must be replaced by $|\boldsymbol{r} - \boldsymbol{r}'|^{-1} \exp(-jk|\boldsymbol{r} - \boldsymbol{r}'|)$. A similar substitution must be effected in (3.49) and (3.50). The boundary condition at the surface of a perfect conductor requires that

$$\lim_{P_1 \to P_0} \boldsymbol{u}_n \times \boldsymbol{H} = \lim_{P_1 \to P_0} (\boldsymbol{u}_n \times \boldsymbol{H}_s) + \boldsymbol{u}_n \times \boldsymbol{H}_i = \boldsymbol{J}_s(P_0). \tag{3.52}$$

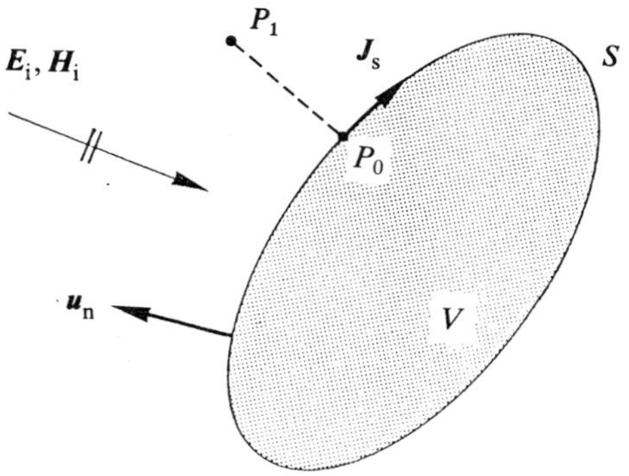

Fig. 3.6. Scattering of an incident electromagnetic wave by a perfect conductor.

Combining (3.49) and (3.52) gives rise to the 'magnetic-field integral equation' (the MFIE)

$$\boldsymbol{J}_s(\boldsymbol{r}) - \frac{1}{2\pi}\boldsymbol{u}_n(\boldsymbol{r}) \times \lim_{\delta \to 0} \iint_{S-\Sigma_\delta} \boldsymbol{J}_s(\boldsymbol{r}') \times \operatorname{grad}' \frac{e^{-jk|\boldsymbol{r}-\boldsymbol{r}'|}}{|\boldsymbol{r}-\boldsymbol{r}'|}\,dS' = 2\boldsymbol{u}_n \times \boldsymbol{H}_i(\boldsymbol{r}) \quad (\boldsymbol{r} \text{ on } S). \tag{3.53}$$

3.9 Second derivatives of the volume potential

The second derivatives of concern are of the form

$$I_1 = \frac{\partial^2}{\partial x_i \partial x_j}\left(\frac{1}{4\pi\varepsilon_0}\iiint_V \frac{\rho(\boldsymbol{r}')}{|\boldsymbol{r}-\boldsymbol{r}'|}\,dV'\right) \quad (i,j = 1, 2, 3). \tag{3.54}$$

These derivatives exist at an interior point P provided ρ satisfies a Hölder condition, i.e. provided there are three positive constants c, A, and α, such that

$$|\rho(Q) - \rho(P)| < A|\overline{PQ}|^\alpha \tag{3.55}$$

for all points Q for which $|\overline{PQ}| < c$. Under these conditions the second derivatives satisfy Poisson's equation

$$\nabla^2\phi = -\rho/\varepsilon_0. \tag{3.56}$$

The second derivatives appearing in (3.54) are very sensitive to small errors in the numerical evaluation of the potential (the function between brackets). It would be preferable, consequently, to let the derivatives operate on analytically known functions. This can be achieved, formally at least, by bringing them behind the integral sign, where they find the $\boldsymbol{r}$-dependent function $1/|\boldsymbol{r} - \boldsymbol{r}'|$ on which to operate. Great care must be exercised, however, when $\boldsymbol{r}$ is an *interior* point of the charge-carrying volume. For such case the second derivative produces a $|\mathbf{r}-\mathbf{r}'|^{-3}$ type of singularity, which is not generally integrable. Methods to get around this difficulty are based on either classical analysis or the theory of generalised functions (Gel'fand *et al.* 1964; Günter 1967; Asvestas 1983). An approach which fits our purpose particularly well consists in splitting the second derivative into three terms. For the radiation problem the splitting is (Lee *et al.* 1980; Asvestas 1983):

$$I_1 = \frac{\partial^2}{\partial x_i \partial x_j} \lim_{\delta \to 0} \iiint_{V-V_\delta} \rho(\mathbf{r}') \frac{e^{-jk|\mathbf{r}-\mathbf{r}'|}}{|\mathbf{r}-\mathbf{r}'|} \, dV' = A_{ij} + B_{ij} + C_{ij}, \quad (3.57)$$

where

$$A_{ij} = \iiint_{V-V^*} \rho(\mathbf{r}') \frac{\partial^2}{\partial x_i' \partial x_j'} \frac{e^{-jk|\mathbf{r}-\mathbf{r}'|}}{|\mathbf{r}-\mathbf{r}'|} \, dV',$$

$$B_{ij} = \rho(\mathbf{r}) \frac{\partial^2}{\partial x_i \partial x_j} \iiint_{V^*} \frac{1}{|\mathbf{r}-\mathbf{r}'|} \, dV',$$

$$C_{ij} = \iiint_{V^*} \left(\rho(\mathbf{r}') \frac{\partial^2}{\partial x_i' \partial x_j'} \frac{e^{-jk|\mathbf{r}-\mathbf{r}'|}}{|\mathbf{r}-\mathbf{r}'|} - \rho(\mathbf{r}) \frac{\partial^2}{\partial x_i' \partial x_j'} \frac{1}{|\mathbf{r}-\mathbf{r}'|} \right) dV'.$$

In the above integrals, V^* is an arbitrary fixed volume, which is only required to contain the field point $\mathbf{r}$. In particular, V^* need not be small, and there is no need to take limits as its size $|V^*|$ tends to zero.

We shall discuss the three terms in (3.57) in sequence. The integral A_{ij} does not give any trouble: it is convergent because the integrand remains finite. In B_{ij}, the second derivative has been carefully kept outside the integral. The latter is proportional to the electrostatic potential due to a charge distribution of unit density in V^*, and should therefore be interpreted in the sense of (3.4). The term B_{ij} can be rewritten in an alternate form by first performing the differentiation with respect to x_j. From Section 3.6,

$$\begin{aligned} \frac{\partial}{\partial x_j} \iiint_{V^*} \frac{1}{|\mathbf{r}-\mathbf{r}'|} \, dV' &= \mathbf{u}_j \cdot \iiint_{V^*} \operatorname{grad} \frac{1}{|\mathbf{r}-\mathbf{r}'|} \, dV' \\ &= -\mathbf{u}_j \cdot \iiint_{V^*} \operatorname{grad}' \frac{1}{|\mathbf{r}-\mathbf{r}'|} \, dV' = -\mathbf{u}_j \cdot \iint_{\Sigma^*} \frac{\mathbf{u}_n(\mathbf{r}')}{|\mathbf{r}-\mathbf{r}'|} \, dS'. \end{aligned} \quad (3.58)$$

In this equation, $\mathbf{u}_j$ is a unit vector in the x_j direction, and Σ^* is the boundary surface of V^* (Fig. 3.7). We now differentiate the integral with respect to x_i. This gives

$$\begin{aligned} \frac{\partial}{\partial x_i} \iint_{\Sigma^*} \frac{\mathbf{u}_n(\mathbf{r}') \, d\Sigma'}{|\mathbf{r}-\mathbf{r}'|} &= \mathbf{u}_i \cdot \iint_{\Sigma^*} \left(\operatorname{grad} \frac{1}{|\mathbf{r}-\mathbf{r}'|} \right) \mathbf{u}_n(\mathbf{r}') \, dS' \\ &= \mathbf{u}_i \cdot \iint_{\Sigma^*} \frac{\mathbf{u}_R(\mathbf{r}') \mathbf{u}_n(\mathbf{r}')}{|\mathbf{r}-\mathbf{r}'|^2} \, dS'. \end{aligned} \quad (3.59)$$

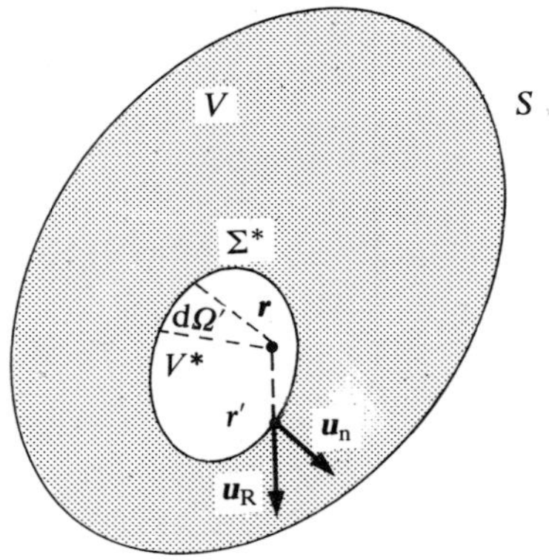

Fig. 3.7. A small volume V^* containing the field point $\boldsymbol{r}$.

The unit vector $\boldsymbol{u}_R$ is in the $\boldsymbol{r}' - \boldsymbol{r}$ direction. Combining (3.58) and (3.59) yields

$$\frac{\partial^2}{\partial x_i \partial x_j} \iiint_{V^*} \frac{1}{|\boldsymbol{r} - \boldsymbol{r}'|} \, \mathrm{d}V' = - \iint_{\Sigma^*} \frac{[\boldsymbol{u}_i \cdot \boldsymbol{u}_\mathrm{R}(\boldsymbol{r}')][\boldsymbol{u}_\mathrm{n}(\boldsymbol{r}') \cdot \boldsymbol{u}_j]}{|\boldsymbol{r} - \boldsymbol{r}'|^2} \, \mathrm{d}S'. \quad (3.60)$$

Finally, the term C_{ij} converges because $\rho(\boldsymbol{r})$ satisfies a Hölder condition, and because the singularities of the kernels neutralize each other. This becomes clear when we write the integrand, in the limit $|\boldsymbol{r} - \boldsymbol{r}'| \to 0$, as

$$\{\rho(\boldsymbol{r}) + [\rho(\boldsymbol{r}') - \rho(\boldsymbol{r})]\} \frac{\partial^2}{\partial x_i' \partial x_j'} \left(\frac{1}{|\boldsymbol{r} - \boldsymbol{r}'|} - \mathrm{j}k + \cdots \right) - \rho(\boldsymbol{r}) \frac{\partial^2}{\partial x_i' \partial x_j'} \frac{1}{|\boldsymbol{r} - \boldsymbol{r}'|}. \quad (3.61)$$

This form shows that the singularities in $|\boldsymbol{r} - \boldsymbol{r}'|^{-3}$ cancel out, leaving only terms in $|\boldsymbol{r} - \boldsymbol{r}'|^{-3+\alpha}$, in view of the Hölder condition (3.55). The integral over V^* therefore approaches zero together with δ, maximum chord of V.

3.10 The L dyadic

The form of (3.57) shows that the second derivative cannot simply be transferred behind the integral sign. Such a procedure would only produce a term such as A_{ij}. The equation also shows that the values of A_{ij}, B_{ij}, and C_{ij} depend on the choice of V^*, but that the sum of these three terms must remain constant. Finally, we notice that the term B_{ij}, which embodies the effect of the fundamental singularity, depends only on the *shape* of V^*, and not on the linear dimensions of the volume. More specifically, B_{ij} remains invariant when V^* and Σ^* are magnified or reduced through a similarity transformation (multiplication) with center at $\boldsymbol{r}$. This becomes clear when we rewrite B_{ij} as

$$B_{ij} = \frac{\partial^2}{\partial x_i \partial x_j} \iiint_{V^*} \frac{1}{|\boldsymbol{r} - \boldsymbol{r}'|} \, \mathrm{d}V' = -4\pi \, \boldsymbol{u}_i \cdot \mathbf{L}_{V^*} \cdot \boldsymbol{u}_j, \tag{3.62}$$

where $\mathbf{L}_{V^*}$ is the dyadic (Fig. 3.7)

$$\mathbf{L}_{V^*} = \frac{1}{4\pi} \iint_{\Sigma^*} \frac{\boldsymbol{u}_\mathrm{R}(\boldsymbol{r}')\boldsymbol{u}_\mathrm{n}(\boldsymbol{r}')}{|\boldsymbol{r} - \boldsymbol{r}'|^2} \, \mathrm{d}S' = \frac{1}{4\pi} \iint_{4\pi} \boldsymbol{u}_\mathrm{R} \boldsymbol{u}_\mathrm{n} (\boldsymbol{u}_\mathrm{R} \cdot \boldsymbol{u}_\mathrm{n}) \, \mathrm{d}\Omega'. \tag{3.63}$$

The formula confirms that, everything else being equal, $\mathbf{L}_{V^*}$ depends only on the shape of V^*, and not on a general scale factor. The explicit value of $\mathbf{L}_{V^*}$ is available for a few simple shapes (Yaghjian 1980; Lee *et al.* 1980). For a sphere centred on $\boldsymbol{r}$, for example,

$$\mathbf{L}_{V^*} = \tfrac{1}{3} \mathbf{I}. \tag{3.64}$$

The same value holds for a cube centred at $\boldsymbol{r}$, and arbitrarily oriented. For the circular cylinder shown in Fig. 3.8a, centred on $\boldsymbol{r}$,

$$\begin{aligned} \mathbf{L}_{V^*} &= \frac{h}{2\sqrt{(a^2 + h^2)}} (\boldsymbol{u}_x \boldsymbol{u}_x + \boldsymbol{u}_y \boldsymbol{u}_y) + \left(1 - \frac{h}{\sqrt{(a^2 + h^2)}}\right) \boldsymbol{u}_z \boldsymbol{u}_z \\ &= (\tfrac{1}{2} \cos\theta_0) \mathbf{I}_{xy} + (1 - \cos\theta_0) \boldsymbol{u}_z \boldsymbol{u}_z, \end{aligned} \tag{3.65}$$

where $\mathbf{I}_{xy} = \boldsymbol{u}_x \boldsymbol{u}_x + \boldsymbol{u}_y \boldsymbol{u}_y$.

For a very thin needle, of circular or square cross-section,

$$\mathbf{L}_{V^*} = \tfrac{1}{2}(\boldsymbol{u}_x \boldsymbol{u}_x + \boldsymbol{u}_y \boldsymbol{u}_y) = \tfrac{1}{2} \mathbf{I}_{xy}. \tag{3.66}$$

For a very flat cylindrical box, of arbitrary cross-section,

$$\mathbf{L}_{V^*} = \boldsymbol{u}_z \boldsymbol{u}_z. \tag{3.67}$$

Notice that expressions (3.66) and (3.67) are only mathematical limits, since they correspond to surfaces enclosing zero volume. For a rectangular box centred at $\boldsymbol{r}$ (Fig. 3.8b),

$$\mathbf{L}_{V^*} = \frac{1}{4\pi} (\Omega_x \boldsymbol{u}_x \boldsymbol{u}_x + \Omega_y \boldsymbol{u}_y \boldsymbol{u}_y + \Omega_z \boldsymbol{u}_z \boldsymbol{u}_z), \tag{3.68}$$

where Ω_x, Ω_y, and Ω_z are twice the solid angle subtended at $\boldsymbol{r}$ by a side perpendicular to the x, y, and z directions, respectively. The sum of the three solid angles is equal to 4π, which implies that the trace of $\mathbf{L}_{V^*}$ is unity. This property, and the symmetric character of the dyadic, are very general, and hold for arbitrary shapes of V^*.

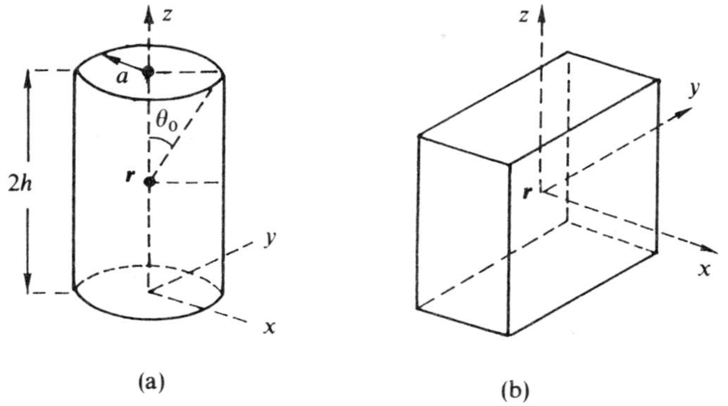

Fig. 3.8. Typical volumes V^*: (a) circular cylinder, (b) rectangular box.

3.11 Green's dyadic for the electric field: exterior points

The electric field in vacuum satisfies the vector Helmholtz equation

$$-\operatorname{curl}\operatorname{curl}\boldsymbol{E}+k^2\boldsymbol{E}=\mathrm{j}\omega\mu_0\boldsymbol{J} \tag{3.69}$$

as well as the radiation condition

$$\lim_{R\to\infty} R(\boldsymbol{u}_R\times\operatorname{curl}\boldsymbol{E}-\mathrm{j}k\boldsymbol{E})=0. \tag{3.70}$$

Our goal is to derive an expression for $\boldsymbol{E}(\boldsymbol{r})$ in terms of $\boldsymbol{J}(\boldsymbol{r})$, the given current density.

A first possible solution is based on the Helmholtz equations satisfied by $\boldsymbol{E}$, viz.

$$\begin{aligned}\nabla^2\boldsymbol{E}+k^2\boldsymbol{E}&=\operatorname{grad}\operatorname{div}\boldsymbol{E}-\operatorname{curl}\operatorname{curl}\boldsymbol{E}+k^2\boldsymbol{E}\\&=-(1/\mathrm{j}\omega\varepsilon_0)\operatorname{grad}\operatorname{div}\boldsymbol{J}+\mathrm{j}\omega\mu_0\boldsymbol{J},\end{aligned} \tag{3.71}$$

$$\nabla^2(\boldsymbol{E}+\boldsymbol{J}/\mathrm{j}\omega\varepsilon_0)+k^2[\boldsymbol{E}+(\boldsymbol{J}/\mathrm{j}\omega\varepsilon_0)]=-(1/\mathrm{j}\omega\varepsilon_0)\operatorname{curl}\operatorname{curl}\boldsymbol{J}. \tag{3.72}$$

The solution of the vector Helmholtz equation

$$\nabla^2\boldsymbol{B}+k^2\boldsymbol{B}=\boldsymbol{C} \tag{3.73}$$

which satisfies the radiation condition is well known. From Section 3.4, it is

$$\boldsymbol{B}(\boldsymbol{r})=-\frac{1}{4\pi}\iiint_V \boldsymbol{C}(\boldsymbol{r}')\,\frac{\mathrm{e}^{-\mathrm{j}k|\boldsymbol{r}-\boldsymbol{r}'|}}{|\boldsymbol{r}-\boldsymbol{r}'|}\,\mathrm{d}V'. \tag{3.74}$$

This expression can be rewritten in terms of a Green's dyadic as

$$\begin{aligned} B(r) &= \iiint_V \left(-\frac{1}{4\pi} \frac{e^{-jk|r-r'|}}{|r-r'|} \mathbf{I} \right) \cdot C(r')\, dV' \\ &= \iiint_V \mathbf{G}(r|r') \cdot C(r')\, dV', \end{aligned} \tag{3.75}$$

where $\mathbf{G}(r|r') = G(r|r')\mathbf{I}$.

It now suffices to replace C by the second members shown in (3.71) and (3.72) to obtain $E(r)$. The solution involves an integration, which was our original purpose. But second derivatives of J are present as well, and this may lead to important errors if J is not known with great accuracy, i.e. if there is non-negligible 'noise' on J. It is easy, however, to get rid of the second derivatives by expressing E in terms of scalar and vector potentials. In the Lorenz gauge,

$$E = -j\omega A - \text{grad}\,\phi = -j\omega A + \frac{1}{j\omega\varepsilon_0\mu_0} \text{grad div}\, A. \tag{3.76}$$

The potentials are specifically (Fig. 3.9a)

$$\begin{aligned} A(r) &= -\mu_0 \iiint_V J(r') G(r|r')\, dV' \\ \phi(r) &= -\frac{1}{j\omega\varepsilon_0\mu_0} \text{div}\, A = \frac{1}{j\omega\varepsilon_0} \iiint_V \text{div}\,[G(r|r') J(r')]\, dV' \\ &= \frac{1}{j\omega\varepsilon_0} \iiint_V \text{grad}\, G(r|r') \cdot J(r')\, dV \\ &= \frac{1}{j\omega\varepsilon_0} \iiint_V G(r|r')\, \text{div}'\, J(r')\, dV' - \frac{1}{j\omega\varepsilon_0} \iint_S G(r|r') u_n \cdot J(r')\, dS'. \end{aligned} \tag{3.77}$$

The first derivatives of J still appear in the last form of ϕ, namely in the term div J, but they are not present in the previous form. From (3.76),

$$\begin{aligned} E(r) = j\omega\mu_0 &\iiint_V G(r|r') J(r')\, dV' \\ &- \frac{1}{j\omega\varepsilon_0} \text{grad} \iiint_V \text{grad}\, G(r|r') \cdot J(r')\, dV'. \end{aligned} \tag{3.78}$$

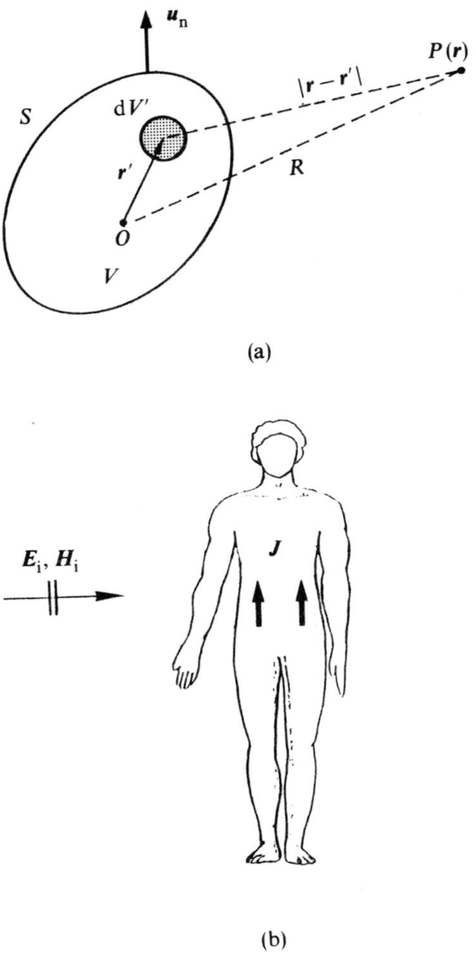

Fig. 3.9. (a) A point $\boldsymbol{P}$ exterior to a current-carrying volume V. (b) Currents induced in a conducting body by an incident wave $\boldsymbol{E}_i$, $\boldsymbol{H}_i$.

When P lies outside the current-carrying volume, the derivatives may safely be brought behind the integral. Thus utilizing definition (B.22) for grad $\boldsymbol{a}$ yields

$$\begin{aligned}\boldsymbol{E}(\boldsymbol{r}) = \mathrm{j}\omega\mu_0 & \iiint_V G(\boldsymbol{r}|\boldsymbol{r}')\boldsymbol{J}(\boldsymbol{r}')\,\mathrm{d}V' \\ & - \frac{1}{\mathrm{j}\omega\Sigma_0}\iiint_V \operatorname{grad}\operatorname{grad} G(\boldsymbol{r}|\boldsymbol{r}')\cdot\boldsymbol{J}(\boldsymbol{r}')\,\mathrm{d}V'.\end{aligned} \tag{3.79}$$

More compactly,

$$\boldsymbol{E}(\boldsymbol{r}) = \mathrm{j}\omega\mu_0 \iiint_V \mathbf{G}_e^0(\boldsymbol{r}|\boldsymbol{r}')\cdot \boldsymbol{J}(\boldsymbol{r}')\,\mathrm{d}V', \tag{3.80}$$

where $\mathbf{G}_e^0$ is the symmetric dyadic

$$\begin{aligned}\mathbf{G}_e^0(\boldsymbol{r}|\boldsymbol{r}') &= \left(\mathbf{I} + \frac{1}{k^2}\operatorname{grad}\operatorname{grad}\right) G(\boldsymbol{r}|\boldsymbol{r}') \\ &= -\frac{1}{4\pi}\left(\mathbf{I} + \frac{1}{k^2}\operatorname{grad}\operatorname{grad}\right)\frac{\mathrm{e}^{-\mathrm{j}k|\boldsymbol{r}-\boldsymbol{r}'|}}{|\boldsymbol{r}-\boldsymbol{r}'|}.\end{aligned} \tag{3.81}$$

Equation (3.80) satisfies our initial goal, which was to express $\boldsymbol{E}$ in terms of $\boldsymbol{J}$, all derivatives of $\boldsymbol{J}$ excluded.

3.12 Interpretation of the elements of $\mathbf{G}_e^0$

The symmetric $\mathbf{G}_e^0$ dyadic can be represented in terms of row and column vectors as

$$\begin{aligned}\mathbf{G}_e^0(\boldsymbol{r}|\boldsymbol{r}') &= \boldsymbol{G}_{ex}^0(\boldsymbol{r}|\boldsymbol{r}')\boldsymbol{u}_x + \boldsymbol{G}_{ey}^0(\boldsymbol{r}|\boldsymbol{r}')\boldsymbol{u}_y + \boldsymbol{G}_{ez}^0(\boldsymbol{r}|\boldsymbol{r}')\boldsymbol{u}_z \\ &= \boldsymbol{u}_x\boldsymbol{G}_{ex}^0(\boldsymbol{r}|\boldsymbol{r}') + \boldsymbol{u}_y\boldsymbol{G}_{ey}^0(\boldsymbol{r}|\boldsymbol{r}') + \boldsymbol{u}_z\boldsymbol{G}_{ez}^0(\boldsymbol{r}|\boldsymbol{r}').\end{aligned} \tag{3.82}$$

From (3.81) and (3.82), it follows that

$$\mathbf{G}_e^0(\boldsymbol{r}|\boldsymbol{r}') = \mathbf{G}_e^0(\boldsymbol{r}'|\boldsymbol{r}) = \mathbf{G}_e^0(\boldsymbol{r}|\boldsymbol{r}')^{\mathrm{T}}. \tag{3.83}$$

The electric field generated by a z-directed current is of the general form

$$\boldsymbol{E}(\boldsymbol{r}) = \mathrm{j}\omega\mu_0 \iiint_V \boldsymbol{G}_{ez}^0(\boldsymbol{r}|\boldsymbol{r}')J_z(\boldsymbol{r}')\,\mathrm{d}V'. \tag{3.84}$$

Let us apply this formula to a z-directed short element of current centred on $\boldsymbol{r}_0$. The corresponding volume current density is

$$\boldsymbol{J}(\boldsymbol{r}) = \mathrm{j}\omega P_e \boldsymbol{u}_z \delta(\boldsymbol{r}-\boldsymbol{r}_0). \tag{3.85}$$

This value, inserted into (3.84), yields

$$\boldsymbol{E}(\boldsymbol{r}) = -\omega^2\mu_0 P_e \boldsymbol{G}_{ez}^0(\boldsymbol{r}|\boldsymbol{r}_0). \tag{3.86}$$

The vector $\boldsymbol{G}_{ez}^0$ is therefore proportional to the electric field of a z-oriented electric dipole. This property is in harmony with the classical interpretation of a Green's function as the response to a 'unit source'. To pursue the identification, let us evaluate $\boldsymbol{G}_{ez}^0$ explicitly. The easiest method consists in introducing spherical coordinates centred on $\boldsymbol{r}'$. If we set $|\boldsymbol{r}-\boldsymbol{r}'| = R$, (B.23) yields

$$\operatorname{grad}\operatorname{grad} G(R) = \frac{\partial^2 G}{\partial R^2}\, \boldsymbol{u}_R \boldsymbol{u}_R + \frac{1}{R}\frac{\partial G}{\partial R}\,(\boldsymbol{u}_\theta \boldsymbol{u}_\theta + \boldsymbol{u}_\varphi \boldsymbol{u}_\varphi)$$

$$= \frac{1}{R}\frac{\mathrm{d}G}{\mathrm{d}R}\,\mathbf{I} + \left(\frac{\mathrm{d}^2 G}{\mathrm{d}R^2} - \frac{1}{R}\frac{\mathrm{d}G}{\mathrm{d}R}\right)\boldsymbol{u}_R \boldsymbol{u}_R. \tag{3.87}$$

Applied to the scalar Green's function (3.20), this formula gives, for $R \neq 0$,

$$\operatorname{grad}\operatorname{grad}\left(-\frac{1}{4\pi}\frac{\mathrm{e}^{-\mathrm{j}kR}}{R}\right) = \frac{\mathrm{e}^{-\mathrm{j}kR}}{4\pi R^3}\,[k^2 R^2 \boldsymbol{u}_R \boldsymbol{u}_R + (1 + \mathrm{j}kR)(\mathbf{I} - 3\,\boldsymbol{u}_R \boldsymbol{u}_R)]. \tag{3.88}$$

If we insert this expression into (3.81), we obtain

$$\mathbf{G}^0_{\mathrm{e}}(\boldsymbol{r}|\boldsymbol{r}') = G(R)\left[\mathbf{I} - \boldsymbol{u}_R \boldsymbol{u}_R - \left(\frac{\mathrm{j}}{kR} + \frac{1}{k^2 R^2}\right)(\mathbf{I} - 3\,\boldsymbol{u}_R \boldsymbol{u}_R)\right]. \tag{3.89}$$

The value of $\boldsymbol{G}^0_{\mathrm{e}z}$ follows immediately from this expression. Inserting this value into (3.86) leads to

$$\boldsymbol{E}(\boldsymbol{r}) = -\frac{k^2 P_{\mathrm{e}}}{\varepsilon_0}\, G(R)\left[\boldsymbol{u}_z\left(1 - \frac{\mathrm{j}}{kR} - \frac{1}{k^2 R^2}\right)\right.$$

$$\left. + (\cos\theta)\boldsymbol{u}_R\left(-1 + \frac{3\mathrm{j}}{kR} + \frac{3}{k^2 R^2}\right)\right]. \tag{3.90}$$

This expression is precisely the electric field of an electric dipole $P_{\mathrm{e}}\boldsymbol{u}_z$, a value which can be found in any elementary textbook. The Green's dyadic therefore consists of the ensemble $\boldsymbol{G}^0_{\mathrm{e}x}$, $\boldsymbol{G}^0_{\mathrm{e}y}$, $\boldsymbol{G}^0_{\mathrm{e}z}$ of the fields produced by elementary dipoles oriented in the x, y, and z directions, respectively.

3.13 Green's dyadic for the electric field: interior points

Equation (3.76), which gives the electric field in terms of the vector potential $\boldsymbol{A}$, contains a term in grad div $\boldsymbol{A}$. When $\boldsymbol{r}$ lies inside the current-carrying volume, this term (which contains nine second derivatives) must be handled in the spirit of Section 3.9. We therefore apply the splitting (3.57) nine times, to obtain

$$E(r) = -j\omega\mu_0 \frac{1}{4\pi} \iiint_V \frac{e^{-jk|r-r'|}}{|r-r'|} J(r')\,dV'$$

$$+ \frac{1}{j\omega\varepsilon_0 4\pi} \iiint_{V-V^*} J(r') \cdot \operatorname{grad}' \operatorname{grad}' \frac{e^{-jk|r-r'|}}{|r-r'|}\,dV'$$

$$- \frac{1}{j\omega\varepsilon_0} \mathbf{L}_{V^*} \cdot J(r) \qquad (3.91)$$

$$+ \frac{1}{j\omega\varepsilon_0 4\pi} \iiint_{V^*} \left(J(r') \cdot \operatorname{grad}' \operatorname{grad}' \frac{e^{-jk|r-r'|}}{|r-r'|} - J(r) \cdot \operatorname{grad}' \operatorname{grad}' \frac{1}{|r-r'|} \right) dV'. \qquad (3.91)$$

This relationship provides a suitable recipe for the evaluation of E.

In (3.91) V^* is an arbitrary volume, not necessarily small, which contains the field point r. When the dimensions of V^* approach zero, the last integral in (3.91), which stems from the term C_{ij} in (3.57), approaches zero at the same time. This behaviour is compensated by an opposite evolution in the integral over $V–V^*$, which ensures that the value of $E(r)$ remains constant (Lee *et al.* 1980). Taking limits as the maximum chord $|V^*|$ of V^* approaches zero in (3.91), we write (Yaghjian 1980)

$$E(r) = j\omega\mu_0 \lim_{|V^*| \to 0} \iiint_{V-V^*} \mathbf{G}_e^0(r|r') \cdot J(r')\,dV' - \frac{1}{j\omega\varepsilon_0} \mathbf{L}_{V^*} \cdot J(r). \qquad (3.92)$$

The limit process must be defined carefully. It requires the linear dimensions of V^* to approach zero, whereas the geometry of V^* (shape, position, orientation with respect to r) is maintained. In other words, V^* contracts to r by a similarity transformation with centre at r. Under these circumstances, $\mathbf{L}_{V^*}$ remains constant. To remind us of this definition we shall replace, as we did in Section 3.1, the notation $\lim_{|V^*| \to 0}$ by $\lim_{\delta \to 0}$, and integrate over $V–V_\delta$, where V_δ is a volume of maximum chord δ.

In numerical practice, V^* is not infinitely small, and the basic approximation leading to (3.92), which was

$$\frac{\partial^2}{\partial x_i \partial x_j} \iiint_{V^*} \frac{e^{-jk|r-r'|}}{|r-r'|}\,dV' \approx \frac{\partial^2}{\partial x_i \partial x_j} \iiint_{V^*} \frac{1}{|r-r'|}\,dV'$$

$$= -4\pi u_i \cdot \mathbf{L}_{V^*} \cdot u_j, \qquad (3.93)$$

becomes unacceptable when the maximum chord δ of V^* exceeds a threshold value, which turns out to be of the order of $\lambda\!\!\!^{-} = \lambda/2\pi$. This transition value is

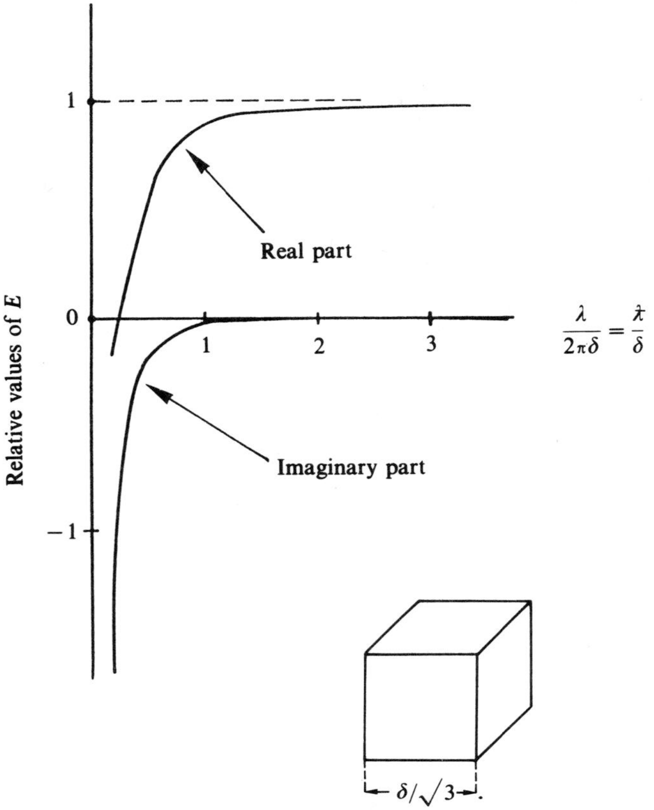

Fig. 3.10. Electric field at the centre of a cube of maximum chord δ (the diagonal). A uniform current flows in the cube. The field is expressed in terms of its static value, obtained for $\delta \ll ƛ$. (from Yaghjian (1981: p. 284) © IEEE)

confirmed by the data appearing in Fig. 3.10, which show that the self-contribution of the cubic cell becomes practically independent of δ once $\delta \leqslant ƛ$ (Yaghjian 1981).

3.14 The full electric Green's dyadic $\mathbf{G}_e$

Equation (3.92) clearly implies that $\boldsymbol{E}(\boldsymbol{r})$ consists of two terms:

The first one is a contribution from the currents outside V, evaluated by means of a dyadic $\mathbf{G}_e^0$. This dyadic was obtained by bringing second derivatives behind the potential integrals in (3.54).

The second term is a contribution from the current in the small volume V^*, derived by carefully *avoiding* the interchange of integral and second derivatives used for $\mathbf{G}_e^0$. This contribution can be attributed to an 'interior' dyadic

$$\mathbf{G}_e^i = (1/k^2)\,\delta(\boldsymbol{r} - \boldsymbol{r}')\,\mathbf{L}_{V^*}. \tag{3.94}$$

Let us introduce the distribution

$$\begin{aligned}\mathbf{G}_e(\boldsymbol{r}|\boldsymbol{r}') &= \mathrm{PV}_{V^*}\mathbf{G}_e^0(\boldsymbol{r}|\boldsymbol{r}') + \mathbf{G}_e^i(\boldsymbol{r}|\boldsymbol{r}') \\ &= \mathrm{PV}_{V^*}\mathbf{G}_e^0(\boldsymbol{r}|\boldsymbol{r}') + (1/k^2)\,\delta(\boldsymbol{r} - \boldsymbol{r}')\,\mathbf{L}_{V^*}.\end{aligned} \tag{3.95}$$

The notation PV_{V^*}, reminiscent of the principal value defined in (1.15), is the generating function for the integral in the right-hand member of (3.92). In terms of $\mathbf{G}_e(\boldsymbol{r}|\boldsymbol{r}')$, the fundamental expression (3.92) becomes simply

$$\boldsymbol{E}(\boldsymbol{r}) = \mathrm{j}\omega\mu_0 \iiint \mathbf{G}_e(\boldsymbol{r}|\boldsymbol{r}')\cdot \boldsymbol{J}(\boldsymbol{r}')\,\mathrm{d}V', \tag{3.96}$$

where the integration is over all space. The $\mathbf{G}_e$ distribution is unique, but its decomposition (3.95) into partial distributions depends on the choice of V^*. It is also clear, from (3.95), that $\mathbf{G}_e$ and $\mathbf{G}_e^0$ have the same value for $\boldsymbol{r} \neq \boldsymbol{r}'$, i.e. anywhere but at the source point (the concept 'value of a distribution' is discussed in Section 1.3).

A form such as (3.96) suggests, just as in the case of (3.80), that $\mathbf{G}_e$ may be conceived as the ensemble of the responses to 'unit' dipole sources. Going one step further, we could also expect $\mathbf{G}_e$ formally to satisfy the differential equation

$$-\operatorname{curl}\operatorname{curl}\mathbf{G}_e(r|r') + k^2\mathbf{G}_e(\boldsymbol{r}|\boldsymbol{r}') = \mathbf{I}\,\delta(\boldsymbol{r} - \boldsymbol{r}'). \tag{3.97}$$

From numerous other examples in mathematical physics, an equation such as (3.97) could serve as a basis to derive the value of $\boldsymbol{E}$ given in (3.96). The point is discussed at some length in Section 3.16. The correct interpretation of (3.97), however, lies in the realm of distribution theory. Distributionally, (3.97) implies that the functionals corresponding to the two members are equal. If we keep in mind the definition (1.125) of the curl of a distribution, (3.97) leads to the functional equality

$$\iiint \mathbf{G}_e(\boldsymbol{r}|\boldsymbol{r}')\cdot(-\operatorname{curl}'\operatorname{curl}'\boldsymbol{\phi} + k^2\boldsymbol{\phi})\,\mathrm{d}V' = \boldsymbol{\phi}(\boldsymbol{r}). \tag{3.98}$$

In this equation, $\boldsymbol{\phi}$ is an arbitrary test vector, and the integration is over all space.

3.15 More on the evaluation of the electric field

The value of $\boldsymbol{E}$ given in (3.92) stems directly from Lee's splitting (3.57). In this splitting, the fundamental singularity is concentrated in the term B_{ij}, which contains the static kernel $|\boldsymbol{r} - \boldsymbol{r}'|^{-1}$. It is also possible to use a splitting based on the *radiation* kernel (3.20). For such a case, the terms B_{ij} and C_{ij} in (3.57) become (Fikioris 1965)

$$B_{ij} = \frac{\partial^2}{\partial x_i \partial x_j} \iiint_{V^*} \frac{e^{-jk|\boldsymbol{r}-\boldsymbol{r}'|}}{|\boldsymbol{r}-\boldsymbol{r}'|} \, dV',$$
$$C_{ij} = \iiint_{V^*} [\rho(\boldsymbol{r}') - \rho(\boldsymbol{r})] \frac{\partial^2}{\partial x'_i \partial x'_y} \frac{e^{-jk|\boldsymbol{r}-\boldsymbol{r}'|}}{|\boldsymbol{r}-\boldsymbol{r}'|} \, dV'. \tag{3.99}$$

Applied to the electric field, and with respect to a spherical volume V^* of radius a centred at $\boldsymbol{r}$, these expressions yield (Fig. 3.11)

$$\boldsymbol{E}(\boldsymbol{r}) = j\omega\mu_0 \iiint_{V-V^*} \mathbf{G}_e^0(\boldsymbol{r}|\boldsymbol{r}') \cdot \boldsymbol{J}(\boldsymbol{r}') \, dV' + \frac{1}{j\omega\varepsilon_0} \boldsymbol{J}(\boldsymbol{r}) [\tfrac{2}{3} e^{-jka}(1 + jka) - 1]$$
$$+ j\omega\mu_0 \iiint_{V^*} \mathbf{G}_e^0(\boldsymbol{r}|\boldsymbol{r}') \cdot [\boldsymbol{J}(\boldsymbol{r}') - \boldsymbol{J}(\boldsymbol{r})] \, dV'. \tag{3.100}$$

The second term in the right-hand member represents the self-field at the centre of a sphere carrying a uniform current $\boldsymbol{J}$. In the limit $ka \to 0$, this field reduces to the value $-\boldsymbol{J}/3j\omega\varepsilon_0$, in agreement with the predictions of (3.64).

The isolation of the singularity by means of a splitting requires evaluation of the fields generated by a small cell carrying a uniform current (Fikioris 1988). Specific results for these fields are available for the sphere (as mentioned above), the circular cylinder with $\boldsymbol{J}$ parallel to the axis, and the rectangular parallelepiped (Van Bladel 1961; Fikioris 1965, 1988; Chen 1977).

In the previous sections, we have consistently sought to express $\boldsymbol{E}$ in terms of $\boldsymbol{J}$ alone, all derivatives of $\boldsymbol{J}$ excluded. It should be noted, however, that an expression such as (3.76), which contains derivatives of $\boldsymbol{J}$, is quite suitable for numerical work when $\boldsymbol{J}$ is known with sufficient accuracy. Furthermore, there

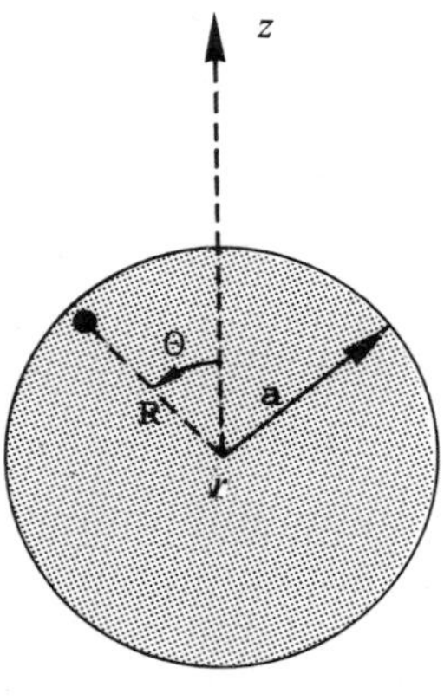

Fig. 3.11. A spherical volume current, with relevant coordinates.

are cases when the derivatives are not even present. In a homogeneous dielectric, for example, div $\boldsymbol{J}$ vanishes, and the scattered field is given by

$$\boldsymbol{E}_s(\boldsymbol{r}) = j\omega\mu_0 \iiint_V G(\boldsymbol{r}|\boldsymbol{r}')\boldsymbol{J}(\boldsymbol{r}')\,dV' + \frac{1}{j\omega\varepsilon_0}\,\text{grad} \iint_S G(\boldsymbol{r}|\boldsymbol{r}')\boldsymbol{u}_n \cdot \boldsymbol{J}(\boldsymbol{r}')\,dS'. \tag{3.101}$$

The currents in the dielectric are the polarization currents $\boldsymbol{J} = j\omega(\varepsilon - \varepsilon_0)\boldsymbol{E}$, hence

$$\boldsymbol{E}_s(\boldsymbol{r}) = \boldsymbol{E}(\boldsymbol{r}) - \boldsymbol{E}_i(\boldsymbol{r}) = \boldsymbol{J}(\boldsymbol{r})/j\omega(\varepsilon - \varepsilon_0) - \boldsymbol{E}_i(\boldsymbol{r}) \quad (\boldsymbol{r} \text{ in } V). \tag{3.102}$$

Substituting this value into (3.101) leads to an integral equation for $\boldsymbol{J}$. In the numerical solution of this equation, the gradient operator in front of the last integral may be usefully replaced by a finite-difference expression (Sarkar *et al.* 1989).

3.16 The source term for $\mathbf{G}_e$

The reader may feel that (3.92) should be obtainable directly from differential equation (3.97). The traditional approach for such derivations is to treat $\mathbf{G}_e$ and $\delta(\boldsymbol{r} - \boldsymbol{r}')$ as 'normal' functions. The essential features of the method become clearer when we consider the example of the clamped string of length L (see e.g. Van Bladel 1985). The displacement of the string satisfies

$$\frac{d^2 y}{dx^2} = f(x), \qquad y = 0 \quad \text{for } x = 0, L. \tag{3.103}$$

The second member $f(x)$ is a time-independent forcing function representing the effect of the vertical force density. The solution of (3.103), expressed in terms of a Green's function, is

$$y(x) = \int_0^L G(x|x')f(x')\,dx'. \tag{3.104}$$

There are two ways to express the basic singularity of $G(x|x')$. In the first approach (method A), $G(x|x')$ is required to satisfy the differential equation

$$\frac{d^2 G}{dx^2} = 0 \quad \text{for } x \neq x', \tag{3.105}$$

with the boundary and 'jump' conditions:

$$G(x|x') = 0 \text{ for } x = 0, L, \quad G(x|x') \text{ is continuous at } x = x',$$
$$\left[\frac{dG}{dx}\right]_{x'+0} - \left[\frac{dG}{dx}\right]_{x'-0} = 1. \tag{3.106}$$

In the second approach (method B), the singularity at $x = x'$ is represented by a δ-function. Thus,

$$\frac{\mathrm{d}^2 G}{\mathrm{d}x^2} = \delta(x - x'). \tag{3.107}$$

We shall seek to derive (3.96) by using both methods, A and B. To apply B, we start from (3.97), dot-multiply this equation with $\boldsymbol{E}(\boldsymbol{r}')$, and integrate over all space, to obtain

$$\iiint [-(\mathrm{curl}'\,\mathrm{curl}'\,\mathbf{G}_e)\cdot\boldsymbol{E}(\boldsymbol{r}') + k^2\mathbf{G}_e\cdot\boldsymbol{E}(\boldsymbol{r}')]\,\mathrm{d}V' = \boldsymbol{E}(\boldsymbol{r}). \tag{3.108}$$

The volume integral can be transformed by means of the vector–dyadic Green's theorem (B.34). This gives

$$\iiint [-(\mathrm{curl}'\,\mathrm{curl}'\,\mathbf{G}_e)\cdot\boldsymbol{E}(\boldsymbol{r}')]\,\mathrm{d}V' = -\iiint \mathbf{G}_e\cdot\mathrm{curl}'\,\mathrm{curl}'\,\boldsymbol{E}\,\mathrm{d}V' \\ + \iint_S [\mathrm{curl}'\,\mathbf{G}_e\cdot(\boldsymbol{u}_R\times\boldsymbol{E}) + \mathbf{G}_e\cdot(\boldsymbol{u}_R\times\mathrm{curl}'\,\boldsymbol{E})]\,\mathrm{d}S', \tag{3.109}$$

where S is a large sphere centred at the origin. The integral over S vanishes because $\boldsymbol{E}$ and $\mathbf{G}_e$ satisfy the radiation condition. Inserting result (3.109) into (3.108) yields

$$\begin{aligned}\boldsymbol{E}(\boldsymbol{r}) &= \iiint [-(\mathrm{curl}'\,\mathrm{curl}'\,\mathbf{G}_e)\cdot\boldsymbol{E} + k^2\mathbf{G}_e\cdot\boldsymbol{E}]\,\mathrm{d}V' \\ &= \iiint \mathbf{G}_e\cdot(-\mathrm{curl}'\,\mathrm{curl}'\,\boldsymbol{E} + k^2\boldsymbol{E})\,\mathrm{d}V' \\ &= \mathrm{j}\omega\mu_0 \iiint \mathbf{G}_e(\boldsymbol{r}|\boldsymbol{r}')\cdot\boldsymbol{J}(\boldsymbol{r}')\,\mathrm{d}V'.\end{aligned} \tag{3.110}$$

This is the desired result. The chosen derivation, based on differential equation (3.97), has the 'a posteriori' merit of leading to the correct value (3.98) in a few simple steps. Such shortcuts are purely formal, however, and must be justified more fully because of the delicate nature of the singularity of $\mathbf{G}_e$. A more acceptable way to derive (3.92) and (3.96)—complementary to the approach discussed in Section 3.13—is to start from differential equation (3.97), avoid the source point $\boldsymbol{r} = \boldsymbol{r}'$, and make use of the known value of $\mathbf{G}_e$ at short distances $|\boldsymbol{r} - \boldsymbol{r}'|$. The steps are similar to those outlined above, but the integration is now over all space minus a small volume V_δ surrounding $\boldsymbol{r}$. In the volume of integration:

$$-\,\mathrm{curl}\,\mathrm{curl}\,\mathbf{G}_e^0 + k^2\mathbf{G}_e^0 = 0. \tag{3.111}$$

With the geometry shown in Fig. 3.7, (3.109) and (3.110) are replaced by

$$j\omega\mu_0 \iiint_{V-V_\delta} \mathbf{G}_e^0(\boldsymbol{r}|\boldsymbol{r}')\cdot\boldsymbol{J}(\boldsymbol{r}')\,dV' - \iint_{\Sigma_\delta} [\text{curl}'\,\mathbf{G}_e^0\cdot(\boldsymbol{u}_n\times\boldsymbol{E}) + \mathbf{G}_e^0\cdot(\boldsymbol{u}_n\times\text{curl}'\,\boldsymbol{E})]\,dS' = 0. \tag{3.112}$$

We now insert into the surface integral the values of $\mathbf{G}_e^0$ and curl $\mathbf{G}_e^0$ which hold at short distances. From (3.89), kR being small,

$$\mathbf{G}_e^0(R) = \frac{1}{4\pi k^2 R^3}(\mathbf{I} - 3\boldsymbol{u}_R\boldsymbol{u}_R) + O\left(\frac{1}{R^2}\right),$$
$$\text{curl}\,\mathbf{G}_e^0(R) = \frac{1}{4\pi R^2}\boldsymbol{u}_R\times\mathbf{I} + O\left(\frac{1}{R}\right). \tag{3.113}$$

Here R stands for $|\boldsymbol{r}-\boldsymbol{r}'|$. When $\boldsymbol{E}$ and $\boldsymbol{J}$ are continuous one obtains, after some algebra (Ball *et al.* 1980; Collin 1986b),

$$\lim_{\delta\to 0}\iint_{\Sigma_\delta} \text{curl}'\,\mathbf{G}_e^0\cdot(\boldsymbol{u}_n\times\boldsymbol{E})\,dS' = \boldsymbol{E}(\boldsymbol{r}) - \mathbf{L}_{V_\delta}\cdot\boldsymbol{E}(\boldsymbol{r}),$$
$$\lim_{\delta\to 0}\iint_{\Sigma_\delta} \mathbf{G}_e^0\cdot(\boldsymbol{u}_n\times\text{curl}'\,\boldsymbol{E})\,dS' = \frac{1}{k^2}\mathbf{L}_{V_\delta}\cdot\text{curl}\,\text{curl}\,\boldsymbol{E}(\boldsymbol{r}). \tag{3.114}$$

Insertion of these values into (3.112) leads to

$$\boldsymbol{E}(\boldsymbol{r}) = j\omega\mu_0\lim_{\delta\to 0}\iiint_{V-V_\delta}\mathbf{G}_e^0(\boldsymbol{r}|\boldsymbol{r}')\cdot\boldsymbol{J}(\boldsymbol{r}')\,dV' + \frac{1}{k^2}\mathbf{L}_{V_\delta}\cdot(k^2\boldsymbol{E} - \text{curl}\,\text{curl}\,\boldsymbol{E}). \tag{3.115}$$

This is exactly (3.92), if we take into account that $\boldsymbol{E}$ satisfies Helmholtz' equation (3.69).

Two additional remarks are of interest here:

(1) The source term for $\mathbf{G}_e^0$ can be derived directly from (3.95) and (3.97). Indeed, these equations give

$$-\text{curl}\,\text{curl}\,\text{PV}_{V_\delta}\mathbf{G}_e^0(\boldsymbol{r}|\boldsymbol{r}') + k^2\,\text{PV}_{V_\delta}\mathbf{G}_e^0(\boldsymbol{r}|\boldsymbol{r}')$$
$$= \mathbf{I}\,\delta(\boldsymbol{r}-\boldsymbol{r}') - \mathbf{L}_{V_\delta}\delta(\boldsymbol{r}-\boldsymbol{r}') + \frac{1}{k^2}\text{curl}\,\text{curl}\,[\delta(\boldsymbol{r}-\boldsymbol{r}')\mathbf{L}_{V_\delta}]. \tag{3.116}$$

Such an equation is of purely formal interest. What really counts in numerical applications is prescription (3.92).

(2) The value of $\boldsymbol{E}$ at an interior point can be found directly from (3.98) when $\boldsymbol{J}$ has the properties of a test vector, in which case $\boldsymbol{E}$ in V is also a test vector. Equation (3.98) now leads directly to (3.96), if Helmholtz' equation

(3.69) is taken into account. Note that $\boldsymbol{J}$ can always be well-approximated by a test vector if it is not a test vector itself.

3.17 The Green's dyadic as a limit

The Green's function of a linear system is the response to an 'infinitely concentrated' forcing function. This response can be obtained by starting from a 'loosely' concentrated source, and letting the concentration increase until it becomes 'infinite'. The forcing function now becomes a δ-function. This procedure can be applied to the Green's dyadic discussed above. We start from a dyadic satisfying (Yaghjian 1982).

$$-\operatorname{curl}\operatorname{curl}\mathbf{G}_n + k^2\mathbf{G}_n = \delta_n(\boldsymbol{r}-\boldsymbol{r}')\mathbf{I}, \tag{3.117}$$

where $\delta_n(\boldsymbol{r}-\boldsymbol{r}')$ is a function 'centred' on $\boldsymbol{r}'$. When δ_n becomes more and more concentrated (i.e. when the δ_n sequence approaches a delta-function) the expression

$$\boldsymbol{E}_n(\boldsymbol{r}) = \mathrm{j}\omega\mu_0 \iiint_V \mathbf{G}_n(\boldsymbol{r}|\boldsymbol{r}')\cdot\boldsymbol{J}(\boldsymbol{r}')\,\mathrm{d}V' \tag{3.118}$$

follows a parallel evolution towards $\boldsymbol{E}_n(\boldsymbol{r})$. To clarify the process, let us start from the specific 'cloud' function

$$\delta_n(R) = \begin{cases} 3n^3/4\pi & \text{if} \quad R = |\boldsymbol{r}-\boldsymbol{r}'| < 1/n, \\ 0 & \text{if} \quad R = |\boldsymbol{r}-\boldsymbol{r}'| \geqslant 1/n, \end{cases} \tag{3.119}$$

for $n = 1, 2, \ldots$. The corresponding $\mathbf{G}_n$ also depends on R only, and is found to be (Boersma, private communication)

$$\mathbf{G}_n(R) = \begin{cases} \dfrac{3n^3}{4\pi k^2}\left[\mathbf{I} - [(1+\mathrm{j}x)e^{-\mathrm{j}x}]_{x=k/n}\left(\mathbf{I} + \dfrac{1}{k^2}\operatorname{grad}\operatorname{grad}\right)\dfrac{\sin kR}{kR}\right] \\ \qquad\qquad\qquad\qquad\qquad\qquad \text{if } R < 1/n, \\ 3\left[\dfrac{\sin x - x\cos x}{x^3}\right]_{x=k/n} \mathbf{G}_{\mathrm{e}}^{0}(R) \qquad \text{if } R \geqslant 1/n. \end{cases} \tag{3.120}$$

In the limit of high values of n (i.e. of small kR), the first line reduces to $(N^3/4\pi k^2)\mathbf{I}$, and the second one to $\mathbf{G}_{\mathrm{e}}^{0}$ (Yaghjian 1982). Substituting (3.120) into (3.118), and letting the radius $\delta = 1/n$ of a small sphere V_0 surrounding $\boldsymbol{r}$ go to zero, yields

$$\boldsymbol{E}(\boldsymbol{r}) = \mathrm{j}\omega\mu_0 \lim_{\delta\to 0} \iiint_{V-V_0} \mathbf{G}_{\mathrm{e}}^{0}(\boldsymbol{r}|\boldsymbol{r}')\cdot\boldsymbol{J}(\boldsymbol{r}')\,\mathrm{d}V' - \frac{1}{3\mathrm{j}\omega\varepsilon_0}\boldsymbol{J}(\boldsymbol{r}). \tag{3.121}$$

This is precisely (3.92) for a sphere. The extension to an arbitrary volume V^* follows by circumscribing V^* with a small spherical volume V_0. From (3.113),

$\mathbf{G}_e^0$ at small kR can be written as

$$\mathbf{G}_e^0 = \frac{1}{4\pi k^2} \operatorname{grad} \frac{\boldsymbol{u}_R}{R^2} + O\left(\frac{1}{R^2}\right). \tag{3.122}$$

Applying (B.33), and remembering that $\mathbf{L}_{V^*}$ is given by (3.63), yields

$$\lim_{\delta \to 0} \iiint_{V_0 - V^*} \mathbf{G}_e^0 \, \mathrm{d}V' = \frac{1}{k^2} (\tfrac{1}{3}\mathbf{I} - \mathbf{L}_{V^*}). \tag{3.123}$$

Combining this relationship with (3.121) reproduces (3.92), which is the desired result.

3.18 Fields in material media

The electric field in a material body immersed in an incident wave can be evaluated from (3.92). In such a body the current density $\boldsymbol{J}$ consists of the combined effects of conduction and polarization. Thus (Fig. 3.9b),

$$\boldsymbol{J} = [\sigma + \mathrm{j}\omega(\varepsilon - \varepsilon_0)]\boldsymbol{E} = [\sigma + \mathrm{j}\omega(\varepsilon - \varepsilon_0)](\boldsymbol{E}_{sc} + \boldsymbol{E}_i), \tag{3.124}$$

where $\boldsymbol{E}_{sc}$ is the (scattered) field (3.92) generated by the induced currents. The unknown $\boldsymbol{J}$ therefore satisfies the integral equation

$$\frac{\boldsymbol{J}(\boldsymbol{r})}{\sigma + \mathrm{j}\omega(\varepsilon - \varepsilon_0)} - \mathrm{j}\omega\mu_0 \lim_{\delta \to 0} \iiint_{V - V_\delta} \mathbf{G}_e^0(\boldsymbol{r}|\boldsymbol{r}') \cdot \boldsymbol{J}(\boldsymbol{r}') \, \mathrm{d}V' + \frac{1}{\mathrm{j}\omega\varepsilon_0} \mathbf{L}_{V_\delta} \boldsymbol{J}(\boldsymbol{r}) = \boldsymbol{E}_i(\boldsymbol{r}) \quad (\boldsymbol{r} \text{ in } V). \tag{3.125}$$

This formulation is well adapted to problems arising in industrial heating and medical hyperthermia, where the unknowns are precisely $\boldsymbol{J}$ and the associated Joule power $|\boldsymbol{J}|^2/2\sigma$ (see e.g. Livesay *et al.* 1974; Casey *et al.* 1988).

The field obtained from the solution of (3.92) is the *Maxwellian* field. An examination of (3.92) shows that the first term on the right-hand side, i.e. the integral over $V - V_\delta$, represents the contribution from the currents outside V_δ, whereas the second term represents the contribution from the uniform currents $\boldsymbol{J}$ flowing in the (very small) volume V_δ. If these currents were suppressed, the electric field $\boldsymbol{E}$ obtained from Maxwell's equations would reduce to its 'cavity value' (Fig. 3.12a)

$$\boldsymbol{E}_c(\boldsymbol{r}) = \mathrm{j}\omega\mu_0 \iiint_{V - V^*} \mathbf{G}_e^0(\boldsymbol{r}|\boldsymbol{r}') \cdot \boldsymbol{J}(\boldsymbol{r}') \, \mathrm{d}V'. \tag{3.126}$$

This is the form (3.80), which may be applied because $\boldsymbol{r}$ is an exterior point of $V - V_\delta$. Consequently,

$$\boldsymbol{E}(\boldsymbol{r}) = \boldsymbol{E}_c(\boldsymbol{r}) - (1/\mathrm{j}\omega\varepsilon_0) \, \mathbf{L}_{V_\delta} \cdot \boldsymbol{J}(\boldsymbol{r}). \tag{3.127}$$

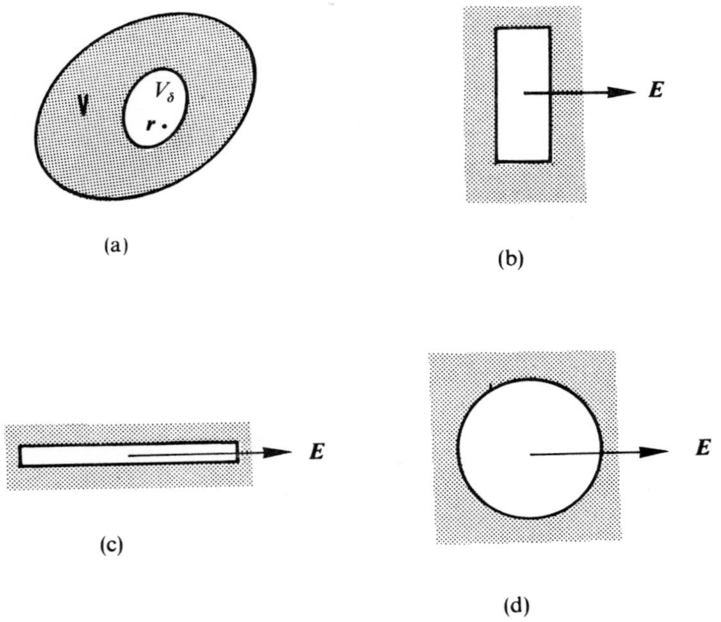

Fig. 3.12. Cavities in a current-carrying volume: (a) general shape, (b) flat box, (c) needle-like cavity, (d) spherical cavity.

The $\mathbf{L}_{V_\delta}$ operator may be interpreted as a *depolarization dyadic*, which serves to represent the nonshrinking contribution from the currents in V_δ. In a nonconducting dielectric, $\boldsymbol{J}$ consists of polarization currents $\mathrm{j}\omega(\varepsilon - \varepsilon_0)\boldsymbol{E} = \mathrm{j}\omega\boldsymbol{M}_\mathrm{e}$. If V_δ is the very flat cavity perpendicular to $\boldsymbol{E}$ shown in Fig. 3.12b, $\mathbf{L}_{V_\delta}$ is given by (3.67), and

$$\boldsymbol{E} = \boldsymbol{E}_\mathrm{c} - \boldsymbol{M}_\mathrm{e}/\varepsilon_0. \tag{3.128}$$

Consequently,

$$\boldsymbol{E}_\mathrm{c} = \varepsilon_\mathrm{r}\boldsymbol{E} = \boldsymbol{D}/\varepsilon_0. \tag{3.129}$$

If V_δ is a very sharp needle parallel to $\boldsymbol{E}$, as illustrated in Fig. 3.12c, we apply (3.66) to obtain

$$\boldsymbol{E}_\mathrm{c} = \boldsymbol{E}. \tag{3.130}$$

In this case, Maxwellian and cavity fields coincide. For a spherical cavity (Fig. 3.12d), equation (3.64) gives

$$\boldsymbol{E} = \boldsymbol{E}_\mathrm{c} - \boldsymbol{M}_\mathrm{e}/3\varepsilon_0. \tag{3.131}$$

The notion of cavity fields was introduced a century ago by Maxwell and Lord Kelvin, in the form of a 'thought experiment'. The operational meaning of the concept has been discussed at length (and critically) in an almost forgotten document published by the American Association of Physics Teachers (Coulomb Law Committee 1950).

3.19 Green's dyadic for the magnetic field

The magnetic field generated by a time-harmonic current is given by the formula

$$\begin{aligned} \boldsymbol{H}(\boldsymbol{r}) &= \frac{1}{4\pi} \iiint_V \boldsymbol{J}(\boldsymbol{r}') \times \operatorname{grad}' \frac{\mathrm{e}^{-\mathrm{j}k|\boldsymbol{r}-\boldsymbol{r}'|}}{|\boldsymbol{r}-\boldsymbol{r}'|} \mathrm{d}V' \\ &= \iiint_V \operatorname{grad}' G(\boldsymbol{r}|\boldsymbol{r}') \times \boldsymbol{J}(\boldsymbol{r}')\mathrm{d}V'. \end{aligned} \tag{3.132}$$

This expression can be written in terms of a dyadic $\mathbf{G}_{\mathrm{m}}$ as

$$\boldsymbol{H}(\boldsymbol{r}) = \iiint_V \mathbf{G}_{\mathrm{m}}(\boldsymbol{r}|\boldsymbol{r}') \cdot \boldsymbol{J}(\boldsymbol{r}')\mathrm{d}V'. \tag{3.133}$$

From (B.20) and (B.29),

$$\begin{aligned} \mathbf{G}_{\mathrm{m}}(\boldsymbol{r}|\boldsymbol{r}') &= \operatorname{grad}' G(\boldsymbol{r}|\boldsymbol{r}') \times \mathbf{I} \\ &= -\operatorname{grad} G(\boldsymbol{r}|\boldsymbol{r}') \times \mathbf{I} = -\operatorname{curl}[G(\boldsymbol{r}|\boldsymbol{r}')\mathbf{I}]. \end{aligned} \tag{3.134}$$

The $\mathbf{G}_{\mathrm{m}}$ dyadic is antisymmetric, and satisfies the relationships

$$\mathbf{G}_{\mathrm{m}}(\boldsymbol{r}'|\boldsymbol{r}) = -\mathbf{G}_{\mathrm{m}}(\boldsymbol{r}|\boldsymbol{r}'), \qquad \mathbf{G}_{\mathrm{m}}^{\mathrm{T}}(\boldsymbol{r}|\boldsymbol{r}') = -\mathbf{G}_{\mathrm{m}}(\boldsymbol{r}|\boldsymbol{r}'). \tag{3.135}$$

From (3.81) and (B.31), we observe that the $\mathbf{G}_{\mathrm{e}}^{0}$ dyadic is related to $\mathbf{G}_{\mathrm{m}}$ by

$$\begin{aligned} \mathbf{G}_{\mathrm{m}}(\boldsymbol{r}|\boldsymbol{r}') &= -\operatorname{curl} \mathbf{G}_{\mathrm{e}}^{0}(\boldsymbol{r}|\boldsymbol{r}'), \\ \mathbf{G}_{\mathrm{e}}^{0}(\boldsymbol{r}|\boldsymbol{r}') &= -(1/k^2)\operatorname{curl} \mathbf{G}_{\mathrm{m}}(\boldsymbol{r}|\boldsymbol{r}') \quad (\boldsymbol{r} \neq \boldsymbol{r}'). \end{aligned} \tag{3.136}$$

At $\boldsymbol{r} = \boldsymbol{r}'$, singular terms appear; they are discussed later in this section.

The electric field is proportional to curl $\boldsymbol{H}$, and should therefore be derivable from (3.132) by taking the curl of both members. A word of caution is necessary here. In (3.132) only first derivatives of $G(R)$ are involved, and the singularity is of the order of R^{-2}. No special precautions are necessary, and the introduction of principal values is not required. But by letting the curl operate on (3.132) we introduce an additional derivative, which requires careful handling of the limit processes involved (Yaghjian 1985). Thus,

$$\boldsymbol{E}(\boldsymbol{r}) = \frac{1}{\mathrm{j}\omega\varepsilon_0} \operatorname{curl} \lim_{|V_\delta| \to 0} \iiint_{V-V_\delta} \operatorname{grad}' G(\boldsymbol{r}|\boldsymbol{r}') \times \boldsymbol{J}(\boldsymbol{r}')\mathrm{d}V' - \frac{1}{\mathrm{j}\omega\varepsilon_0}\boldsymbol{J}(\boldsymbol{r}). \tag{3.137}$$

To transform this expression into (3.92) it is essential to note that V_δ, which contains $\boldsymbol{r}$, shifts position with $\boldsymbol{r}$. In other words, the limits of the integral in

(3.137) are variable. The difficulty can be resolved by using a three-dimensional version of the Leibniz rule for differentiating integrals with variable limits (Silberstein 1991).

The fields generated by magnetic currents $\boldsymbol{K}$ can be obtained by methods analogous to those used for $\boldsymbol{J}$. Starting from Maxwell's equations (2.85), (2.86), and the expression for the vector potential (2.137), we find

$$\boldsymbol{E}(\boldsymbol{r}) = \mathrm{j}\omega\mu_0 \lim_{\delta\to 0} \iiint_{V-V_\delta} \mathbf{G}^0_{\mathrm{e}}(\boldsymbol{r}|\boldsymbol{r}')\cdot\boldsymbol{J}(\boldsymbol{r}')\,\mathrm{d}V' + \iiint_V \mathbf{G}^{\mathrm{T}}_{\mathrm{m}}(\boldsymbol{r}|\boldsymbol{r}')\cdot\boldsymbol{K}(\boldsymbol{r}')\,\mathrm{d}V' - \frac{1}{\mathrm{j}\omega\varepsilon_0}\mathbf{L}_{V_\delta}\cdot\boldsymbol{J}(\boldsymbol{r}) \tag{3.138}$$

$$\boldsymbol{H}(\boldsymbol{r}) = \iiint_V \mathbf{G}_{\mathrm{m}}(\boldsymbol{r}|\boldsymbol{r}')\cdot\boldsymbol{J}(\boldsymbol{r}')\,\mathrm{d}V' + \mathrm{j}\omega\varepsilon_0 \lim_{\delta\to 0}\iiint_{V-V_\delta} \mathbf{G}^0_{\mathrm{h}}(\boldsymbol{r}|\boldsymbol{r}')\cdot\boldsymbol{K}(\boldsymbol{r}')\,\mathrm{d}V' - \frac{1}{\mathrm{j}\omega\mu_0}\mathbf{L}_{V_\delta}\cdot\boldsymbol{K}(\boldsymbol{r}). \tag{3.139}$$

The dyadic $\mathbf{G}^0_{\mathrm{h}}$ is equal to $\mathbf{G}^0_{\mathrm{e}}$, but is given a separate notation in order to respect the symmetry of the equations. Equations (3.138) and (3.139) have the nature of circuit equations. The impedance is provided by the term $\mathrm{j}\omega\mu_0\mathbf{G}^0_{\mathrm{e}}/k^2$, and the admittance by $\mathrm{j}\omega\varepsilon_0\mathbf{G}^0_{\mathrm{h}}/k^2$ (Daniele 1982; Daniele *et al.* 1984).

The 'mixed' dyadic $\mathbf{G}_{\mathrm{m}}$ formally satisfies the differential equation

$$-\operatorname{curl}\operatorname{curl}\mathbf{G}_{\mathrm{m}}(\boldsymbol{r}|\boldsymbol{r}') + k^2\mathbf{G}_{\mathrm{m}}(\boldsymbol{r}|\boldsymbol{r}') = -\operatorname{curl}[\delta(\boldsymbol{r}-\boldsymbol{r}')\mathbf{I}]. \tag{3.140}$$

We observe that both members of this equation are antisymmetric solenoidal dyadics. From the relationship

$$\iiint \operatorname{curl}[\delta(\boldsymbol{r}-\boldsymbol{r}')\mathbf{I}]\,\boldsymbol{\phi}(\boldsymbol{r})\,\mathrm{d}V = -\operatorname{curl}'\boldsymbol{\phi}(\boldsymbol{r}'), \tag{3.141}$$

we can deduce the distributional meaning of (3.140), which is

$$\iiint \mathbf{G}_{\mathrm{m}}(\boldsymbol{r}|\boldsymbol{r}')\cdot(-\operatorname{curl}\operatorname{curl}\boldsymbol{\phi} + k^2\boldsymbol{\phi})\,\mathrm{d}V = \operatorname{curl}'\boldsymbol{\phi}(\boldsymbol{r}'). \tag{3.142}$$

By taking the curl of both members of (3.140) we similarly obtain

$$\iiint \operatorname{curl}\mathbf{G}_{\mathrm{m}}(\boldsymbol{r}|\boldsymbol{r}')\cdot(-\operatorname{curl}\operatorname{curl}\boldsymbol{\phi} + k^2\boldsymbol{\phi})\,\mathrm{d}V = -\operatorname{curl}'\operatorname{curl}'\boldsymbol{\phi}(\boldsymbol{r}'). \tag{3.143}$$

But we have shown in (3.98) that

$$\iiint \mathbf{G}_{\mathrm{e}}(\boldsymbol{r}'|\boldsymbol{r})\cdot(-\operatorname{curl}\operatorname{curl}\boldsymbol{\phi} + k^2\boldsymbol{\phi})\,\mathrm{d}V = \boldsymbol{\phi}(\boldsymbol{r}'). \tag{3.144}$$

Combining (3.143) and (3.144) gives

$$\iiint [\operatorname{curl} \mathbf{G}_\mathrm{m}(\boldsymbol{r}|\boldsymbol{r}') + k^2 \mathbf{G}_\mathrm{e}(\boldsymbol{r}'|\boldsymbol{r})] \cdot (-\operatorname{curl}\operatorname{curl}\boldsymbol{\phi} + k^2\boldsymbol{\phi})\,\mathrm{d}V \tag{3.145}$$

$$= \iiint \mathbf{I}\,\delta(\boldsymbol{r}-\boldsymbol{r}')\cdot(-\operatorname{curl}\operatorname{curl}\boldsymbol{\phi} + k^2\boldsymbol{\phi})\,\mathrm{d}V.$$

Noting that $-\operatorname{curl}\operatorname{curl}\boldsymbol{\phi} + k^2\boldsymbol{\phi}$ is a test vector, we observe that (3.145) is satisfied if the following distributional equation holds:

$$\operatorname{curl} \mathbf{G}_\mathrm{m}(\boldsymbol{r}|\boldsymbol{r}') + k^2 \mathbf{G}_\mathrm{e}(\boldsymbol{r}|\boldsymbol{r}') = \mathbf{I}\,\delta(\boldsymbol{r}-\boldsymbol{r}'). \tag{3.146}$$

By analogous arguments,

$$\mathbf{G}_\mathrm{m}(\boldsymbol{r}|\boldsymbol{r}') = -\operatorname{curl} \mathbf{G}_\mathrm{e}(\boldsymbol{r}|\boldsymbol{r}'). \tag{3.147}$$

3.20 Helmholtz' theorem in unbounded space

Helmholtz' theorem states that a vector function $\boldsymbol{a}(\boldsymbol{r})$ may be split according to the formula

$$\boldsymbol{a}(\boldsymbol{r}) = \operatorname{grad}\phi + \operatorname{curl}\boldsymbol{v}. \tag{3.148}$$

The first term, grad ϕ, is the irrotational (lamellar or longitudinal) component. The second term, curl $\boldsymbol{v}$, is the solenoidal (transverse) component. The splitting shown in (3.148) is obviously not unique, because harmonic vectors may always be added to, and subtracted from, the right-hand side. Harmonic vectors are both solenoidal and irrotational, and can therefore be expressed either as a gradient or a curl. The ambiguity inherent in (3.148) may be lifted by imposing additional conditions on ϕ or $\boldsymbol{v}$, for example, regularity at large distances.

In our search for ϕ and $\boldsymbol{v}$, we first notice that a possible splitting is generated by any solution of the equation

$$\nabla^2\boldsymbol{f} = \operatorname{grad}\operatorname{div}\boldsymbol{f} - \operatorname{curl}\operatorname{curl}\boldsymbol{f} = \boldsymbol{a}. \tag{3.149}$$

The splitting obtains by setting $\operatorname{div}\boldsymbol{f} = \phi$ and $-\operatorname{curl}\boldsymbol{f} = \boldsymbol{v}$. Let us assume that $\boldsymbol{a}$ is concentrated in a volume V bounded by S. If we require $\boldsymbol{f}$ to be regular at infinity, i.e. to be $O(R^{-1})$, with derivatives $O(R^{-2})$, the solution of (3.149) is the unique vector

$$\boldsymbol{f}(\boldsymbol{r}) = -\frac{1}{4\pi}\iiint_V \frac{\boldsymbol{a}(\boldsymbol{r}')}{|\boldsymbol{r}-\boldsymbol{r}'|}\,\mathrm{d}V'. \tag{3.150}$$

This results in the unambiguous splitting

$$\begin{aligned}\boldsymbol{a}(\boldsymbol{r}) = \operatorname{grad}\left(\frac{1}{4\pi}\iint_S \frac{\boldsymbol{u}_n\cdot\boldsymbol{a}(\boldsymbol{r}')}{|\boldsymbol{r}-\boldsymbol{r}'|}\,dS' - \frac{1}{4\pi}\iiint_V \frac{\operatorname{div}'\boldsymbol{a}(\boldsymbol{r}')}{|\boldsymbol{r}-\boldsymbol{r}'|}\,dV'\right)\\ + \operatorname{curl}\left(\frac{1}{4\pi}\iiint_V \frac{\operatorname{curl}'\boldsymbol{a}(\boldsymbol{r}')}{|\boldsymbol{r}-\boldsymbol{r}'|}\,dV' - \frac{1}{4\pi}\iint_S \frac{\boldsymbol{u}_n\times\boldsymbol{a}(\boldsymbol{r}')}{|\boldsymbol{r}-\boldsymbol{r}'|}\,dS'\right).\end{aligned} \tag{3.151}$$

An alternate form is

$$\begin{aligned}\boldsymbol{a}(\boldsymbol{r}) = \operatorname{grad}\frac{1}{4\pi}\iiint_V \boldsymbol{a}(\boldsymbol{r}')\cdot\operatorname{grad}'\frac{1}{|\boldsymbol{r}-\boldsymbol{r}'|}\,dS'\\ + \operatorname{curl}\frac{1}{4\pi}\iiint_V \boldsymbol{a}(\boldsymbol{r}')\times\operatorname{grad}'\frac{1}{|\boldsymbol{r}-\boldsymbol{r}'|}\,dS'.\end{aligned} \tag{3.152}$$

Formal introduction of the gradient and curl operators behind the integral leads to a splitting in terms of longitudinal and transverse dyadics (Morse *et al.* 1953; Howard, 1974; Howard *et al.* 1978; Johnson *et al.* 1979). Thus,

$$\boldsymbol{a}(\boldsymbol{r}) = \iiint [\boldsymbol{\delta}_\ell(\boldsymbol{r}|\boldsymbol{r}') + \boldsymbol{\delta}_t(\boldsymbol{r}|\boldsymbol{r}')]\cdot\boldsymbol{a}(\boldsymbol{r}')\,dV', \tag{3.153}$$

where

$$\boldsymbol{\delta}_\ell(\boldsymbol{r}|\boldsymbol{r}') = -\frac{1}{4\pi}\operatorname{grad}'\operatorname{grad}'\frac{1}{|\boldsymbol{r}-\boldsymbol{r}'|}, \qquad \boldsymbol{\delta}_t(\boldsymbol{r}|\boldsymbol{r}') = \frac{1}{4\pi}\operatorname{curl}'\operatorname{curl}'\frac{\mathbf{I}}{|\boldsymbol{r}-\boldsymbol{r}'|}.$$

The longitudinal and transverse dyadics $\boldsymbol{\delta}_\ell$ and $\boldsymbol{\delta}_t$ clearly satisfy

$$\boldsymbol{\delta}_\ell(\boldsymbol{r}|\boldsymbol{r}') + \boldsymbol{\delta}_t(\boldsymbol{r}|\boldsymbol{r}') = \mathbf{I}\,\delta(\boldsymbol{r}-\boldsymbol{r}'). \tag{3.154}$$

It is obvious, from their form, that $\boldsymbol{\delta}_\ell$ and $\boldsymbol{\delta}_t$ are not 'concentrated at one point'. In the language of Section 1.3, their support does not reduce to $\boldsymbol{r} = \boldsymbol{r}'$, i.e. they do not vanish identically away from the source. This is confirmed by the actual value of $\operatorname{grad}\operatorname{grad} R^{-1}$ obtained from (2.11) or (3.88). More precisely (Belinfante 1946; Weiglhofer, 1989b):

$$\operatorname{grad}\operatorname{grad}\frac{1}{R} = \frac{1}{R^3}(3\boldsymbol{u}_R\boldsymbol{u}_R - \mathbf{I}) \quad R \neq 0. \tag{3.155}$$

3.21 Green's dyadics for a cavity

Figure 3.13 shows a cavity energized by electric and magnetic currents. The walls of the cavity are perfectly conducting, but their surface may be interrupted by an aperture S_a. The tangential component of the electric field vanishes on the wall, but is different from zero in the aperture.

The problem of determining $\boldsymbol{E}$ and $\boldsymbol{H}$ is classically solved by means of expansions in electric and magnetic eigenvectors. The essentials of the

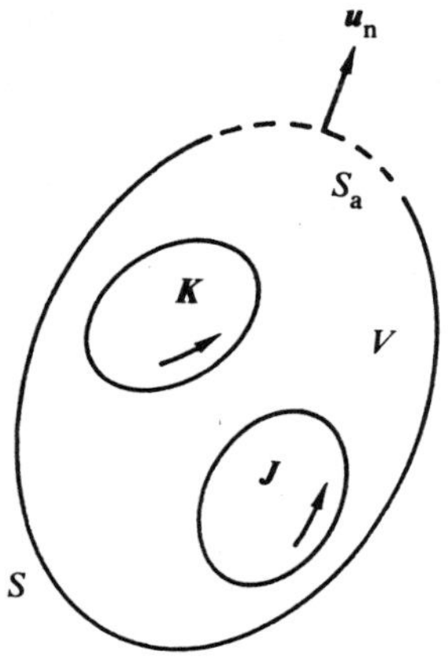

Fig. 3.13. A cavity with electric and magnetic sources.

method are summarized in Appendix C. The expansions, which are given in detail in (C.12), represent a perfectly valid method of determining $\boldsymbol{E}$ and $\boldsymbol{H}$. Consequently, the introduction of suitable Green's dyadics is by no means necessary. Dyadics, however, have the advantage of endowing the solution with symbolic simplicity. To determine these dyadics, we note that the integrations in (C.12) extend over the total current-carrying volume, and that the expansions hold for $\boldsymbol{r}$ both inside or outside the currents. The expansions therefore define a $\mathbf{G}_e$ and a $\mathbf{G}_h$. The electric field generated by electric currents $\boldsymbol{J}$, for example, is

$$\boldsymbol{E}(\boldsymbol{r}) = \mathrm{j}\omega\mu_0 \iiint_V \mathbf{G}_e(\boldsymbol{r}|\boldsymbol{r}') \cdot \boldsymbol{J}(\boldsymbol{r}')\, \mathrm{d}V', \tag{3.156a}$$

where

$$\mathbf{G}_e(\boldsymbol{r}|\boldsymbol{r}') = \sum_m \frac{1}{k^2} \boldsymbol{f}_m(\boldsymbol{r}) \boldsymbol{f}_m(\boldsymbol{r}') + \sum_m \frac{1}{k^2 - k_m^2} \boldsymbol{e}_m(\boldsymbol{r}) \boldsymbol{e}_m(\boldsymbol{r}'). \tag{3.156b}$$

As in free space, the Green's dyadic $\mathbf{G}_e$ satisfies the differential equation

$$-\operatorname{curl}\operatorname{curl} \mathbf{G}_e + k^2 \mathbf{G}_e = \mathbf{I}\delta(\boldsymbol{r} - \boldsymbol{r}'). \tag{3.157}$$

The boundary condition is now

$$\boldsymbol{u}_n \times \mathbf{G}_e = 0 \quad (\boldsymbol{r} \text{ on } S). \tag{3.158}$$

From (3.156b), it is clear that

$$\mathbf{G}_e(\boldsymbol{r}|\boldsymbol{r}') = \mathbf{G}_e^{\mathrm{T}}(\boldsymbol{r}'|\boldsymbol{r}). \tag{3.159}$$

This property is characteristic of self-adjoint vector transformations in real space (Van Bladel 1985).

The $\mathbf{G}_e$ dyadic for the cavity has the same kind of singularity as its counterpart in unbounded space (Yaghjian 1980; Daniele *et al.* 1984). When $\boldsymbol{r} \to \boldsymbol{r}'$, i.e. close to the source point, $\mathbf{G}_e(\boldsymbol{r}|\boldsymbol{r}')$ behaves as $\mathbf{G}_e^0(\boldsymbol{r}|\boldsymbol{r}')$, and the expansion is found to converge very slowly. At $\boldsymbol{r} = \boldsymbol{r}'$ itself, the series for $\mathbf{G}_e$ diverges.

The solution for the magnetic field, given in (C.12), can again be expressed in terms of a Green's dyadic. Thus,

$$\boldsymbol{H}(\boldsymbol{r}) = \iiint_V \mathbf{G}_m(\boldsymbol{r}|\boldsymbol{r}') \cdot \boldsymbol{J}(\boldsymbol{r}')\,\mathrm{d}V', \tag{3.160}$$

where

$$\mathbf{G}_m(\boldsymbol{r}|\boldsymbol{r}') = -\sum_m \frac{k_m}{k^2 - k_m^2} \boldsymbol{h}_m(\boldsymbol{r})\boldsymbol{e}_m(\boldsymbol{r}').$$

At a regular point (i.e. for $\boldsymbol{r} \neq \boldsymbol{r}'$),

$$G_m(\boldsymbol{r}|\boldsymbol{r}') = -\operatorname{curl} \mathbf{G}_e(\boldsymbol{r}|\boldsymbol{r}'). \tag{3.161}$$

In a similar vein, the fields produced by volume magnetic currents $\boldsymbol{K}$ and surface currents $\boldsymbol{K}_s = \boldsymbol{u}_n \times \boldsymbol{E}$ are given by

$$\begin{aligned}\boldsymbol{E}(\boldsymbol{r}) = &\iiint_V [-\mathbf{G}_m^T(\boldsymbol{r}'|\boldsymbol{r})] \cdot \boldsymbol{K}(\boldsymbol{r}')\,\mathrm{d}V' \\ &+ \iint_S [-\mathbf{G}_m^T(\boldsymbol{r}'|\boldsymbol{r})] \cdot \boldsymbol{K}_s(\boldsymbol{r}')\,\mathrm{d}S',\end{aligned} \tag{3.162}$$

where

$$\mathbf{G}_m^T(\boldsymbol{r}'|\boldsymbol{r}) = -\sum_m \frac{k_m}{k^2 - k_m^2} \boldsymbol{e}_m(\boldsymbol{r})\boldsymbol{h}_m(\boldsymbol{r}').$$

Also,

$$\begin{aligned}\boldsymbol{H}(\boldsymbol{r}) = &\,\mathrm{j}\omega\varepsilon_0 \iiint_V \mathbf{G}_h(\boldsymbol{r}|\boldsymbol{r}') \cdot \boldsymbol{K}(\boldsymbol{r}')\,\mathrm{d}V' \\ &+ \mathrm{j}\omega\varepsilon_0 \iint_S \mathbf{G}_h(\boldsymbol{r}|\boldsymbol{r}') \cdot \boldsymbol{K}_s(\boldsymbol{r}')\,\mathrm{d}S',\end{aligned} \tag{3.163}$$

where

$$\mathbf{G}_h(\boldsymbol{r}|\boldsymbol{r}') = \sum_m \left(\frac{1}{k^2} \boldsymbol{g}_m(\boldsymbol{r})\boldsymbol{g}_m(\boldsymbol{r}') + \frac{1}{k^2 - k_m^2} \boldsymbol{h}_m(\boldsymbol{r})\boldsymbol{h}_m(\boldsymbol{r}') \right).$$

The symmetry properties of $\mathbf{G}_h$ are those satisfied by $\mathbf{G}_e$, viz.

$$\mathbf{G}_h(\boldsymbol{r}|\boldsymbol{r}') = \mathbf{G}_h^T(\boldsymbol{r}'|\boldsymbol{r}). \tag{3.164}$$

Explicit forms of the dyadics for a few simple shapes—for example the sphere, the circular cylinder, or the rectangular box—are available in literature (see e.g. Tai *et al.* 1976; Bressan *et al.* 1985; Hadidi *et al.* 1988).

The operational meaning of the dyadic is clear from the preceding analysis. Expressed in a nutshell: when the dyadic $\mathbf{G}_e$, given in (3.156b), is introduced in (3.96), it generates a series which is precisely the original modal solution of our field problem, i.e. (3.156a).

3.22 Helmholtz' theorem in a bounded volume

The Helmholtz splitting in the volume V shown in Fig. 3.13 has the same form as in free space, namely

$$\boldsymbol{a}(\boldsymbol{r}) = \operatorname{grad} \phi + \operatorname{curl} \boldsymbol{v}. \tag{3.165}$$

To determine ϕ and $\boldsymbol{v}$, we take the divergence and the curl of both members of the equation. The divergence yields

$$\nabla^2 \phi_D = \operatorname{div} \boldsymbol{a}. \tag{3.166}$$

The splitting becomes unambiguous when we add conditions on ϕ and $\boldsymbol{v}$ which determine these functions uniquely. In the *electric* (or Dirichlet) splitting, the added requirement is

$$\phi_D = 0 \quad \text{on } S. \tag{3.167}$$

For that choice, the irrotational part, grad ϕ_D, is perpendicular to S, and ϕ_D is obtained by solving the Dirichlet problem consisting of (3.166) and (3.167). The solution can be achieved with the help of a Green's function satisfying

$$\nabla^2 G_D(\boldsymbol{r}|\boldsymbol{r}') = \delta(\boldsymbol{r} - \boldsymbol{r}'), \qquad G_D(\boldsymbol{r}|\boldsymbol{r}') = 0 \quad (\boldsymbol{r} \text{ on } S). \tag{3.168}$$

Elementary steps, based on Green's theorem, show that

$$\begin{aligned} \phi_D(\boldsymbol{r}) &= \iiint_V (\operatorname{div}' \boldsymbol{a}) G_D(\boldsymbol{r}|\boldsymbol{r}') \, dV' \\ &= -\iiint_V \boldsymbol{a}(\boldsymbol{r}') \cdot \operatorname{grad}' G_D(\boldsymbol{r}|\boldsymbol{r}') \, dV'. \end{aligned} \tag{3.169}$$

The two terms of the splitting satisfy the orthogonality relationship

$$\iiint_V \operatorname{grad} \phi_D \cdot \operatorname{curl} \boldsymbol{v}_D \, dV = 0. \tag{3.170}$$

In a second splitting, of the *magnetic* (or Neumann) type, $\boldsymbol{v}$ is required to satisfy

$$\operatorname{curl} \operatorname{curl} \boldsymbol{v}_N = \operatorname{curl} \boldsymbol{a}, \qquad \boldsymbol{u}_n \times \boldsymbol{v}_N = 0 \quad \text{on } S. \tag{3.171}$$

The curl of a vector perpendicular to S is tangent to S. It follows that curl $\boldsymbol{v}$ is tangent to S, hence that ϕ_N satisfies

$$\nabla^2 \phi_N = \operatorname{div} \boldsymbol{a}, \qquad \frac{\partial \phi_N}{\partial n} = \boldsymbol{u}_n \cdot \boldsymbol{a} \quad \text{on } S. \tag{3.172}$$

The determination of ϕ_N is a Neumann problem, which in turn may be solved with the help of a suitable Green's function.

A third Helmholtz decomposition is the *mixed* splitting

$$\boldsymbol{a}(\boldsymbol{r}) = \operatorname{grad} \phi + \operatorname{curl} \boldsymbol{v} + \boldsymbol{a}_0. \tag{3.173}$$

Here, ϕ vanishes at the boundary, $\boldsymbol{v}$ is perpendicular to the latter, and $\boldsymbol{a}_0$ is a suitable sourceless (harmonic) vector. To obtain the various terms in (3.173), we first evaluate grad ϕ, which is a term furnished by the electric splitting, and then apply the magnetic splitting to what is left after grad ϕ is subtracted from $\boldsymbol{a}$.

In the previous analysis V is tacitly assumed to be simply connected and simply bounded. When V is doubly bounded or doubly connected, a third term must be added to the original splitting (3.165). In a doubly bounded region (e.g. in the space between concentric spheres), the *electric* splitting must include a harmonic vector perpendicular to both boundary surfaces S_1 and S_2. This vector is proportional to the electrostatic field which arises when a difference of potential is applied between the (metallized) S_1 and S_2. In a doubly connected region (such as the ringlike volume of a circular particle accelerator) the magnetic splitting must include a harmonic vector tangent to the boundary. In fluid dynamics, this vector is the velocity of an incompressible fluid flowing irrotationally in the ringlike volume (Van Bladel 1960).

3.23 Helmholtz' theorem in eigenvector form

The Helmholtz splitting can be generated with the help of eigenfunctions and eigenvectors of the type discussed in Appendix C. Because the electric eigenvectors $\boldsymbol{f}_m$ and $\boldsymbol{e}_m$ form a complete orthonormal set, we write, in the volume V of the cavity,

$$\begin{aligned}\boldsymbol{a}(\boldsymbol{r}) &= \sum_{m=0}^{\infty} \boldsymbol{f}_m(\boldsymbol{r}) \iiint_V \boldsymbol{a}(\boldsymbol{r}')\cdot \boldsymbol{f}_m(\boldsymbol{r}')\,\mathrm{d}V', + \sum_{m=0}^{\infty} \boldsymbol{e}_m(\boldsymbol{r}) \iiint_V \boldsymbol{a}(\boldsymbol{r}')\cdot \boldsymbol{e}_m(\boldsymbol{r}')\,\mathrm{d}V' \\ &= \boldsymbol{a}_{\ell\mathrm{D}} + \boldsymbol{a}_{\mathrm{tD}},\end{aligned} \tag{3.174}$$

where

$$\boldsymbol{a}_{\ell\mathrm{D}} = \operatorname{grad} \phi_{\mathrm{D}} = \iiint_V \left(\sum_{m=0}^{\infty} \boldsymbol{f}_m(\boldsymbol{r})\, \boldsymbol{f}_m(\boldsymbol{r}') \right) \cdot \boldsymbol{a}(\boldsymbol{r}')\,\mathrm{d}V',$$

$$\boldsymbol{a}_{\mathrm{tD}} = \operatorname{curl} \boldsymbol{v}_{\mathrm{D}} = \iiint_V \left(\sum_{m=0}^{\infty} \boldsymbol{e}_m(\boldsymbol{r}) \boldsymbol{e}_m(\boldsymbol{r}') \right) \cdot \boldsymbol{a}(\boldsymbol{r}')\,\mathrm{d}V'.$$

Here $\boldsymbol{m}$ stands for an integer triple (m_1, m_2, m_3), and the sum is a triply infinite sum. The following expansions are of interest. From (C.3),

$$G_{\mathrm{D}}(\boldsymbol{r}|\boldsymbol{r}') = -\sum_{\boldsymbol{m}=0}^{\infty} \phi_{\boldsymbol{m}}(\boldsymbol{r})\phi_{\boldsymbol{m}}(\boldsymbol{r}'), \qquad \delta(\boldsymbol{r}-\boldsymbol{r}') = \sum_{\boldsymbol{m}=0}^{\infty} \mu_{\boldsymbol{m}}^2\phi_{\boldsymbol{m}}(\boldsymbol{r})\phi_{\boldsymbol{m}}(\boldsymbol{r}'). \tag{3.175}$$

Also

$$\mathbf{I}\delta(\boldsymbol{r}-\boldsymbol{r}') = \boldsymbol{\delta}_{\ell\mathrm{D}}(\boldsymbol{r}|\boldsymbol{r}') + \boldsymbol{\delta}_{\mathrm{tD}}(\boldsymbol{r}|\boldsymbol{r}'), \tag{3.176}$$

where

$$\boldsymbol{\delta}_{\ell\mathrm{D}}(\boldsymbol{r}|\boldsymbol{r}') = \sum_{\boldsymbol{m}=0}^{\infty} \boldsymbol{f}_{\boldsymbol{m}}(\boldsymbol{r})\,\boldsymbol{f}_{\boldsymbol{m}}(\boldsymbol{r}'), \qquad \boldsymbol{\delta}_{\mathrm{tD}}(\boldsymbol{r}|\boldsymbol{r}') = \sum_{\boldsymbol{m}=0}^{\infty} \boldsymbol{e}_{\boldsymbol{m}}(\boldsymbol{r})\boldsymbol{e}_{\boldsymbol{m}}(\boldsymbol{r}').$$

We have provided the longitudinal and transverse $\boldsymbol{\delta}$ dyadics with a subscript D, to emphasize that the splitting into a $\boldsymbol{\delta}_\ell$ and a $\boldsymbol{\delta}_{\mathrm{t}}$ is not unique but depends on the choice made for the Helmholtz splitting itself (in the present case, we chose the Dirichlet type). Thus,

$$\boldsymbol{a}_{\ell\mathrm{D}}(\boldsymbol{r}) = \iiint_V \boldsymbol{\delta}_{\ell\mathrm{D}}(\boldsymbol{r}|\boldsymbol{r}')\cdot\boldsymbol{a}(\boldsymbol{r}')\,\mathrm{d}V', \qquad \boldsymbol{a}_{\mathrm{tD}}(\boldsymbol{r}) = \iiint_V \boldsymbol{\delta}_{\mathrm{tD}}(\boldsymbol{r}|\boldsymbol{r}')\cdot\boldsymbol{a}(\boldsymbol{r}')\,\mathrm{d}V'. \tag{3.177}$$

Similar considerations hold for the magnetic splitting, which is

$$\boldsymbol{a}(\boldsymbol{r}) = \boldsymbol{a}_{\ell\mathrm{N}}(\boldsymbol{r}|\boldsymbol{r}') + \boldsymbol{a}_{\mathrm{tN}}(\boldsymbol{r}|\boldsymbol{r}'), \tag{3.178}$$

where

$$\boldsymbol{a}_{\ell\mathrm{N}}(\boldsymbol{r}|\boldsymbol{r}') = \iiint_V \left(\sum_{\boldsymbol{n}=0}^{\infty} \boldsymbol{g}_{\boldsymbol{n}}(\boldsymbol{r})\boldsymbol{g}_{\boldsymbol{n}}(\boldsymbol{r}')\right)\cdot\boldsymbol{a}(\boldsymbol{r}')\,\mathrm{d}V',$$

$$\boldsymbol{a}_{\mathrm{tN}}(\boldsymbol{r}|\boldsymbol{r}') = \iiint_V \left(\sum_{\boldsymbol{n}=0}^{\infty} \boldsymbol{h}_{\boldsymbol{n}}(\boldsymbol{r})\boldsymbol{h}_{\boldsymbol{n}}(\boldsymbol{r}')\right)\cdot\boldsymbol{a}(\boldsymbol{r}')\,\mathrm{d}V'.$$

When we particularize $\boldsymbol{a}$ to be the electric field, the electric splitting becomes, from (3.156a),

$$\boldsymbol{E}_\ell(\boldsymbol{r}) = \mathrm{j}\omega\mu_0 \iiint_V \mathbf{G}_{\mathrm{e}\ell}(\boldsymbol{r}|\boldsymbol{r}')\cdot\boldsymbol{J}(\boldsymbol{r}')\,\mathrm{d}V', \tag{3.179}$$

$$\boldsymbol{E}_{\mathrm{t}}(\boldsymbol{r}) = \mathrm{j}\omega\mu_0 \iiint_V \mathbf{G}_{\mathrm{et}}(\boldsymbol{r}|\boldsymbol{r}')\cdot\boldsymbol{J}(\boldsymbol{r}')\,\mathrm{d}V', \tag{3.180}$$

where

$$\mathbf{G}_{\mathrm{e}\ell}(\boldsymbol{r}|\boldsymbol{r}') = \frac{1}{k^2}\sum_{\boldsymbol{m}=0}^{\infty} \boldsymbol{f}_{\boldsymbol{m}}(\boldsymbol{r})\,\boldsymbol{f}_{\boldsymbol{m}}(\boldsymbol{r}'),$$

$$\mathbf{G}_{\mathrm{et}}(\boldsymbol{r}|\boldsymbol{r}') = \sum_{\boldsymbol{m}=0}^{\infty} \frac{1}{k^2 - k_{\boldsymbol{m}}^2}\boldsymbol{e}_{\boldsymbol{m}}(\boldsymbol{r})\boldsymbol{e}_{\boldsymbol{m}}(\boldsymbol{r}').$$

Comparison with (3.153) and (3.176) shows that

$$\mathbf{G}_{e\ell}(\boldsymbol{r}|\boldsymbol{r}') = (1/k^2)\boldsymbol{\delta}_\ell(\boldsymbol{r}|\boldsymbol{r}') = -(1/k^2)\,\mathrm{grad}\,\mathrm{grad}'\,G_\mathrm{D}(\boldsymbol{r}|\boldsymbol{r}'). \tag{3.181}$$

The essential singularity of G_D is

$$\lim_{\boldsymbol{r}\to\boldsymbol{r}'} G_\mathrm{D}(\boldsymbol{r}|\boldsymbol{r}') = -\frac{1}{4\pi|\boldsymbol{r}-\boldsymbol{r}'|}. \tag{3.182}$$

It follows that $\mathbf{G}_\ell$ has the same singularity as the dyadic $\boldsymbol{\delta}_\ell$ relative to unbounded space, a behaviour which is described by (3.155). From (3.81), (3.87), and (3.88), we observe that $\mathbf{G}_\mathrm{e}^0$ is dominated, at small R, by the terms

$$\mathbf{G}_\mathrm{e}^0(R) = \frac{1}{4\pi k^2 R^3}(\mathbf{I} - 3\boldsymbol{u}_R\boldsymbol{u}_R) - \frac{1}{8\pi R}(\mathbf{I} + \boldsymbol{u}_R\boldsymbol{u}_R). \tag{3.183}$$

The main singularity of $\mathbf{G}_\mathrm{e}$, of order R^{-3}, may therefore be attributed to the quasi-static term grad grad R^{-1} (Howard *et al.* 1978; Daniele *et al.* 1984). Expression (3.183) suggests that the convergence of the eigenvector expansion of $\mathbf{G}_\mathrm{e}$ can be accelerated by extracting the term in R^{-3} from $\mathbf{G}_\mathrm{e}^0$. It is even possible to go one step further, and extract the term in R^{-1} as well, leaving a residue which remains finite when $\boldsymbol{r}\to\boldsymbol{r}'$ (Bressan *et al.* 1985). The technique of extracting the strongest singularity, and thus obtaining a rest term which is smoother and converges faster, is of very general use in mathematical physics. It has been applied, for example, to the evaluation of the fields generated by surface currents (Arcioni *et al.* 1988).

The discussion given above for the *electric* splitting can be extended, mutatis mutandis, to the *magnetic* one. The details are trivial, and are left to the reader.

3.24 Magnetic field generated by electric currents in a waveguide

The mathematical background needed to determine $\boldsymbol{E}$ and $\boldsymbol{H}$ from a knowledge of $\boldsymbol{J}$ is given in Appendix D. A few elementary steps, based on (D.3) and (D.4), yield the following modal expansion for the magnetic field (Fig. 3.14):

$$\begin{aligned}\boldsymbol{H}(\boldsymbol{r}) = \sum_{m=0}^{\infty}\Bigg(&-\int_{-\infty}^{\infty} G_2(z\,|\,z'\,|\,\gamma_m)f_{m1}(z')\,\mathrm{d}z'\\ &+\mu_m^2\int_{-\infty}^{\infty} G_1(z\,|\,z'\,|\,\gamma_m)f_{m2}(z')\,\mathrm{d}z'\Bigg)\boldsymbol{u}_z\times\mathrm{grad}\,\phi_m\\ &+\sum_{n\neq 0}^{\infty}\Bigg(-\int_{-\infty}^{\infty} G_2(z\,|\,z'|\gamma_n)f_n(z')\,\mathrm{d}z'\Bigg)\mathrm{grad}\,\psi_n\\ &+\sum_{n\neq 0}^{\infty}\Bigg(-\nu_n^2\int_{-\infty}^{\infty} G_1(z\,|\,z'|\gamma_n)f_n(z')\,\mathrm{d}z'\Bigg)\psi_n\boldsymbol{u}_z.\end{aligned} \tag{3.184}$$

The symbols $\boldsymbol{m}$ and $\boldsymbol{n}$ stand for integer pairs, and the sum is a doubly infinite sum. The eigenfunction $\psi_{\boldsymbol{n}}$ corresponding to (0, 0) will be denoted by ψ_0. It is a constant. The notation $\boldsymbol{n} \neq 0$ means that the $\psi_0 = 1$ mode is not involved in the summation. The scalar Green's functions appearing in (3.184) are

$$G_1(z\,|\,z'\,|\,\gamma) = -(1/2\gamma)\mathrm{e}^{-\gamma|z-z'|},$$
$$G_2(z\,|\,z'\,|\,\gamma) = \frac{\mathrm{d}G_1(z\,|\,z'\,|\,\gamma)}{\mathrm{d}z} = \tfrac{1}{2}\operatorname{sgn}(z-z')\mathrm{e}^{-\gamma|z-z'|}, \tag{3.185}$$

with

$$\frac{\mathrm{d}G_2(z\,|\,z'\,|\,\gamma)}{\mathrm{d}z} = \gamma^2 G_1(z\,|\,z'\,|\,\gamma) + \delta(z-z'). \tag{3.186}$$

The presence of a δ-function in the second member of (3.186) is of great importance for the evaluation of the electric field. The point is further discussed in Section 3.26.

The γs appearing in (3.184) are, in an '$\boldsymbol{m}$' (or TM) mode, given by

$$\gamma_{\boldsymbol{m}} = \begin{cases} \sqrt{(\mu_{\boldsymbol{m}}^2 - k^2)} & \text{for } k \leqslant \mu_{\boldsymbol{m}} \text{ (damped mode)}, \\ \mathrm{j}\sqrt{(k^2 - \mu_{\boldsymbol{m}}^2)} & \text{for } k \geqslant \mu_{\boldsymbol{m}} \text{ (propagated mode)}. \end{cases} \tag{3.187}$$

In an '$\boldsymbol{n}$' (or TE) mode,

$$\gamma_{\boldsymbol{n}} = \begin{cases} \sqrt{(\nu_{\boldsymbol{n}}^2 - k^2)} & \text{for } k \leqslant \nu_{\boldsymbol{n}} \text{ (damped mode)}, \\ \mathrm{j}\sqrt{(k^2 - \nu_{\boldsymbol{n}}^2)} & \text{for } k \geqslant \nu_{\boldsymbol{n}} \text{ (propagated mode)}. \end{cases} \tag{3.188}$$

The functions $f_{\boldsymbol{m}}$, $f_{\boldsymbol{m}2}$, and $f_{\boldsymbol{n}}$ express the coupling of $\boldsymbol{J}$ to the various mode components. Thus,

$$f_{\boldsymbol{m}1}(z) = \iint_S \boldsymbol{J}\cdot \operatorname{grad}\phi_{\boldsymbol{m}}\,\mathrm{d}S, \qquad f_{\boldsymbol{m}2}(z) = \iint_S \boldsymbol{J}\cdot(\phi_{\boldsymbol{m}}\boldsymbol{u}_z)\,\mathrm{d}S, \tag{3.189}$$

$$f_{\boldsymbol{n}}(z) = \iint_S \boldsymbol{J}\cdot(\operatorname{grad}\psi_{\boldsymbol{n}} \times \boldsymbol{u}_z)\,\mathrm{d}S.$$

Equation (3.184) is the fundamental solution for $\boldsymbol{H}$. As in any other volume, however, the fields may be expressed in terms of Green's dyadics by the relationships

$$\boldsymbol{E}(\boldsymbol{r}) = \mathrm{j}\omega\mu_0 \iiint \mathbf{G}_{\mathrm{e}}(\boldsymbol{r}|\boldsymbol{r}')\cdot \boldsymbol{J}(\boldsymbol{r}')\,\mathrm{d}V',$$
$$\boldsymbol{H}(\boldsymbol{r}) = \iiint \mathbf{G}_{\mathrm{m}}(\boldsymbol{r}|\boldsymbol{r}')\cdot \boldsymbol{J}(\boldsymbol{r}')\,\mathrm{d}V'. \tag{3.190}$$

In the waveguide, $\mathrm{d}V = \mathrm{d}S\,\mathrm{d}z$. The value of $\mathbf{G}_\mathrm{m}$, for example, can easily be derived from the modal expansion (3.184). By inspection,

$$\begin{aligned}\mathbf{G}_\mathrm{m}(\boldsymbol{r}|\boldsymbol{r}') = &-\sum_{m=0}^{\infty} G_2(z\,|\,z'\,|\,\gamma_m)\boldsymbol{u}_z \times \operatorname{grad}\phi_m \operatorname{grad}'\phi'_m \\ &+\sum_{m=0}^{\infty} [\mu_m^2 G_1(z\,|\,z'\,|\,\gamma_m)\boldsymbol{u}_z \times \operatorname{grad}\phi_m]\phi'_m\boldsymbol{u}_z \\ &-\sum_{n\neq 0}^{\infty} G_2(z\,|\,z'\,|\,\gamma_n)\operatorname{grad}\psi_n \operatorname{grad}'\psi'_n \times \boldsymbol{u}_z \\ &-\sum_{n\neq 0}^{\infty} \nu_n^2 G_1(z\,|\,z'\,|\,\gamma_n)\psi_n\boldsymbol{u}_z \operatorname{grad}'\psi'_n \times \boldsymbol{u}_z.\end{aligned} \tag{3.191}$$

In this formula, ϕ_m stands for $\phi_m(x, y)$, and ϕ'_m for $\phi_m(x', y')$. Similar conventions are used for ψ_n and the gradients. When inserted into (3.190), the series for $\mathbf{G}_\mathrm{m}$ reproduces the fundamental modal expansion (3.184). This series is often rewritten as

$$\begin{aligned}\mathbf{G}_\mathrm{m}(\boldsymbol{r}|\boldsymbol{r}') = &\sum_{m=0}^{\infty}\frac{1}{2\gamma_m}\boldsymbol{M}_m(\mp\gamma_m)\operatorname{curl}'\boldsymbol{M}'_m(\pm\gamma_m) \\ &+\sum_{n\neq 0}^{\infty}\frac{1}{2\gamma_n}\boldsymbol{N}_n(\mp\gamma_n)\operatorname{curl}'\boldsymbol{N}'_n(\pm\gamma_n),\end{aligned} \tag{3.192}$$

where

$$\begin{aligned}\boldsymbol{M}_m(\gamma) &= \operatorname{curl}[\phi_m(x, y)\mathrm{e}^{\gamma z}\boldsymbol{u}_z] = \mathrm{e}^{\gamma z}\operatorname{grad}\phi_m \times \boldsymbol{u}_z, \\ \boldsymbol{N}_n(\gamma) &= \frac{1}{k}\operatorname{curl}\operatorname{curl}[\psi_n(x, y)\mathrm{e}^{\gamma z}\boldsymbol{u}_z] = \frac{\gamma}{k}\mathrm{e}^{\gamma z}\operatorname{grad}\psi_n + \frac{\nu_n^2}{k}\mathrm{e}^{\gamma z}\psi_n\boldsymbol{u}_z, \\ \operatorname{curl}\boldsymbol{M}_m(\gamma) &= \operatorname{curl}\operatorname{curl}[\phi_m(x, y)\mathrm{e}^{\gamma z}\boldsymbol{u}_z] = \gamma\mathrm{e}^{\gamma z}\operatorname{grad}\phi_m + \mu_m^2\mathrm{e}^{\gamma z}\phi_m\boldsymbol{u}_z, \\ \operatorname{curl}\boldsymbol{N}_n(\gamma) &= \frac{\nu_n^2-\gamma^2}{k}\mathrm{e}^{\gamma z}\operatorname{grad}\psi_n \times \boldsymbol{u}_z.\end{aligned} \tag{3.193}$$

In (3.192), the upper and lower signs correspond to $z' < z$ and $z' > z$, respectively. It is easy to see that a term such as $\boldsymbol{M}_m(-\gamma_m)$ stands for the magnetic field in a sourceless TM mode, propagated (or damped) in the direction of increasing z. The term $\boldsymbol{M}_m(\gamma_m)$ can be given a similar interpretation, but now with respect to decreasing z. The curl $\boldsymbol{M}_m$ and curl $\boldsymbol{N}_n$ terms represent the electric field of the modes.

With the help of (D.17) to (D.19), the modal expansion (3.184) can also be written in the form

$$\boldsymbol{H}(\boldsymbol{r}) = \operatorname{curl}(\pi_1\boldsymbol{u}_z) + \operatorname{curl}\operatorname{curl}(\pi_2\boldsymbol{u}_z), \tag{3.194}$$

where π_1 and π_2 are the Hertz potentials

$$\pi_1(\boldsymbol{r}) = \sum_{m=0}^{\infty}\left(\int_{-\infty}^{\infty} G_2(z\,|\,z'\,|\,\gamma_m) f_{m1}(z')\,\mathrm{d}z' - \mu_m^2 \int_{-\infty}^{\infty} G_1(z\,|\,z'\,|\,\gamma_m) f_{m2}(z')\,\mathrm{d}z'\right)\phi_m(\boldsymbol{\rho}), \tag{3.195}$$

$$\pi_2(\boldsymbol{r}) = -\sum_{n\neq 0}^{\infty}\left(\int_{-\infty}^{\infty} G_1(z\,|\,z'\,|\,\gamma_n) f_n(z')\,\mathrm{d}z'\right)\psi_n(\boldsymbol{\rho}).$$

The symbol $\boldsymbol{\rho} = x\boldsymbol{u}_x + y\boldsymbol{u}_y$ denotes the radius vector in the transverse plane.

To illustrate the use of the modal expansion, we shall evaluate the magnetic field generated by a very simple source: a z-oriented electric dipole located at $\boldsymbol{r}_0$. The corresponding current density is

$$\boldsymbol{J} = \mathrm{j}\omega P_{\mathrm{e}1}\,\delta(\boldsymbol{r} - \boldsymbol{r}_0)\boldsymbol{u}_z. \tag{3.196}$$

For this current, the modal expansion becomes the series

$$\boldsymbol{H}(\boldsymbol{r}) = -\frac{\mathrm{j}\omega P_{\mathrm{e}1}}{2}\sum_{m=0}^{\infty}\frac{\mu_m^2}{\gamma_m}\mathrm{e}^{-\gamma_m|z-z_0|}\phi_m(x_0, y_0)\boldsymbol{u}_z \times \operatorname{grad}\phi_m(x, y). \tag{3.197}$$

At sufficiently large distances (i.e. at large $|z - z_0|$), the sum reduces to a finite number of terms, namely those which are propagated. This reduction represents a major simplification in the actual evaluation of the series.

3.25 Electric field generated by electric currents in a waveguide

From basic theory outlined in Appendix D, the electric field in an infinite guide can be written as

$$\begin{aligned}\boldsymbol{E}(\boldsymbol{r}) = \frac{1}{\mathrm{j}\omega\varepsilon_0}\sum_{m=0}^{\infty}&\left(\gamma_m^2\int_{-\infty}^{\infty} G_1(z\,|\,z'\,|\,\gamma_m) f_{m1}(z')\,\mathrm{d}z' + \mu_m^2\int_{-\infty}^{\infty} G_2(z\,|\,z'\,|\,\gamma_m) f_{m2}(z')\,\mathrm{d}z'\right)\operatorname{grad}\phi_m \\ &+ \frac{1}{\mathrm{j}\omega\varepsilon_0}\sum_{m=0}^{\infty}\left(\mu_m^2\int_{-\infty}^{\infty} G_2(z\,|\,z'\,|\,\gamma_m) f_{m1}(z')\,\mathrm{d}z' - \mu_m^4\int_{-\infty}^{\infty} G_1(z\,|\,z'\,|\,\gamma_m) f_{m2}(z')\,\mathrm{d}z' - \mu_m^2 f_{m2}(z)\right)\phi_m\boldsymbol{u}_z \\ &+ \mathrm{j}\omega\mu_0\sum_{n\neq 0}\int_{-\infty}^{\infty} G_1(z\,|\,z'\,|\,\gamma_n) f_n(z')\,\mathrm{d}z'\operatorname{grad}\psi_n \times \boldsymbol{u}_z.\end{aligned} \tag{3.198}$$

From (3.189) and (3.190), $\boldsymbol{E}$ can also be expressed in terms of the electric Green's dyadic

$$\mathbf{G}_e(\boldsymbol{r}|\boldsymbol{r}') = -\frac{1}{k^2}\sum_{m=0}^{\infty} \gamma_m^2 G_1(z\,|\,z'\,|\,\gamma_m)\,\mathrm{grad}\,\phi_m\,\mathrm{grad}'\,\phi_m$$
$$+\frac{1}{k^2}\sum_{m=0}^{\infty} [\mu_m^2 G_2(z\,|\,z'\,|\,\gamma_m)\,\mathrm{grad}\,\phi_m]\,\phi_m'\boldsymbol{u}_z$$
$$-\frac{1}{k^2}\sum_{m=0}^{\infty} \mu_m^2 G_2(z\,|\,z'\,|\,\gamma_m)\,\phi_m\,\boldsymbol{u}_z\,\mathrm{grad}'\,\phi_m$$
$$+\frac{1}{k^2}\left(\sum_{m=0}^{\infty} [\mu_m^4 G_1(z\,|\,z'\,|\,\gamma_m) + \mu_m^2\delta(z-z')]\,\phi_m\phi_m'\right)\boldsymbol{u}_z\boldsymbol{u}_z$$
$$+\sum_{n\neq 0}^{\infty} G_1(z\,|\,z'\,|\,\gamma_n)(\mathrm{grad}\,\psi_n \times \boldsymbol{u}_z)(\mathrm{grad}'\,\psi_n \times \boldsymbol{u}_z). \tag{3.199}$$

Specific expressions for $\mathbf{G}_e$ and $\mathbf{G}_m$ are available in the literature for rectangular, circular and coaxial waveguides (see e.g. Tai 1971, 1983; Rahmat-Samii 1975; Pan Shenggen 1986). The electric field can also be expressed in terms of the previously defined Hertz potentials. Indeed, combining (3.195) and (3.197) yields

$$\boldsymbol{E}(\boldsymbol{r}) = \frac{1}{\mathrm{j}\omega\varepsilon_0}\mathrm{curl}\,\mathrm{curl}\,(\pi_1\boldsymbol{u}_z) + \frac{1}{\mathrm{j}\omega\varepsilon_0}\mathrm{curl}\,\mathrm{curl}\,\mathrm{curl}\,(\pi_2\boldsymbol{u}_z)$$
$$-\frac{1}{\mathrm{j}\omega\varepsilon_0}\left(\sum_{m=0}^{\infty} f_{m1}(z)\,\mathrm{grad}\,\phi_m + \sum_{m=0}^{\infty}\mu_m^2 f_{m2}(z)\,\phi_m\boldsymbol{u}_z \right. \tag{3.200}$$
$$\left. + \sum_{n\neq 0}^{\infty} f_n(z)\,\mathrm{grad}\,\psi_n \times \boldsymbol{u}_z\right).$$

The first two terms on the right-hand side clearly stand for $\mathrm{curl}\,\boldsymbol{H}/\mathrm{j}\omega\varepsilon_0$. According to Maxwell's equations, the term in large parentheses must therefore be $\boldsymbol{J}$. This prediction can be verified by inserting the actual values of f_{m1}, f_{m2}, and f_n into (3.199), and making use of the relationships

$$\sum_{m=0}^{\infty} \mu_m^2\phi_m(\boldsymbol{\rho})\phi_m(\boldsymbol{\rho}') = \delta(\boldsymbol{\rho}-\boldsymbol{\rho}'), \tag{3.201}$$

$$\sum_{m=0}^{\infty} \mathrm{grad}\,\phi_m\,\mathrm{grad}'\,\phi_m + \sum_{n\neq 0}^{\infty} \mathrm{grad}\,\psi_n \times \boldsymbol{u}_z\,\mathrm{grad}'\,\psi_n \times \boldsymbol{u}_z = \mathbf{I}_{xy}\delta(\boldsymbol{\rho}-\boldsymbol{\rho}'), \tag{3.202}$$

where $\mathbf{I}_{xy} = \boldsymbol{u}_x\boldsymbol{u}_x + \boldsymbol{u}_y\boldsymbol{u}_y$. Both expansions are based on the orthogonality and normalization properties of the two complete sets formed respectively by the eigenfunctions ϕ_m and the eigenvectors $(\mathrm{grad}\,\phi_m,\ \mathrm{grad}\,\psi_n \times \boldsymbol{u}_z)$.

The value of $\boldsymbol{E}$ given in (3.200) shows that the electric field consists of a solenoidal part, complemented by a nonsolenoidal term $-\boldsymbol{J}/\mathrm{j}\omega\varepsilon_0$. More explicitly,

$$\boldsymbol{E} = \operatorname{curl}\left(\frac{1}{\mathrm{j}\omega\varepsilon_0}\operatorname{curl}(\pi_1 \boldsymbol{u}_z) + \frac{1}{\mathrm{j}\omega\varepsilon_0}\operatorname{curl}\operatorname{curl}(\pi_2 \boldsymbol{u}_z)\right) - \frac{1}{\mathrm{j}\omega\varepsilon_0}\boldsymbol{J}. \tag{3.203}$$

The electric field is nonsolenoidal in the source region, since

$$\operatorname{div}\boldsymbol{E} = -\frac{1}{\mathrm{j}\omega\varepsilon_0}\operatorname{div}\boldsymbol{J} = \frac{\rho}{\varepsilon_0}. \tag{3.204}$$

Expanding $\boldsymbol{E}$ exclusively in solenoidal terms must therefore lead to erroneous results (Tai 1971, 1973a). The point is further discussed in Section 3.26.

3.26 Splitting the electric dyadic

The presence of a $\delta(z - z')$ function in (3.199) is noteworthy. This function, combined with (3.201), allows us to write the Green's dyadic in the form

$$\mathbf{G}_\mathrm{e}(\boldsymbol{r}|\boldsymbol{r}') = \mathbf{S}_\mathrm{e}(\boldsymbol{r}|\boldsymbol{r}') + (1/k^2)\delta(\boldsymbol{r} - \boldsymbol{r}')\boldsymbol{u}_z\boldsymbol{u}_z. \tag{3.205}$$

From the value of $\mathbf{G}_\mathrm{e}$ given in (3.199),

$$\begin{aligned}\mathbf{S}_\mathrm{e}(\boldsymbol{r}|\boldsymbol{r}') = &-\frac{1}{2k^2}\sum_{m=0}^{\infty}\frac{1}{\gamma_m}\operatorname{curl}\boldsymbol{M}_m(\mp\gamma_m)\operatorname{curl}'\boldsymbol{M}'_m(\pm\gamma_m)\\ &-\frac{1}{2k^2}\sum_{n\neq 0}^{\infty}\frac{1}{\gamma_n}\operatorname{curl}\boldsymbol{N}_n(\mp\gamma_n)\operatorname{curl}'\boldsymbol{N}'_n(\pm\gamma_n).\end{aligned} \tag{3.206}$$

Inserting (3.205) into (3.190) leads to

$$\boldsymbol{E}(\boldsymbol{r}) = \mathrm{j}\omega\mu_0\iiint \mathbf{S}_\mathrm{e}(\boldsymbol{r}|\boldsymbol{r}')\cdot(\boldsymbol{J}_t + \boldsymbol{J}_z)\,\mathrm{d}S'\,\mathrm{d}z' - \frac{1}{\mathrm{j}\omega\varepsilon_0}\boldsymbol{J}_z(\boldsymbol{r}). \tag{3.207}$$

We observe that the term in $\delta(\boldsymbol{r} - \boldsymbol{r}')$ in (3.205) (henceforth called the 'second' term) contributes to the z component of $\boldsymbol{E}$ by means of the z component of $\boldsymbol{J}$. In the series for $\mathbf{S}_\mathrm{e}$, the rapidly oscillatory part can be written in closed form; this property is of great interest for numerical applications, since it leads to improved convergence (Wang 1982).

In the source volume, the divergence of $\boldsymbol{E}$ does not vanish: it has the value given in (3.204). Since the solenoidal basis vectors defined in (3.193) are not complete with respect to nonsolenoidal fields, the expansion for $\mathbf{G}_\mathrm{e}$ must include the irrotational basis vectors

$$\boldsymbol{L}_m(\gamma) = \operatorname{grad}[\phi_m(x, y)\mathrm{e}^{\gamma z}] = \mathrm{e}^{\gamma z}\operatorname{grad}\phi_m + \gamma\mathrm{e}^{\gamma z}\phi_m\boldsymbol{u}_z. \tag{3.208}$$

In (3.205), the 'second' term corresponds to the contribution of a principal volume in the shape of a flat cylindrical box, which has an $\mathbf{L}_{V*}$ equal to $\boldsymbol{u}_z\boldsymbol{u}_z$.

In the spirit of (3.95), the dyadic $\mathbf{S}_e$ may therefore be interpreted as the corresponding PV $\mathbf{G}_e^0(\boldsymbol{r}|\boldsymbol{r}')$.

The need to include the 'second' term has generated a flurry of articles. The term always surfaces, whatever the method used to evaluate $\boldsymbol{E}$. It can be obtained, for example, by means of a Fourier transform of the field equations with respect to z (Collin 1973). More generally, it is revealed by a Fourier transform with respect to the $\boldsymbol{r}$ variable (Chew 1989; Wang 1990). The transform of $\mathbf{G}_e$ is

$$\mathbf{G}_e(\boldsymbol{\kappa}\,|\,\boldsymbol{r}') = \iiint \mathbf{G}_e(\boldsymbol{r}|\boldsymbol{r}')\mathrm{e}^{\mathrm{j}\boldsymbol{\kappa}\cdot\boldsymbol{r}}\,\mathrm{d}V. \tag{3.209}$$

Fourier transforming (3.97) leads to

$$\boldsymbol{\kappa} \times [\boldsymbol{\kappa} \times \mathbf{G}_e(\boldsymbol{\kappa}\,|\,\boldsymbol{r}')] + k^2\mathbf{G}_e(\boldsymbol{\kappa}\,|\,\boldsymbol{r}') = \mathbf{I}\mathrm{e}^{\mathrm{j}\boldsymbol{\kappa}\cdot\boldsymbol{r}'}. \tag{3.210}$$

The solution of (3.210) is clearly

$$\mathbf{G}_e(\boldsymbol{\kappa}\,|\,\boldsymbol{r}') = \frac{\boldsymbol{\kappa}\boldsymbol{\kappa} - k^2\,\mathbf{I}}{k^2(\kappa^2 - k^2)}\,\mathrm{e}^{\mathrm{j}\boldsymbol{\kappa}\cdot\boldsymbol{r}'}, \tag{3.211}$$

where $\kappa^2 = \kappa_x^2 + \kappa_y^2 + \kappa_z^2$. The value of $\mathbf{G}_e$ in $\boldsymbol{r}$ space follows by inversion. Thus,

$$\mathbf{G}_e(\boldsymbol{r}|\boldsymbol{r}') = \frac{1}{8\pi^3}\iiint_{-\infty}^{\infty} \frac{\boldsymbol{\kappa}\boldsymbol{\kappa} - k^2\mathbf{I}}{k^2(\kappa^2 - k^2)}\,\mathrm{e}^{-\mathrm{j}\boldsymbol{\kappa}\cdot(\boldsymbol{r} - \boldsymbol{r}')}\,\mathrm{d}\kappa_x\,\mathrm{d}\kappa_y\,\mathrm{d}\kappa_z. \tag{3.212}$$

This generalized dyadic contains distributions such as the delta function. It can be rewritten as

$$\mathbf{G}_e(\boldsymbol{r}|\boldsymbol{r}') = \frac{1}{8\pi^3}\iiint_{-\infty}^{\infty} \left(\frac{\boldsymbol{\kappa}\boldsymbol{\kappa} - k^2\mathbf{I}}{k^2(\kappa^2 - k^2)} - \frac{\boldsymbol{u}_z\boldsymbol{u}_z}{k^2}\right)\mathrm{e}^{-\mathrm{j}\boldsymbol{\kappa}\cdot(\boldsymbol{r} - \boldsymbol{r}')}\,\mathrm{d}\kappa_x\,\mathrm{d}\kappa_y\,\mathrm{d}\kappa_z.$$
$$+ \frac{\boldsymbol{u}_z\boldsymbol{u}_z}{k^2}\iiint_{-\infty}^{\infty} \frac{1}{8\pi^3}\,\mathrm{e}^{-\mathrm{j}\boldsymbol{\kappa}\cdot(\boldsymbol{r} - \boldsymbol{r}')}\,\mathrm{d}\kappa_x\,\mathrm{d}\kappa_y\,\mathrm{d}\kappa_z. \tag{3.213}$$

Contour integration can be performed on the first term of this equation, whereas the integral in the second term is equal to $\delta(\boldsymbol{r} - \boldsymbol{r}')$. It follows that this term is also the 'second' term of (3.205).

The series $\mathbf{S}_e$ is discontinuous at $z = z'$, and is also singular at $\boldsymbol{r} = \boldsymbol{r}'$. This is apparent from the relationship

$$\mathbf{S}_e(\boldsymbol{r}|\boldsymbol{r}') = (1/k^2)[\delta(\boldsymbol{r} - \boldsymbol{r}')\mathbf{I}_{xy} - \operatorname{curl}\mathbf{G}_m(\boldsymbol{r}|\boldsymbol{r}')], \tag{3.214}$$

which is an immediate consequence of (3.146). The discontinuous behaviour across $z = z'$ may be interpreted in terms of equivalent layers of surface charge, current, and polarization, all of these being located in the $z = z'$ plane (Tai 1981, 1983).

The splitting (3.205) may be extended to cases when there is a 'preferred' direction, corresponding to a coordinate p (instead of z). The coordinate p is part of an orthogonal curvilinear coordinate system (v_1, v_2, p). The splitting of the Green's dyadic is now (Pathak 1983; Wang 1990)

$$\mathbf{G}_e(\boldsymbol{r}|\boldsymbol{r}') = \mathbf{S}_e(\boldsymbol{r}|\boldsymbol{r}') + (1/k^2)\delta(\boldsymbol{r} - \boldsymbol{r}')\boldsymbol{u}_p\boldsymbol{u}_p. \tag{3.215}$$

3.27 Magnetic currents in a waveguide

The magnetic currents consist of volume currents $\boldsymbol{K}$ and surface currents $\boldsymbol{K}_s = \boldsymbol{u}_n \times \boldsymbol{E}$. Maxwell's equations in the presence of these currents become

$$\operatorname{curl}\boldsymbol{E} = -j\omega\mu_0\boldsymbol{H} - \boldsymbol{K} \qquad \operatorname{curl}\boldsymbol{H} = j\omega\varepsilon_0\boldsymbol{E}. \tag{3.216}$$

They yield the divergence equations

$$\operatorname{div}\boldsymbol{E} = 0 \qquad \operatorname{div}\boldsymbol{H} = -(1/j\omega\mu_0)\operatorname{div}\boldsymbol{K} = \rho_m/\mu_0, \tag{3.217}$$

where ρ_m is the (formal) magnetic charge density. A comparison of these equations with those relative to electric sources reveals that the roles of $(\boldsymbol{E}, \boldsymbol{H}, \boldsymbol{J})$ are interchanged with those of $(\boldsymbol{H}, -\boldsymbol{E}, \boldsymbol{K})$. It is therefore elementary to derive, for the $\boldsymbol{K}$ sources, the modal expansions and Green's dyadics which correspond to (3.184), (3.191), (3.198), and (3.199). We shall not go into the details of this exercise, except to underline the need to include the eigenfunction $\psi_0 = 1$ in the Neumann expansion for H_z. Without ψ_0, the set of the ψ_n would not be complete. The function ψ_0 does not generate a transverse eigenvector (since $\operatorname{grad}\psi_0 = 0$), and for that reason is sometimes forgotten in the analysis. The term in ψ_0 actually represents the average value of a function $f(\boldsymbol{\rho})$ over the cross-section S. Thus, if s is the area of S, then

$$(f)_{\text{ave}} = \frac{1}{s}\iint_S f(\boldsymbol{\rho})\,dS. \tag{3.218}$$

The full expansion of $f(\boldsymbol{\rho})$ in terms of Neumann eigenfunctions is now

$$f(\boldsymbol{\rho}) = \sum_{n\neq 0}^{\infty} \nu_n^2\psi_n(\boldsymbol{\rho})\iint_S f(\boldsymbol{\rho}')\psi_n(\boldsymbol{\rho}')\,dS' + \frac{1}{s}\iint_S f(\boldsymbol{\rho}')\,dS'. \tag{3.219}$$

This expression shows that

$$\delta(\boldsymbol{\rho} - \boldsymbol{\rho}') = \sum_{n\neq 0}^{\infty} \nu_n^2\psi_n(\boldsymbol{\rho}')\psi_n(\boldsymbol{\rho}') + \frac{1}{s}. \tag{3.220}$$

Also,

$$\begin{aligned} \mathbf{I}\delta(\boldsymbol{r} - \boldsymbol{r}') = &\sum_{m=0}^{\infty} (\boldsymbol{u}_z \times \operatorname{grad}\phi_m)(\boldsymbol{u}_z \times \operatorname{grad}'\phi_m) \\ &+ \sum_{n\neq 0}^{\infty} \operatorname{grad}\psi_n \operatorname{grad}'\psi_n + \left(\sum_{n\neq 0}^{\infty} \nu_n^2\psi_n\psi_n' + \frac{1}{s}\right)\boldsymbol{u}_z\boldsymbol{u}_z. \end{aligned} \tag{3.221}$$

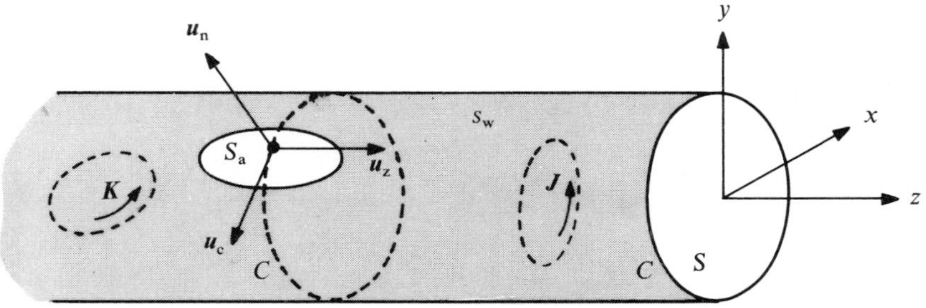

Fig. 3.14. A waveguide with electric and magnetic sources.

It is easy to understand why the $\psi_0 = 1$ term may be omitted in the expansion of H_z when the sources are electric. In this case, Maxwell's equation $\operatorname{curl} \boldsymbol{E} = -\mathrm{j}\omega\mu_0 \boldsymbol{H}$ leads to (Fig. 3.14)

$$\mathrm{j}\omega\phi = \mathrm{j}\omega\mu_0 s[H_z]_{\mathrm{ave}} = -\int_C \boldsymbol{E}\cdot\boldsymbol{u}_\mathrm{c}\,\mathrm{d}C, \tag{3.222}$$

where ϕ is the flux of $\boldsymbol{B}$ through the cross-section. However, the line integral along C vanishes because $\boldsymbol{E}$ is perpendicular to S_w. The integral would, of course, be different from zero in the presence of apertures, but these are excluded because an aperture and its electric field represent a *magnetic* source. We conclude that the term in $\psi_0 = 1$ disappears from the expansion. In the case of magnetic currents, however, (3.216) shows that

$$\mathrm{j}\omega\phi = \mathrm{j}\omega\mu_0 s[H_z]_{\mathrm{ave}} = -\int_C (\boldsymbol{u}_\mathrm{n}\times\boldsymbol{E})\cdot\boldsymbol{u}_z\,\mathrm{d}C - s[K_z]_{\mathrm{ave}}. \tag{3.223}$$

As a result,

$$[H_z]_{\mathrm{ave}} = -\frac{1}{\mathrm{j}\omega\mu_0}[K_z]_{\mathrm{ave}} - \frac{1}{\mathrm{j}\omega\mu_0 s}\int_C \boldsymbol{K}_s\cdot\boldsymbol{u}_z\,\mathrm{d}C. \tag{3.224}$$

The right-hand member is normally different from zero, which implies that $(H_z)_{\mathrm{ave}}$ does not vanish, hence that the term in ψ_0 *must* be present.

Another consequence of (3.216) is that $\boldsymbol{E}$ has become solenoidal. This is confirmed by the relationship

$$\boldsymbol{E}(\boldsymbol{r}) = \operatorname{curl}(\pi_1\boldsymbol{u}_z) + \operatorname{curl}\operatorname{curl}(\pi_2\boldsymbol{u}_z), \tag{3.225}$$

which corresponds to (3.194). The Hertz potentials are

$$\pi_1(\mathbf{r}) = \sum_{n \neq 0}^{\infty} \left(- \int_{-\infty}^{\infty} G_2(z \mid z' \mid \gamma_n) g_{n1}(z') \, dz' + \nu_n^2 \int_{-\infty}^{\infty} G_1(z \mid z' \mid \gamma_n) g_{n2}(z') \, dz' \right) \psi_n(\boldsymbol{\rho}) \tag{3.226}$$

$$\pi_2(\mathbf{r}) = - \sum_{m=0}^{\infty} \left(\int_{-\infty}^{\infty} G_1(z \mid z' \mid \gamma_m) g_m(z') \, dz' \right) \phi_m(\boldsymbol{\rho}).$$

The functions g_{n1}, g_{n2}, and g_m are defined in (D.3) and (D.4). The magnetic field is correspondingly

$$\mathbf{H}(\mathbf{r}) = -\frac{1}{j\omega\mu_0} \operatorname{curl} \operatorname{curl} (\pi_1 \mathbf{u}_z) - \frac{1}{j\omega\mu_0} \operatorname{curl} \operatorname{curl} \operatorname{curl} (\pi_2 \mathbf{u}_z) - \frac{1}{j\omega\mu_0} \mathbf{K}. \tag{3.227}$$

Suitable Green's dyadics can easily be deduced from these expressions.

3.28 The importance of the ψ_0 term

To illustrate the necessity of including the $\psi_0 = 1$ term in the expansion for H_z, let us discuss two examples involving aperture fields. The first one concerns a slot in a rectangular waveguide (Fig. 3.15a). The main unknown here is the $\mathbf{u}_n \times \mathbf{E}$ field in the slot. A classical method of solution consists in assuming $\mathbf{u}_n \times \mathbf{E}$ as known, evaluating the resulting $\mathbf{H}$ inside and outside the guide, and finally requiring the tangential component of $\mathbf{H}$ to be continuous across the slot. The procedure yields an integral equation for $\mathbf{u}_n \times \mathbf{E}$. Because this equation must be satisfied in the slot, the $\psi_0 = 1$ contribution to $\mathbf{H}$ must be included, lest erroneous results obtain (Lyon *et al.* 1983; Sangster *et al.* 1987). Figure 3.15b shows results pertaining to the frequency dependence of the impedance of a certain dielectric-covered slot. The dots are experimental points. The theoretical results stem from the 'moment' solution of the integral equation mentioned above. Omission of the ψ_0 term clearly leads to major errors in the value of the reactance X, whereas the value of R is not influenced. The reason is simple: the ψ_0 term contributes to the reactive power, but does not affect the radiated power. The errors in X are sufficiently important to cause unacceptable shifts in the value of the slot resonant frequency, which is obtained by setting $X(\omega) = 0$.

The second example concerns a small loop of current radiating into a circular waveguide (Fig. 3.16a). It is convenient to express the fields as the sum of the primary fields $(\mathbf{E}_i, \mathbf{H}_i)$ generated by the loop in free space, and the secondary fields $(\mathbf{E}_s, \mathbf{H}_s)$ produced by the presence of the walls. The latter fields result from the existence of a magnetic current

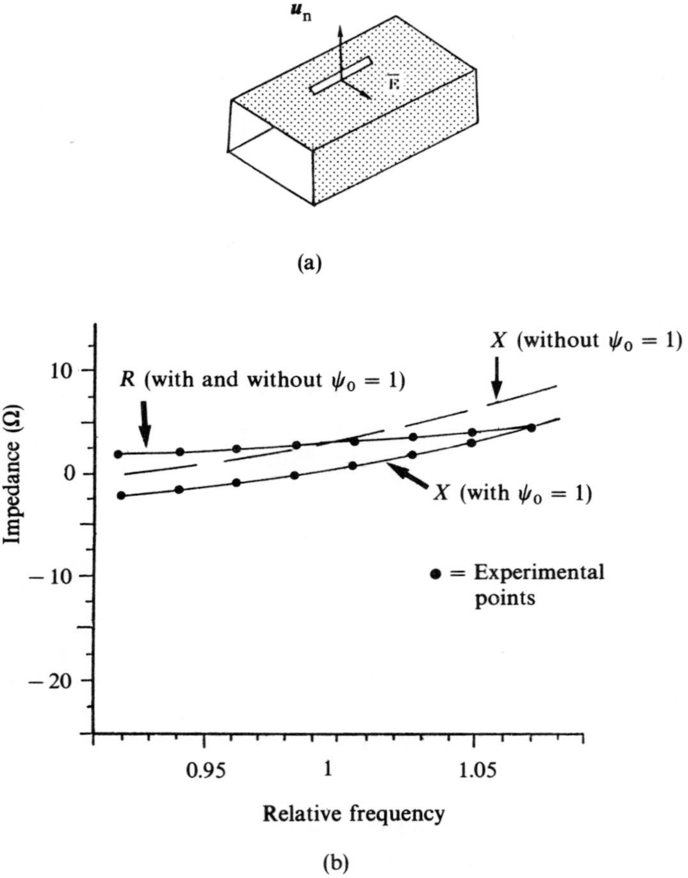

Fig. 3.15. (a) A slot in a rectangular waveguide. (b) Theoretical and experimental results for shunt impedance of a rectangular slot. (Communicated by R. W. Lyon.)

$\boldsymbol{K}_s = \boldsymbol{u}_n \times \boldsymbol{E}_s = -\boldsymbol{u}_n \times \boldsymbol{E}_i$ along the whole length of the wall. The $\boldsymbol{H}_s$ field *must* therefore include a $\psi_0 = 1$ contribution. The consequences of omitting this contribution appear clearly in Fig. 3.16b (Van Bladel 1981).

3.29 Derivation of the dyadics from a differential equation

Because $\boldsymbol{E}$ (and therefore curl $\boldsymbol{H}$) are perpendicular to the wall S_w, the differential equation and boundary condition to be satisfied by $\mathbf{G}_m$ are

$$\begin{aligned} -\operatorname{curl}\operatorname{curl}\mathbf{G}_m + k^2\mathbf{G}_m &= \operatorname{curl}[\delta(\boldsymbol{r}-\boldsymbol{r}')\mathbf{I}], \\ \boldsymbol{u}_n \times \operatorname{curl}\mathbf{G}_m &= 0 \quad \text{on } S_w. \end{aligned} \tag{3.228}$$

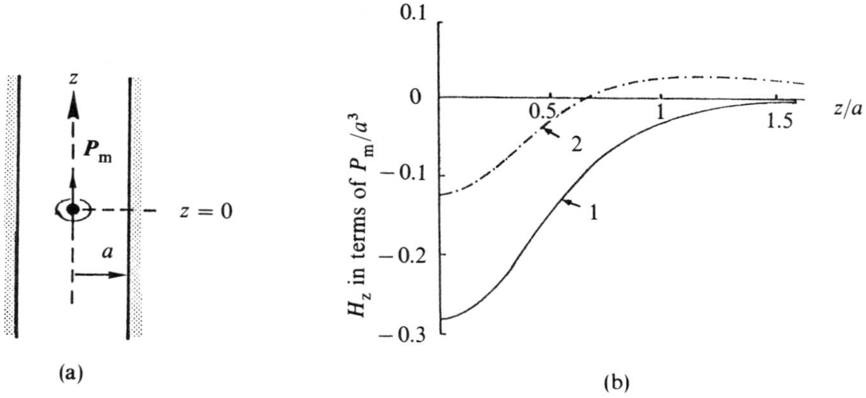

Fig. 3.16. (a) A small loop (dipole moment $P_m \boldsymbol{u}_z$) in a circular waveguide. (b) H_z along the wall in the low-frequency limit ($ka \to 0$): exact results in curve 1; ψ_0 contribution omitted in curve 2. (from Van Bladel (1981: p. 250) © IEE).

The $\mathbf{G}_m$ dyadic is clearly solenoidal (in the sense that div $\mathbf{G}_m = 0$). The first step in obtaining $\mathbf{G}_m$ consists in expanding both $\mathbf{G}_m$ and the right-hand member of (3.228) in the eigenvectors defined in (3.193) (Tai 1973a, b, 1983). Because $\mathbf{G}_m$ is solenoidal, only the $\boldsymbol{M}$ and $\boldsymbol{N}$ vectors are needed. We omit the details but show, as a matter of illustration, the form of the expansion appropriate for the source term in (3.228). This is

$$\operatorname{curl}\,[\delta(\boldsymbol{r}-\boldsymbol{r}')\mathbf{I}] = \int_{-\infty}^{\infty}\left(\sum_{m=0}^{\infty} \boldsymbol{M}_m(\gamma)\boldsymbol{A}_m(\gamma) + \sum_{n=0}^{\infty} \boldsymbol{N}_n(\gamma)\boldsymbol{B}_n(\gamma)\right)\mathrm{d}\gamma. \qquad (3.229)$$

The vectors $\boldsymbol{A}_m$ and $\boldsymbol{B}_n$ are proportional to curl$'\,\boldsymbol{M}_m(-h)$ and curl $\boldsymbol{N}'_n(-h)$, with proportionality coefficients easily determined. Solution of (3.228) yields $\mathbf{G}_m$ in the form of an integral of type (3.229), which must be evaluated by the techniques of complex integration. The end result reproduces (3.192). Similar steps confirm the relationships (3.205) and (3.206). Specific values of $\mathbf{G}_e$ and $\mathbf{G}_m$ are available for the rectangular, circular, elliptic, and coaxial waveguides (Tai 1973a, b, 1983), as well as for the cavity obtained by shorting the end planes of a rectangular waveguide (Rahmat-Samii 1975; Tai *et al.* 1976).

Derivation of the Green's dyadic from the differential equation is also possible for conical waveguides, and even for free space, considered as a 'spherical' waveguide. In the latter case, the fields can be written in terms of Hertz potentials as

$$\boldsymbol{E}(\boldsymbol{r}) = \frac{1}{\mathrm{j}\omega\varepsilon_0}\operatorname{curl}\operatorname{curl}(\pi_1 R\boldsymbol{u}_R) + \frac{1}{\mathrm{j}\omega\varepsilon_0}\operatorname{curl}\operatorname{curl}\operatorname{curl}(\pi_2 R\boldsymbol{u}_R) - \frac{1}{\mathrm{j}\omega\varepsilon_0}\boldsymbol{J},$$
$$\boldsymbol{H}(\boldsymbol{r}) = \operatorname{curl}(\pi_1 R\boldsymbol{u}_R) + \operatorname{curl}\operatorname{curl}(\pi_2 R\boldsymbol{u}_R). \qquad (3.230)$$

The πs are conveniently expressed in terms of spherical harmonics. For example,

$$\pi_1(R, \varphi, \theta) = \sum_{n=0}^{\infty} \sum_{m=1}^{n} [A_{nm}(R)\sin m\varphi + B_{nm}(R)\cos m\varphi]\,\mathrm{P}_n^m(\cos\theta), \tag{3.231}$$

with a similar expression for π_2. The function P_n^m is an associated Legendre function of the first kind. The determination of the R-dependent coefficients in terms of the current $\boldsymbol{J}(\boldsymbol{r})$ is a classical problem (see e.g. Van Bladel 1985). Once it is solved, the form of $\mathbf{G}_e$ and $\mathbf{G}_m$ in spherical coordinates follows immediately. Here, again, it is useful to express the dyadics in terms of $\boldsymbol{M}$, $\boldsymbol{N}$, $\boldsymbol{L}$ eigenvectors (Stratton 1941; Tai 1971; Collin 1986a). In the present case these are

$$\boldsymbol{M}_{nm}(\lambda, \boldsymbol{r}) = \operatorname{curl}(\lambda\psi_{nm}R\boldsymbol{u}_R), \qquad \boldsymbol{N}_{nm}(\lambda, \boldsymbol{r}) = (1/\lambda)\operatorname{curl}\operatorname{curl}(\lambda\psi_{nm}R\boldsymbol{u}_R),$$
$$\boldsymbol{L}_{nm}(\lambda, \boldsymbol{r}) = \operatorname{grad}\psi_{nm}, \tag{3.232}$$

where

$$\psi_{nm} = [Z_n^1(\lambda R)\sin m\varphi + Z_n^2(\lambda R)\cos m\varphi]\,\mathrm{P}_n^m(\cos\theta) \tag{3.233}$$

and the Z_n are solutions of the spherical Bessel equation. By use of spectral methods, one arrives at the expression

$$\mathbf{G}_e = \int_0^{\infty} \sum_{n,m} \left(\frac{\boldsymbol{M}_{nm}(\lambda, \boldsymbol{r})\boldsymbol{M}_{nm}(\lambda, \boldsymbol{r}') + \boldsymbol{N}_{nm}(\lambda, \boldsymbol{r})\boldsymbol{N}_{nm}(\lambda, \boldsymbol{r}')}{n(n+1)Q_{nm}(k^2 - \lambda^2)} + \frac{\boldsymbol{L}_{nm}(\lambda, \boldsymbol{r})\boldsymbol{L}_{nm}(\lambda, \boldsymbol{r}')}{k^2 Q_{nm}} \right) \mathrm{d}\lambda, \tag{3.234}$$

where Q_{nm} is the normalization factor

$$Q_{nm} = \frac{\varepsilon_{0m}\pi^2}{(2n+1)} \frac{(n+m)!}{(n-m)!}. \tag{3.235}$$

In this formula, $\varepsilon_{00} = 2$, and $\varepsilon_{0m} = 1$ for $m > 0$. The dyadic has the alternate form

$$\mathbf{G}_e = (-1/k^2)\operatorname{curl}\mathbf{G}_m + (1/k^2)\mathbf{I}\,\delta(\boldsymbol{r} - \boldsymbol{r}'), \tag{3.236}$$

where

$$-\frac{1}{k^2}\operatorname{curl}\mathbf{G}_m = \int_0^{\infty} \sum_{n,m} \frac{\lambda^2(\boldsymbol{M}_{nm}\boldsymbol{M}'_{nm} + \boldsymbol{N}_{nm}\boldsymbol{N}'_{nm})}{n(n+1)k^2(k^2 - \lambda^2)Q_{nm}}\,\mathrm{d}\lambda.$$

References

Arcioni, P., Bressan, M., and Conciauro, G. (1988). A new algorithm for the wide-band analysis of arbitrarily shaped planar circuits. *IEEE Transactions on Microwave Theory and Techniques*, **36**, 1426–37.

Asvestas, J.S. (1983). Comments on "Singularity in Green's function and its numerical evaluation". *IEEE Transactions on Antennas and Propagation*, **31**, 174–7.

Ball, J.A.R., and Khan, P.J. (1980). Source region electric field derivation by a dyadic Green's function approach. *IEE Proceedings*, **H127**, 301–4.

Belinfante, F.J. (1946). On the longitudinal and the transversal delta-function, with some applications. *Physica*, **XII**, 1–16.

Bressan, M., and Conciauro, G. (1985). Singularity extraction from the electric Green's function for a spherical resonator. *IEEE Transactions on Microwave Theory and Techniques*, **33**, 407–14.

Casey, J.P., and Bansal, R. (1988). Square helical antenna with a dielectric core. *IEEE Transactions on Electromagnetic Compatibility*, **30**, 429–36.

Chen, K.M. (1977). A simple physical picture of tensor Green's function in source region. *Proceedings of the IEEE*, **65**, 1201–4.

Chew, W.C. (1989). Some observations on the spatial and eigenfunction representations of dyadic Green's functions. *IEEE Transactions on Antennas and Propagation*, **37**, 1322–7.

Collin, R.E. (1973). On the incompleteness of E and H modes in waveguides. *Canadian Journal of Physics*, **51**, 1135–40.

Collin, R.E. (1986*a*). Dyadic Green's function expansions in spherical coordinates. *Electromagnetics*, **6**, 183–207.

Collin, R.E. (1986*b*). The dyadic Green's function as an inverse operator. *Radio Science*, **21**, 883–90.

Collin, R.E. (1990). *Field theory of guided waves*, IEEE Press, New York. Revised and expanded version of a text first published by McGraw-Hill in 1960, New York.

Coulomb's Law Committee (1950). The teaching of electricity and magnetism at the college level. *American Journal of Physics*, **18**, 1–25.

Daniele, V.G. (1982). New expressions of dyadic Green's functions in uniform waveguides with perfectly conducting walls. *IEEE Transactions on Antennas and Propagation*, **30**, 497–9.

Daniele, V.G., and Orefice, M. (1984). Dyadic Green's functions in bounded media. *IEEE Transactions on Antennas and Propagation*, **32**, 193–6.

De Zutter, D. (1982). The dyadic Green's function for the Fourier spectra of the fields from harmonic sources in uniform motion. *Electromagnetics*, **2**, 221–37.

Durand, E. (1964). *Electrostatique, I. Les distributions*, pp. 243–6,300,304. Masson, Paris.

Faché, N., Van Hese, J., and De Zutter, D. (1989). Generalized space domain Green's dyadic for multilayered media with special application to microwave interconnections. *Journal of Electromagnetic Waves and Applications*, **3**, 651–69.

Fikioris, J.G. (1965). Electromagnetic field inside a current-carrying region. *Journal of Mathematical Physics*, **6**, 1617–20.

Fikioris, J.G. (1988). The electromagnetic field of constant-density distributions in finite regions. *Journal of Electromagnetic Waves and Applications*, **2**, 141–53.

Gel'fand, I.M., and Shilov, G.E. (1964). *Generalized Functions*, Vol. 1. Academic Press, New York.

Günter, N.M. (1967). *Potential theory and its applications to basic problems of mathematical physics*. F. Ungar, New York.

Hadidi, A., and Hamid, M. (1988). Electric and magnetic dyadic Green's functions of bounded regions. *Canadian Journal of Physics*, **66**, 249–57.

Howard, A.Q. (1974). On the longitudinal component of the Green's function dyadic. *Proceedings of the IEEE*, **62**, 1704–5.

Howard, A.Q., and Seidel, D.B. (1978). Singularity extraction in kernel functions in closed region problems. *Radio Science*, **13**, 425–9.

Johnson, W.A., Howard, A.Q., and Dudley, D.G. (1979). On the irrotational component of the electric Green's dyadic. *Radio Science*, **14**, 961–7.

Kastner, R. (1987). On the singularity of the full spectral Green's dyadic. *IEEE Transactions on Antennas and Propagation*, **35**, 1303–5.

Kellogg, O.D. (1967). *Foundations of Potential Theory*, pp. 18, 152. Springer Verlag, New York.

Lakhtakia, A., Varadan, V.V., and Varadan, V.K. (1989). Time-harmonic and time-dependent dyadic Green's functions for some uniaxial gyro-electromagnetic media. *Applied Optics*, **28**, 1049–52.

Lee, S.W., Boersma, J., Law, C.L., and Deschamps, G.A. (1980). Singularity in Green's function and its numerical evaluation. *IEEE Transactions on Antennas and Propagation*, **28**, 311–7.

Livesay, D.E., and Chen, K.M. (1974). Electromagnetic fields induced inside arbitrarily shaped biological bodies. *IEEE Transactions on Microwave Theory and Techniques*, **22**, 1273–80.

Lyon, R.W., and Hizal, A. (1983). A moment method analysis of narrow dielectric covered rectangular slots in the broad wall of a rectangular waveguide. *Proceedings of the IEE International Conference on Antennas and Propagation*, 150–3.

MacMillan, W.D. (1958). *The Theory of the Potential*. p. 227, Dover Publications, New York.

Martensen, E. (1968). *Potentialtheorie*, p. 206. Teubner, Stuttgart.

Mikhlin, S.G. (1970). *Mathematical Physics, an Advanced Course*, pp. 344–370, North-Holland, Amsterdam.

Morse, P.M., and Feshbach, H. (1953). *Methods of Theoretical Physics*, Chapter 13. McGraw Hill, New York.

Müller, C. (1969). *Foundations of the Mathematical Theory of Electromagnetic Waves*, Springer, Berlin.

Pan Shenggen (1986). On the question of the dyadic Green's function at the source region. *Kexue Tongbao*, **31**, 725–30.

Pathak, P.H. (1983). On the eigenfunction expansion of electromagnetic dyadic Green's functions. *IEEE Transactions on Antennas and propagation*, **31**, 837–46.

Rahmat-Samii, Y. (1975). On the question of computation of the dyadic Green's function at the source region in waveguides and cavities. *IEEE Transactions on Microwave Theory and Techniques*, **23**, 762–5.

Sangster, A.J., and McCormick, A.H.I. (1987). Some observations on the computer-aided design/synthesis of slotted-waveguide antennas. *International Journal of Electronics*, **62**, 641–62.

Sarkar, T.K., Arvas, E., and Ponnapalli, S. (1989). Electromagnetic scattering from dielectric bodies. *IEEE Transactions on Antennas and Propagation*, **37**, 673–6.

Silberstein, M.E. (1991). Application of a generalized Leibniz rule for calculating electromagnetic fields within continuous source regions. To appear in *Radio Science*, **26**.

Sternberg, W.J., and Smith, T.L. (1952). *The Theory of Potential and Spherical Harmonics*. The University of Toronto Press, Toronto.

STRATTON, J.A. (1941). *Electromagnetic Theory*, p. 418. McGraw-Hill, New York.

TAI, C.T. (1971). *Dyadic Green's Functions in Electromagnetic Theory*, Intext, Scranton, Pa.

TAI, C.T. (1973*a*). On the eigenfunction expansion of dyadic Green's functions. *Proceedings of the IEEE*, **61**, 480–1.

TAI, C.T. (1973*b*). *Eigen-function expansion of dyadic Green's functions*. Mathematic Note 28. Weapons Systems Laboratory, Kirtland Air force Base, Albuquerque, N.M., U.S.A.

TAI, C.T., and Rozenfeld, D. (1976). Different representations of dyadic Green's functions for a rectangular cavity. *IEEE Transactions on Microwave Theory and Techniques*, **24**, 597–601.

TAI, C.T. (1981). Equivalent layers of surface charge, current sheet, and polarization in the eigenfunction expansions of Green's functions in electromagnetic theory. *IEEE Transactions on Antennas and Propagation*, **29**, 733–9.

TAI, C.T. (1983). Dyadic Green's functions for a coaxial line. *IEEE Transactions on Antennas and Propagation*, **31**, 355–8.

VAN BLADEL, J. (1960). On Helmholtz' theorem in multiply-bounded and multiply-connected regions. *Journal of the Franklin Institute*, **269**, 445–62.

VAN BLADEL, J. (1961). Some remarks on Green's dyadic for infinite space. *IRE Transactions on Antennas and Propagation*, **9**, 503–6.

VAN BLADEL, J. (1981). Contribution of the ψ = constant mode to the modal expansion in a waveguide. *IEE Proceedings*, **H128**, 247–51.

VAN BLADEL, J. (1985). *Electromagnetic Fields*, pp. 10, 16, 194, 205. Hemisphere, Washington. Reprinted, with corrections, from a text published in 1964 by McGraw-Hill, New York.

WANG, J.J.H. (1982). A unified and consistent view on the singularities of the electric dyadic Green's function in the source region. *IEEE Transactions on Antennas and Propagation*, **30**, 463–8.

WANG, J.J.H. (1990). *Generalized Moment Methods in Electromagnetics—Formulation and Computer Solution of Integral Equations*. J. Wiley, New York.

WEIGLHOFER, W. (1989*a*). A simple and straightforward derivation of the dyadic Green's function of an isotropic chiral medium. *Archiv für Elektronik und Übertragungstechnik*, **43**, 51–2.

WEIGLHOFER, W. (1989*b*). Delta-function identities and electromagnetic field singularities. *American Journal of Physics*, **57**, 455–6.

YAGHJIAN, A.D. (1980). Electric dyadic Green's functions in the source region. *Proceedings of the IEEE*, **68**, 248–63.

YAGHJIAN, A.D. (1981). Reply to Tai, C.T., Comments on electric dyadic Green's functions in the source region, *Proceedings of the IEEE*, **69**, 282–5.

YAGHJIAN, A.D. (1982). A delta-distribution derivation of the electric field in the source region. *Electromagnetics*, **2**, 161–7.

YAGHJIAN, A.D. (1985). Maxwellian and cavity electromagnetic fields within continuous sources. *American Journal of Physics*, **53**, 859–63.

4

SINGULARITIES AT AN EDGE

Wedges consisting of one or more materials are frequently encountered in practical structures (Fig. 4.1). The wedges are often quite sharp; in that case, their sharpness is routinely assumed to be infinite. This idealized model, which is only valid when the radius of curvature of the edge is negligible with respect to other relevant lengths, leads to the presence of 'infinities' in fields and source densities. It is important, for electromagnetic practice, to know the mathematical nature of these singularities. One of our goals is to show why this is so. Our second purpose is to give detailed information on the singular fields, and in particular on their components and singularity exponents. The mathematics involved are quite elementary.

4.1 The importance of understanding field behaviour at an edge

The rationale for including the present Chapter is well illustrated by a study of the fields near a charged strip. Assume first that the surface charge density ρ_s is uniform (Fig. 4.2a). The x component of the resulting electric field is given by (Durand 1964)

$$e_x = (\rho_s/2\pi\varepsilon_0)\log_e(r_1/r_2). \tag{4.1}$$

On the strip, this component is directed towards the end points, where it becomes infinite. Assume now that the strip is perfectly conducting. The boundary condition at the metal requires e_x to vanish between $x = a$ and $x = -a$. Such a requirement implies the presence of a strong charge density near the end points, strong enough to create a centre-directed e'_x which precisely cancels out the contribution e''_x from the central charges (Fig. 4.2b). This expectation is confirmed by the actual analytical form of ρ_s, which is (Durand 1964)

$$\rho_s = \rho_L/\pi a\sqrt{[1-(x/a)^2]}. \tag{4.2}$$

Here ρ_L is the linear charge density on the strip (in $\mathrm{C\,m^{-1}}$). We observe that, near the end points, ρ_s is proportional to the inverse of the square root of the distance d to these points. Knowing the nature of this behaviour is of great interest, because it reveals the following interesting features:

(1) Although ρ_s becomes infinite for $x = \pm a$, the integrated charge in the vicinity of each end point is proportional to $\sqrt{d}$, and therefore approaches zero with d. We conclude that no 'linear charge density' accumulates at the edge.

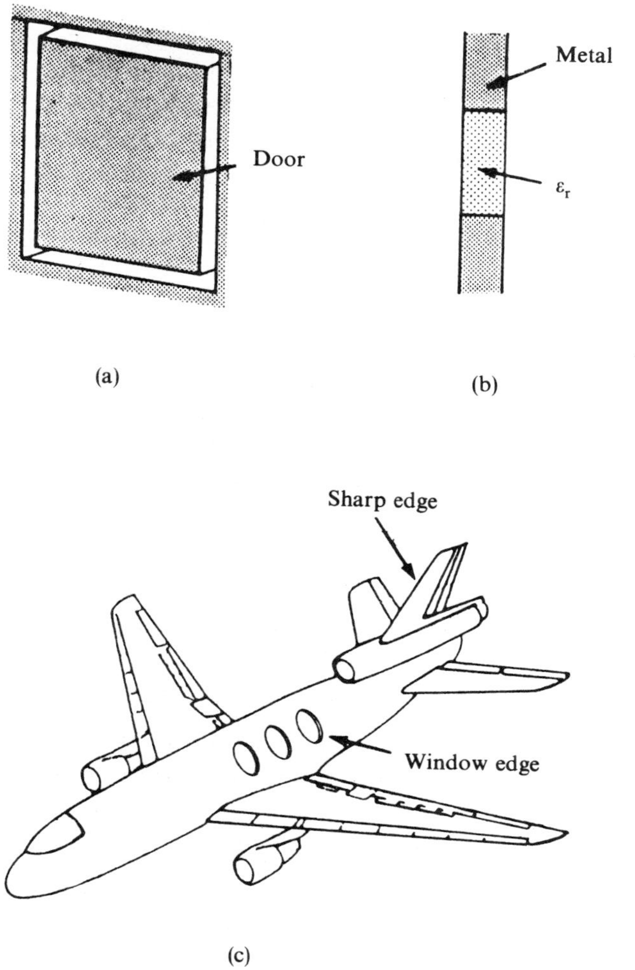

Fig. 4.1. Examples of sharp edges in practical structures. (a) Cracks around the door of a Faraday cage. (b) Perfectly conducting wall with dielectric window. (c) Edges present in an aircraft frame.

(2) The electric field near the end points is proportional to $1/\sqrt{d}$. This property makes it possible to estimate the extent of the region in which electric breakdown may occur, an important factor in high-voltage applications.

The results obtained for the strip may be extrapolated, mutatis mutandis, to other structures presenting sharp edges. In all these cases, the knowledge of

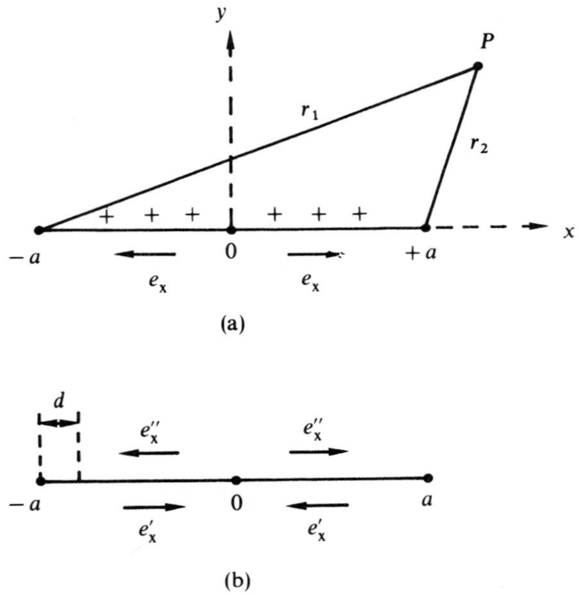

Fig. 4.2. (a) A uniformly charged strip. (b) A charged metallic strip and tangential electric fields.

the singular behaviour presents the following advantages:

(1) The singularity may be incorporated into the numerical algorithm used to solve the problem. Such a move increases the speed of convergence of the numerical process.
(2) When the singularity is *not* incorporated into the algorithm, the accuracy of the chosen numerical method may be checked by verifying whether fields and sources behave as predicted in the vicinity of 'singular' points (or zones).
(3) The same checks apply to analytical solutions.

These various advantages will be discussed in more detail in later sections.

4.2 The singularity exponent near a perfectly conducting wedge

A typical wedge is shown in Fig. 4.3a. Its boundary is often curved, as at the rim of a circular aperture. At distances d much smaller than the main radius of curvature of the edge, the latter may be considered as 'straight', and the wedge becomes two-dimensional (Poincelot 1963; Jones 1964; Ruck *et al.* 1970).

The singular behaviour at an edge follows from the requirement that the energy per m be finite along the edge, or, equivalently, that the energy density

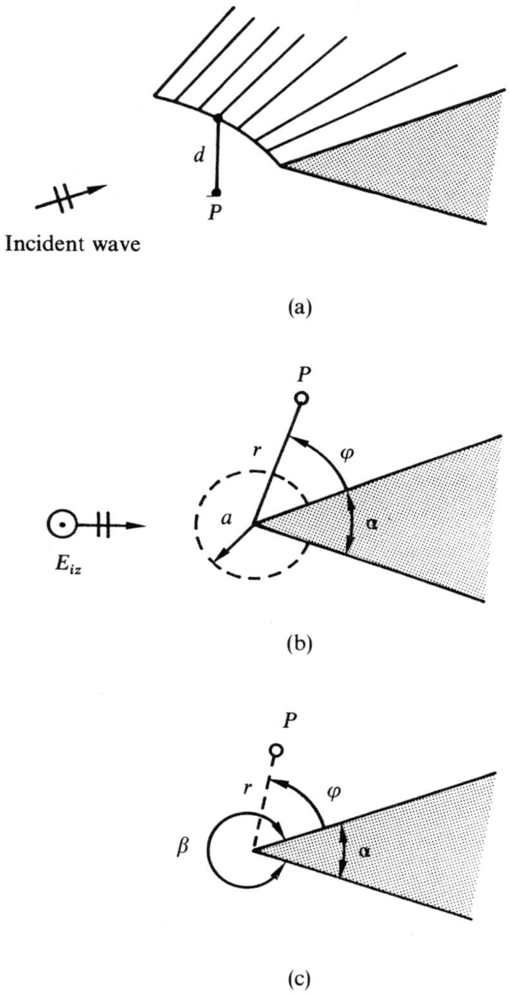

Fig. 4.3. (a) A curved metallic wedge. (b) A perfectly conducting wedge immersed in an E wave. (c) Coordinates relevant for the determination of ν.

near the edge remain integrable (Meixner 1949). To illustrate the point, consider a wave with an electric field parallel to the edge (an E wave: Fig. 4.3b). In a cylinder of radius a, and in application of the energy requirement, the integral

$$E = \tfrac{1}{2}\varepsilon_0 \int_0^{2\pi-\alpha} \int_0^a |E_z|^2 \, r \, \mathrm{d}r \, \mathrm{d}\varphi \quad (\mathrm{J\,m^{-1}}) \tag{4.3}$$

must remain finite. Assume now that E_z becomes proportional to r^ν at small distances r. From (4.3), the finiteness of E requires ν to be larger than -1. The magnetic field corresponding to E_z, on the other hand, is proportional to curl $\mathbf{E}$, hence to $r^{\nu-1}$. Since the magnetic energy must remain finite as well, we conclude that ν must be positive. A positive ν implies that E_z, component parallel to the edge, vanishes at the edge. This property is in agreement with the known boundary conditions at a perfect conductor.

The next step in our search for the actual value of ν is to assume that the singularity of E_z (and that of its derivatives) results from a factor r^ν. We write

$$E_z = r^\nu[a_0(\varphi, z) + r\, a_1(\varphi, z) + r^2 a_2(\varphi, z) + \cdots]. \tag{4.4}$$

The E_z component must satisfy the Helmholtz equation

$$\frac{\partial^2 E_z}{\partial r^2} + \frac{1}{r}\frac{\partial E_z}{\partial r} + \frac{1}{r^2}\frac{\partial^2 E_z}{\partial \varphi^2} + \frac{\partial^2 E_z}{\partial z^2} + k^2 E_z = 0. \tag{4.5}$$

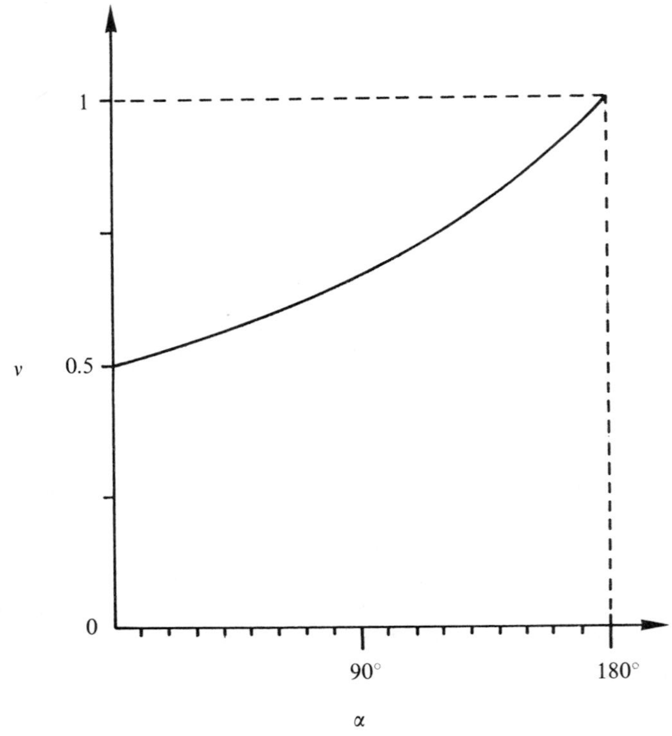

Fig. 4.4. The singularity exponent ν as a function of the opening angle α.

Introducing (4.4) into this equation yields

$$r^{\nu-2}\left(\frac{d^2a_0}{d\varphi^2}+\nu^2a_0\right)+r^{\nu-1}\left(\frac{d^2a_1}{d\varphi^2}+(\nu+1)^2a_1\right)$$
$$+r^{\nu}\left(\frac{d^2a_2}{d\varphi^2}+(\nu+2)^2a_2+\frac{d^2a_0}{dz^2}+k^2a_0\right)+\cdots=0. \tag{4.6}$$

The coefficient of the term in $r^{\nu-2}$ must vanish, which implies that a_0 must be a linear combination of $\sin\nu\varphi$ and $\cos\nu\varphi$. Since E_z vanishes at the surface of the wedge, i.e. for $\varphi=0$ and $\varphi=\beta$ (Fig. 4.3c), the solution is (Motz 1947):

$$a_0=A\sin\nu_n\varphi,\qquad \nu_n=n\pi/(2\pi-\alpha)=n\pi/\beta\quad(n=1,2,\dots). \tag{4.7}$$

A similar kind of argument can be used for an H wave. The boundary condition now requires $\partial H_z/\partial\varphi$ to vanish at the surface of the wedge, which yields

$$a_0=B\cos\tau_n\varphi,\qquad \tau_n=n\pi/(2\pi-\alpha)=n\pi/\beta\quad(n=1,2,\dots). \tag{4.8}$$

We observe that both E and H waves have the same singularity exponent. The main singularity corresponds to the lowest value of ν_n (or τ_n), obtained for $n=1$. This value will henceforth be denoted by ν. The minimum value of ν, equal to $\frac{1}{2}$, occurs for $\alpha=0$, i.e. for a half-infinite plane. For higher values of α, the exponent increases, reaches unity for a plane boundary ($\alpha=\pi$), and becomes larger than one for a reentrant wedge. The law of variation of ν in terms of α is shown in Fig. 4.4.

4.3 Fields, currents, and charges associated with a perfectly conducting wedge

The various components of the singular field are obtained by inserting the known singular values of E_z and H_z into Maxwell's equations. The result is

$$E_z=j\omega\mu_0 A\,r^{\nu}\sin\nu\varphi,\qquad H_z=j\omega\varepsilon_0 B\,r^{\nu}\cos\nu\varphi+\text{constant},$$
$$\boldsymbol{E}_t=-\frac{\nu B}{r^{1-\nu}}(\boldsymbol{u}_r\sin\nu\varphi+\boldsymbol{u}_\varphi\cos\nu\varphi),\qquad \boldsymbol{H}_t=\frac{\nu A}{r^{1-\nu}}(-\boldsymbol{u}_r\cos\nu\varphi+\boldsymbol{u}_\varphi\sin\nu\varphi). \tag{4.9}$$

The subscript t denotes a transverse component, i.e. a component in a plane perpendicular to the edge. The symbols A and B are arbitrary coefficients. We observe that only transverse components can become infinite. The (possible) infinity of $\boldsymbol{E}_t$, in particular, is not surprising. This component, which must remain perpendicular to the conductor, suddenly changes its orientation at the edge. Such a discontinuity implies that $\boldsymbol{E}_t$ is either zero at the tip, or infinite there. The discussion in Section 4.4 confirms the correctness of this simplistic argument.

The components (4.9) are the singular parts, on which more regular parts are superimposed. We observe that infinities occur only for sharp wedges, and that their 'strength' is determined by the smallest value of ν_n, viz.

$$\nu = \pi/(2\pi - \alpha) = \pi/\beta. \tag{4.10}$$

Charges and currents at the surface of the wedge follow from the value of the field components. The charge density near the edge, for example, is

$$\rho_s = -\frac{\nu B \varepsilon_0}{r^{1-\nu}} \quad (\mathrm{C\,m^{-2}}). \tag{4.11}$$

This density, which has the same value on both sides of the wedge, becomes infinite for $r = 0$. The total charge contained between 0 and r, however, is

$$\rho_L = -2\varepsilon_0 B r^\nu \quad (\mathrm{C\,m^{-1}}). \tag{4.12}$$

It approaches zero with r. Consequently, no charge accumulation appears at the edge. Near the latter, the current density has the value

$$\mathbf{J}_s = \frac{\nu A}{r^{1-\nu}} \mathbf{u}_z + \mathrm{j}\omega\varepsilon_0 B r^\nu \mathbf{u}_r \quad (\mathrm{A\,m^{-1}}). \tag{4.13}$$

This current flows on both sides of the wedge. On a semi-infinite plane, the two contributions 'melt' into a single layer of value twice that given in (4.13). The formula shows that the component of $\mathbf{J}_s$ parallel to the edge becomes infinite. The total current flowing between 0 and r, however, is proportional to r^ν, which implies that no current concentration occurs at the edge. We also observe that the component of $\mathbf{J}_s$ perpendicular to the edge remains finite.

4.4 On the possible absence of singularities

The field singularities are only *potentially* present, and cases are known where they are not 'excited'. The symmetry of the configuration plays an important role in these examples. With respect to the symmetry axis $\varphi = \varphi_0$, the singular part of E_z may be written in the form (Fig. 4.5)

$$E_z = \mathrm{j}\omega\mu_0 A r^\nu \cos \nu(\varphi - \varphi_0). \tag{4.14}$$

This component is symmetric with respect to $\varphi = \varphi_0$. If the wedge is illuminated by a wave with an *antisymmetric* E_z, the field pattern (4.14) will not be excited. In such a case, the fields near the edge do *not* exhibit the expected singular behaviour, and remain finite. A typical example is shown in Fig. 4.6, which displays a cylinder illuminated by two plane waves, one incident along the direction $\mathbf{u}_1 = \alpha\mathbf{u}_x - \beta\mathbf{u}_y$, the other along $\mathbf{u}_2 = \alpha\mathbf{u}_x + \beta\mathbf{u}_y$. The total incident field is proportional to $\sin k\beta y$, and is therefore antisymmetric. As a consequence the fields remain finite at the edge of the wing (Vassallo 1985).

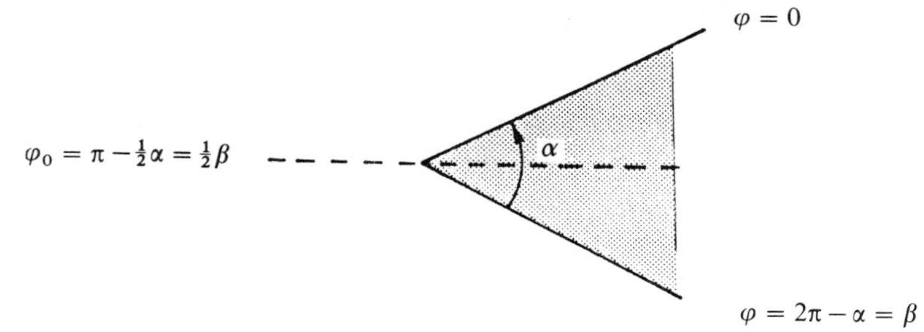

Fig. 4.5. The symmetry axis of a conducting wedge.

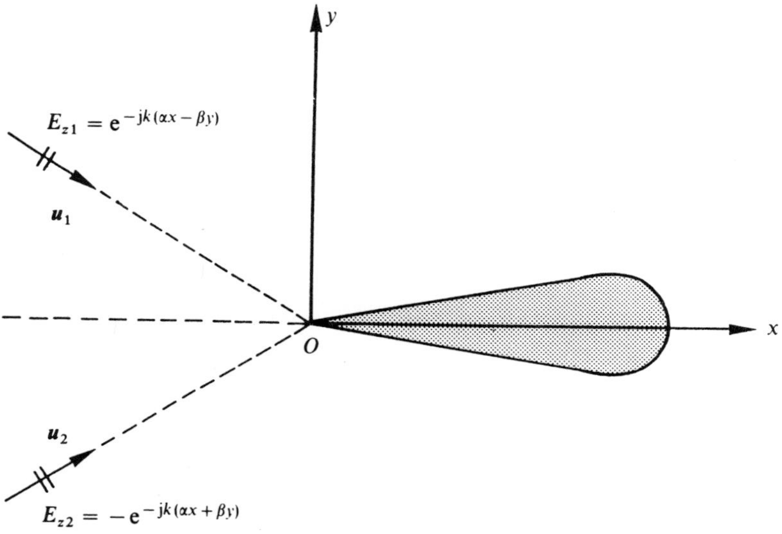

Fig. 4.6. An 'aircraft wing' in an antisymmetric incident wave.

Similar conclusions hold for an H wave, but in this case the singular H_z pattern is antisymmetric with respect to the ($\varphi = \varphi_0$) axis. The singularity is not excited when the incident H_z is *symmetric* with respect to that axis.

Special *dimensional* properties may also be responsible for the absence of infinities. A well-known example is the rectangular groove illuminated by a linearly polarized wave (Fig. 4.7). The groove must be $\frac{1}{2}\lambda$ deep. The total fields, sum of incident and scattered parts, are easily found to be

$$E_y = \mathrm{e}^{-\mathrm{j}kx} - \mathrm{e}^{\mathrm{j}kx} = -2\mathrm{j}\sin kx, \qquad H_z = \frac{1}{R_c}(\mathrm{e}^{-\mathrm{j}kx} + \mathrm{e}^{\mathrm{j}kx}) = \frac{2}{R_c}\cos kx. \tag{4.15}$$

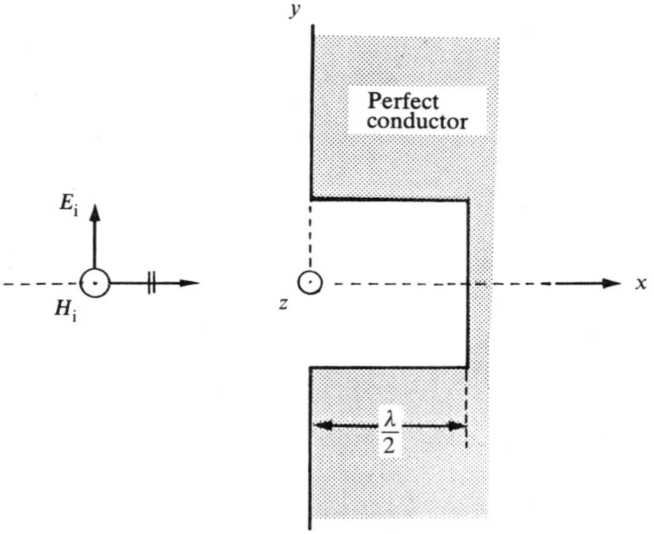

Fig. 4.7. An H wave incident on a rectangular groove.

We observe that these fields satisfy the boundary conditions on the perfect conductor, both in the groove and on the $x = 0$ plane. It is clear that neither E_z nor H_z are singular (Bolomey *et al.* 1971; Vassallo 1985).

To summarize: except under special symmetry conditions, the field components perpendicular to a sharp edge, together with the charge density and the current density parallel to the edge, become infinite like $r^{\nu-1}$. The value of ν lies between $\frac{1}{2}$ (semi-infinite plane) and 1 (full plane). The field components parallel to the edge, and the current density perpendicular to it, remain finite. The electric field parallel to the edge, in particular, must vanish like r^{ν}. These results are sketched graphically in Fig. 4.8.

4.5 Further comments on the edge condition

The E_z and H_z components must satisfy Laplace's equation at distances short with respect to λ. This requirement can serve as a basis for the evaluation of ν by an alternate method, consisting in Fourier-expanding E_z in terms of $\sin n\nu\varphi$, H_z in terms of $\cos n\nu\varphi$, and inserting these expansions into Laplace's equation. It should be remembered, in that respect, that Laplace's equation also has a solution of the form $(A + B \log_e r)(C\varphi + D)$. The logarithmic terms generate transverse fields proportional to $1/r$, an r-dependence which violates the 'bounded energy' requirement. The coefficient B must therefore be zero (Jones 1986). Furthermore, since E_z must vanish for $\varphi = 0$

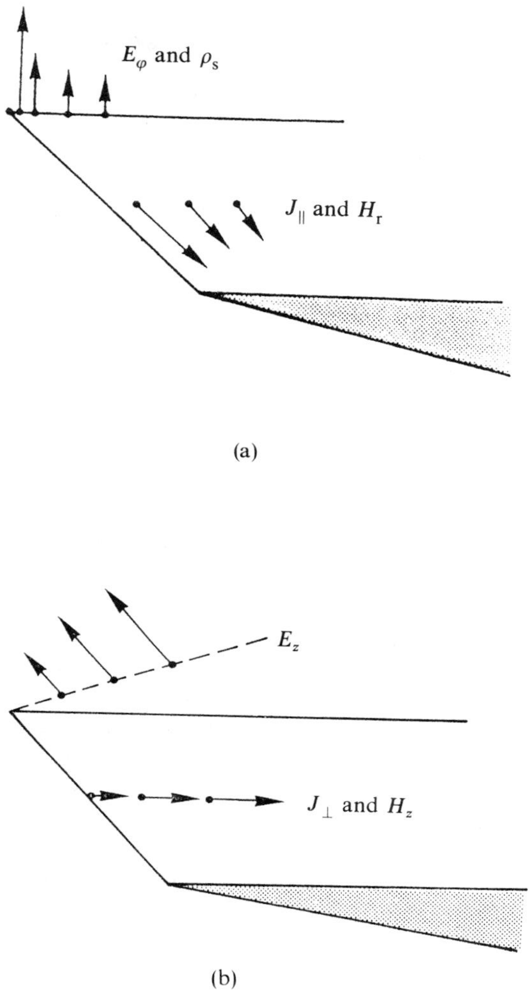

Fig. 4.8. The behaviour of fields, currents, and charges near a sharp edge.

and $\varphi = 2\pi - \alpha$, the coefficients C and D must vanish when the wave is of the E type. In an H wave, the term in D survives; it gives rise to the already mentioned term H_z = constant.

The predominance of the singular fields extends only to sufficiently short distances. We should remember, in our search for the limit, that the fields (4.9) were obtained by neglecting the $k^2r^2a_0$ term in (4.6) with respect to a_0. The singular behaviour is therefore limited to the region $r \ll \lambda$. This criterion,

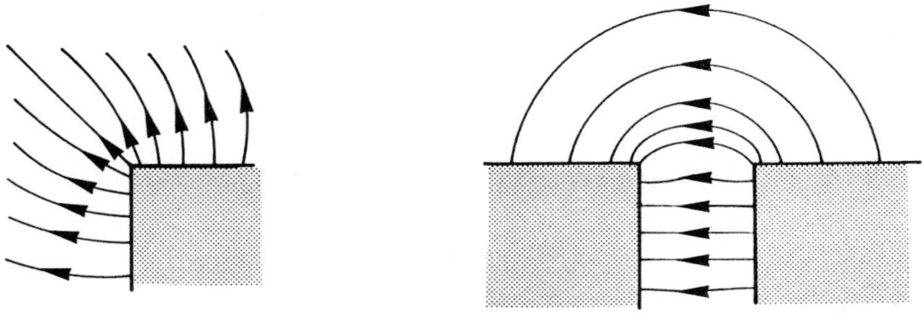

Fig. 4.9. (a) Lines of force in the vicinity of a 90° perfectly conducting wedge. (b) Lines of force between two 90° wedges.

which is confirmed by results obtained in Section 4.6, has obvious practical consequences, for example in the evaluation of a field component such as E_z at a set of discrete points (a 'net'). These points must be separated by distances small with respect to λ (typically less than $\frac{1}{8}\lambda$) to follow the phase variation of the field with sufficient accuracy. The predicted singular behaviour, therefore, is only appropriate for the first few points close to the edge. At these points, the knowledge of the behaviour can be exploited by a variety of schemes (Motz 1947). In the case of a semi-infinite plane, for example, E_z varies as $\sqrt{r}$; hence, if the first two points of the net are at respective distances $r_1 = d$ and $r_2 = 2d$, the ratio of the values of E_z at these points must be $1:\sqrt{2}$. The edge condition may therefore be enforced by setting $E_{z1} = E_{z2}/\sqrt{2}$.

The extent of the singularity zone is limited, not only by the distance to the apex, but also by changes in the shape of the wedge (specifically by variations in the curvature), or by the presence of other bodies. Figure 4.9 shows how the lines of force of $\boldsymbol{E}$ near a 90° metallic wedge ($\nu = \frac{2}{3}$) are distorted by the presence of a neighbouring wedge.

4.6 Scattering by a semi-infinite perfectly conducting plane

Only a few detailed analytical solutions involving bodies with wedges are available in the literature. They all confirm the singular behaviour given in (4.9). One such solution concerns a semi-infinite plane illuminated by a plane wave. This classical configuration was analysed by Sommerfeld at the turn of the century. The incident waves, of the E or H type, have z-oriented fields of the form (Fig. 4.10)

$$\begin{Bmatrix} E_{zi} \\ H_{zi} \end{Bmatrix} = e^{jkr\cos(\varphi - \varphi_i)}. \tag{4.16}$$

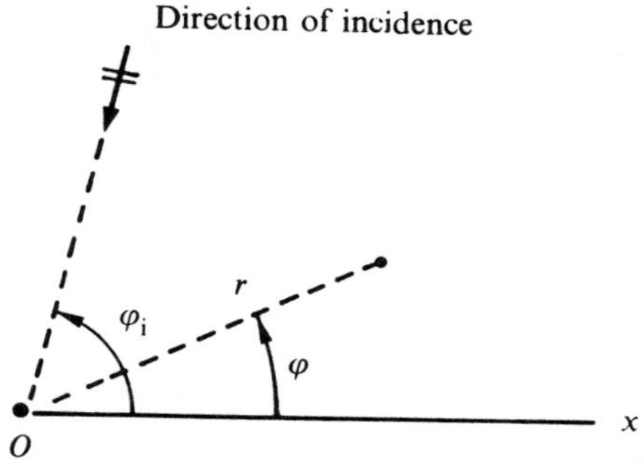

Fig. 4.10. A plane wave incident on a half-infinite conducting plane.

Dimensional factors have been omitted for the sake of simplicity. The general solution of the problem is classical (Sommerfeld 1964; Van Bladel 1985b). It yields

$$\begin{Bmatrix} E_z \\ H_z \end{Bmatrix} = \frac{1+\mathrm{j}}{2}\,\mathrm{e}^{\mathrm{j}kr\cos(\varphi-\varphi_i)}\int_{-\infty}^{2\sqrt{(kr/\pi)}\cos\frac{1}{2}(\varphi-\varphi_i)} \mathrm{e}^{-\frac{1}{2}\mathrm{j}\pi\tau^2}\,\mathrm{d}\tau$$

$$\mp\frac{1+\mathrm{j}}{2}\,\mathrm{e}^{\mathrm{j}kr\cos(\varphi+\varphi_i)}\int_{-\infty}^{2\sqrt{(kr/\pi)}\cos\frac{1}{2}(\varphi+\varphi_i)} \mathrm{e}^{-\frac{1}{2}\pi\tau^2}\,\mathrm{d}\tau. \tag{4.17}$$

The integrals are Fresnel integrals, which may be written as

$$\int_{-\infty}^{w} \mathrm{e}^{-\frac{1}{2}\mathrm{j}\pi\tau^2}\,\mathrm{d}\tau = \int_{-\infty}^{0} \mathrm{e}^{-\frac{1}{2}\mathrm{j}\pi\tau^2}\,\mathrm{d}\tau + \int_{0}^{w} \mathrm{e}^{-\frac{1}{2}\mathrm{j}\pi\tau^2}\,\mathrm{d}\tau$$

$$= \frac{1-\mathrm{j}}{2} + \int_{0}^{w} \mathrm{e}^{-\frac{1}{2}\mathrm{j}\pi\tau^2}\,\mathrm{d}\tau = \frac{1-\mathrm{j}}{2} + \mathrm{F}(w). \tag{4.18}$$

To investigate the fields at short distances, we use the expansion

$$\mathrm{F}(w) = w(1 - \tfrac{1}{6}\mathrm{j}\,\pi w^2 + \cdots). \tag{4.19}$$

Introducing this expansion into (4.17) and (4.18) yields

$$E_z = 2(1+\mathrm{j})\sqrt{(kr/\pi)}\,\sin\tfrac{1}{2}\varphi\,\sin\tfrac{1}{2}\varphi_i + \mathrm{j}kr\sin\varphi\sin\varphi_i$$
$$+ \text{ terms in } (kr)^{\frac{3}{2}} + \cdots, \tag{4.20}$$
$$H_z = 1 + 2(1+\mathrm{j})\sqrt{(kr/\pi)}\,\cos\tfrac{1}{2}\varphi\,\cos\tfrac{1}{2}\varphi_i + \mathrm{j}kr\cos\varphi\cos\varphi_i$$
$$+ \text{ terms in } (kr)^{\frac{3}{2}} + \cdots.$$

The predicted behaviour is confirmed: E_z vanishes like $\sqrt{r}$, and H_z remains finite, albeit with a 'singular' part proportional to $\sqrt{r}$. In the aperture plane (i.e. for $\varphi = \pi$):

$$
\begin{aligned}
E_z &= 2(1+\mathrm{j})\sqrt{(kr/\pi)}\,\sin\tfrac{1}{2}\varphi_i + \text{terms in } (kr)^{\frac{3}{2}} + \cdots, \\
H_z &= 1 - \mathrm{j}kr\cos\varphi_i + \cdots,
\end{aligned}
\tag{4.21}
$$

The absence of the $\sqrt{r}$ term in H_z was expected, since the singular H_z field given in (4.8) vanishes in the $(\varphi = \pi)$ plane. On top of the screen (i.e. for $\varphi = 0$),

$$
\begin{aligned}
E_z &= 0, \qquad H_r = -\frac{1}{R_c}\left(2(1-\mathrm{j})\frac{1}{\sqrt{(\pi kr)}}\sin\tfrac{1}{2}\varphi_i + \sin\varphi_i + \cdots\right), \\
H_z &= 1 + 2(1+\mathrm{j})\sqrt{(kr/\pi)}\,\cos\tfrac{1}{2}\varphi_i + \mathrm{j}\,kr\cos\varphi_i + \cdots, \\
E_\varphi &= -R_c\left((1-\mathrm{j})\frac{1}{\sqrt{(\pi kr)}}\cos\tfrac{1}{2}\varphi_i + \cos\varphi_i + \cdots\right).
\end{aligned}
\tag{4.22}
$$

On the lower side of the screen (i.e. for $\varphi = 2\pi$),

$$
\begin{aligned}
E_z &= 0, \qquad H_r = \frac{1}{R_c}\left(2(1-\mathrm{j})\frac{1}{\sqrt{(\pi kr)}}\sin\tfrac{1}{2}\varphi_i - \sin\varphi_i + \cdots\right), \\
H_z &= 1 - 2(1+\mathrm{j})\sqrt{(kr/\pi)}\,\cos\tfrac{1}{2}\varphi_i + \mathrm{j}\,kr\cos\varphi_i + \cdots, \\
E_\varphi &= R_c\left((1-\mathrm{j})\frac{1}{\sqrt{(\pi kr)}}\cos\tfrac{1}{2}\varphi_i - \cos\varphi_i + \cdots\right).
\end{aligned}
\tag{4.23}
$$

The charge density on the metal follows from the value of E_φ. Thus,

$$
\begin{aligned}
&\rho_s = 0 \quad \text{in an } E \text{ wave}, \\
&\rho_s = \varepsilon_0[E_\varphi]_0 - \varepsilon_0[E_\varphi]_{2\pi} = \frac{2}{c}(\mathrm{j}-1)\frac{1}{\sqrt{(\pi kr)}}\cos\tfrac{1}{2}\varphi_i + \cdots \quad \text{in an } H \text{ wave}.
\end{aligned}
\tag{4.24}
$$

The current density is similarly

$$
\begin{aligned}
J_{sz} &= -[H_r]_0 + [H_r]_{2\pi} = \frac{4}{R_c}(1-\mathrm{j})\frac{1}{\sqrt{(\pi kr)}}\sin\tfrac{1}{2}\varphi_i \quad \text{in an } E \text{ wave}, \\
J_{sr} &= [H_z]_0 - [H_z]_{2\pi} = 4(1+\mathrm{j})\sqrt{(kr/\pi)}\,\cos\tfrac{1}{2}\varphi_i \quad \text{in an } H \text{ wave}.
\end{aligned}
\tag{4.25}
$$

The general formulas of Section 4.3 are therefore confirmed.

We have mentioned, in Section 4.4, that the singular fields are not expected to remain dominant when r, the distance to the edge, becomes comparable to λ. The transition is shown with particular clarity in Figs 4.11 and 4.12, where it is observed to occur for $r \approx \lambda/2\pi$ (Morris 1982). In Fig. 4.11, which concerns an E wave, the current density $|J_{sz}|$ is expressed in terms of E_{zi}/R_c. Figures 4.11a,b represent $|J_{sz}|$ on the upper part of the screen, whereas Fig. 4.11c gives the same information for the lower part. The singular behaviour clearly

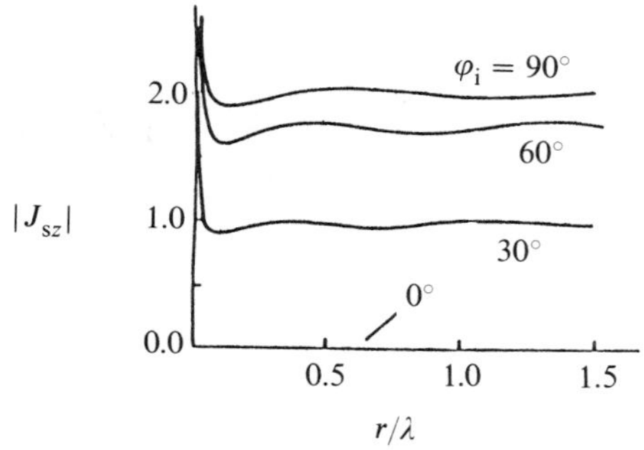

(a)

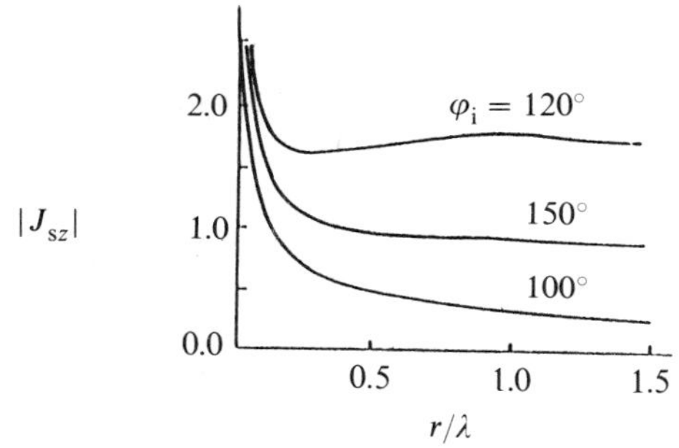

(b)

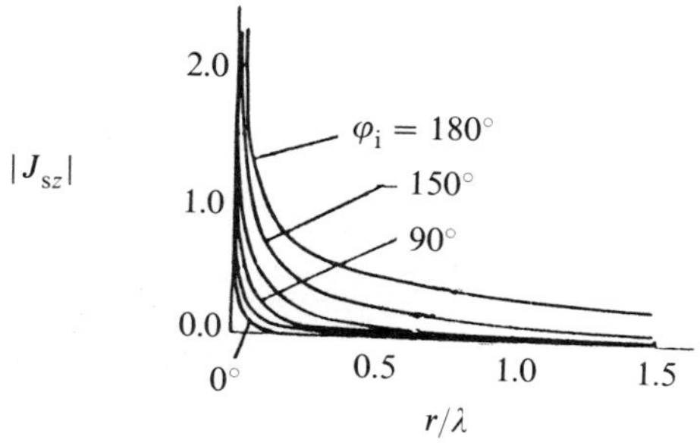

(c)

Fig. 4.11. Current density on a half-plane in an E wave: (a, b) upper part of the screen, (c) lower part of the screen. From Morris (1982: p. 611) © IEE.

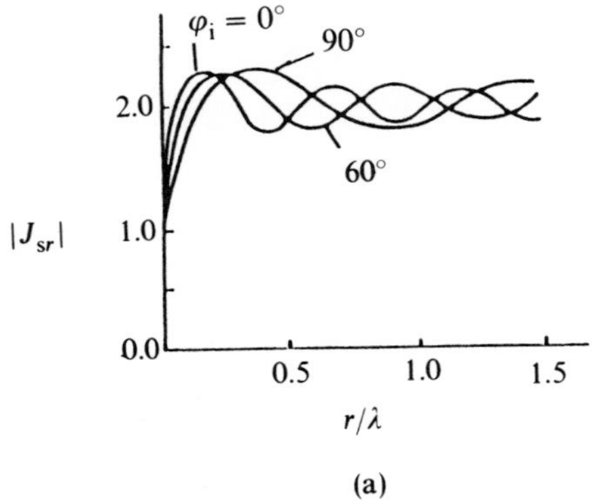

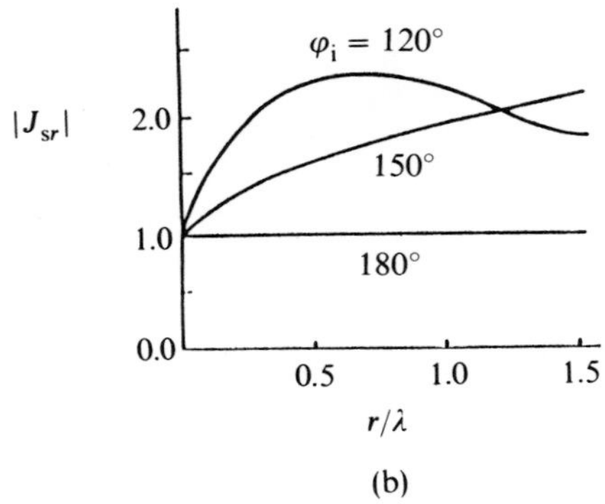

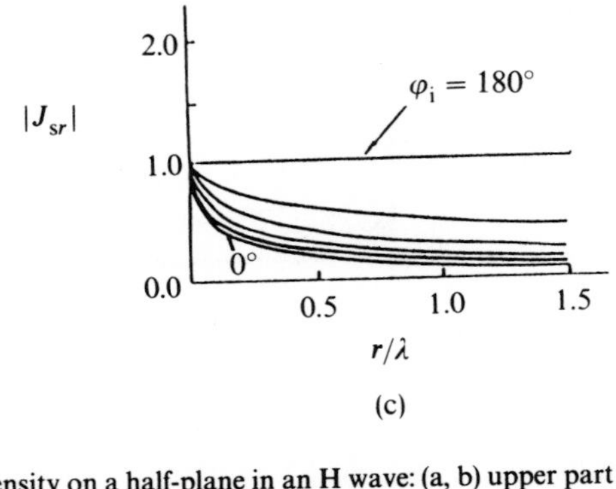

Fig. 4.12. Current density on a half-plane in an H wave: (a, b) upper part of the screen, (c) lower part of the screen. From Morris (1982: p. 612) © IEE.

dominates at small r. As r increases, the current progressively approaches the values which hold for an infinite plane, i.e. $(2E_i/R_c)\sin\varphi_i$ on the upper side, and zero on the lower one. Similar data appear in Fig. 4.12 for an H wave, where $|J_{sr}| = |H_z|$ is expressed in terms of H_{zi}. The values of $|J_{sr}|$ at large r tend asymptotically towards two on the upper side, and zero on the lower one.

4.7 More confirmations of the edge behaviour

Numerous additional confirmations of the validity of equations (4.9), both theoretical and experimental, are available in literature. On the analytical side, we already discussed, in Section 4.1, the charge distribution on a conducting strip. Another analytical check is afforded by the charge density on a conducting circular disk of radius a, raised to a potential V. This is (Durand 1964)

$$\rho_s = \frac{4\varepsilon_0 V}{\pi\sqrt{(a^2 - r^2)}} \tag{4.26}$$

We observe that ρ_s becomes infinite like $1/\sqrt{d} = 1/\sqrt{(a - r)}$, thereby confirming that the curvature of the edge does not influence the singularity when d is sufficiently small. Similar results hold for an elliptic disk (Smythe 1950). Finally, the charge density on a hollow conducting circular cylinder, raised to a potential V, is given by (Butler 1980)

$$\rho_s = \frac{2\varepsilon_0 V}{h\log_e(16a/h)} \frac{1}{\sqrt{[1 - (z/h)^2]}} \quad (h \ll a). \tag{4.27}$$

The expected singularity occurs at $z = \pm h$ (Fig. 4.13a).

Confirmation of the edge singularity is also afforded by 'numerical experiments', specifically those associated with the 'brute force' solution of integral equations. In such a solution, the edge behaviour is *not* incorporated into the algorithm, but the singularity shows up automatically in the final results. An early example (Mei *et al.* 1963) concerns the static charge density on a conducting rectangular cylinder. Detailed numerical results confirm that ρ_s becomes infinite like $d^{-\frac{1}{3}}$ at the 90° corners (Fig. 4.14).

Two more examples are worth mentioning:

(1) the charge density ρ_s on two parallel metallic strips, carried to respective potentials V and $-V$. The results confirm the $1/\sqrt{d}$ singularity (Fig. 4.15).

(2) the axial current on a hollow circular cylinder, acting as a shield for a radiating electric dipole (Fig. 4.13). The sharp edges are immersed in an H type of wave, therefore the current may be expected to be axial, and to vanish like $\sqrt{d}$ at the edge. The numerical results are in agreement with this prediction (Williams *et al.* 1987).

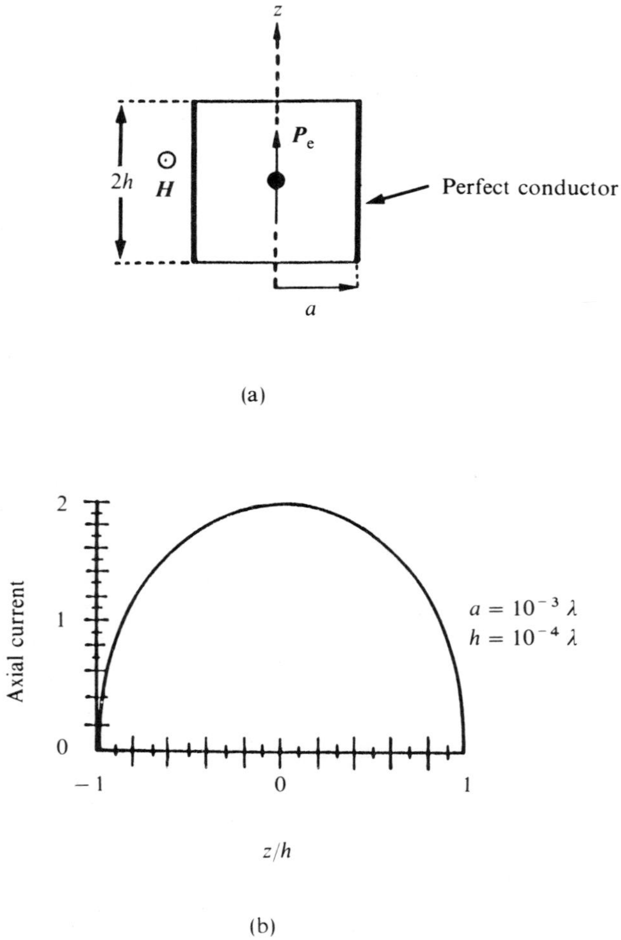

Fig. 4.13. (a) A metallic sleeve and radiating electric dipole. (b) The relative variation of the z-directed current on the sleeve. From Williams *et al.* (1987: p. 222) © Amer. Geoph. Union.

As a last check, we show a few *experimental* values relative to the surface current on an equilateral triangular cylinder (Iizuka *et al.* 1967). The experimental points confirm that J_s becomes infinite in an E wave (for which $\mathbf{J}_s$ is parallel to the edge, Fig. 4.16b), but remains finite in an H wave (for which $\mathbf{J}_s$ is perpendicular to the edge, Fig. 4.16c). In the latter case the theory predicts a vertical tangent to the curve at the edge points. The experimental data are in harmony with this prediction.

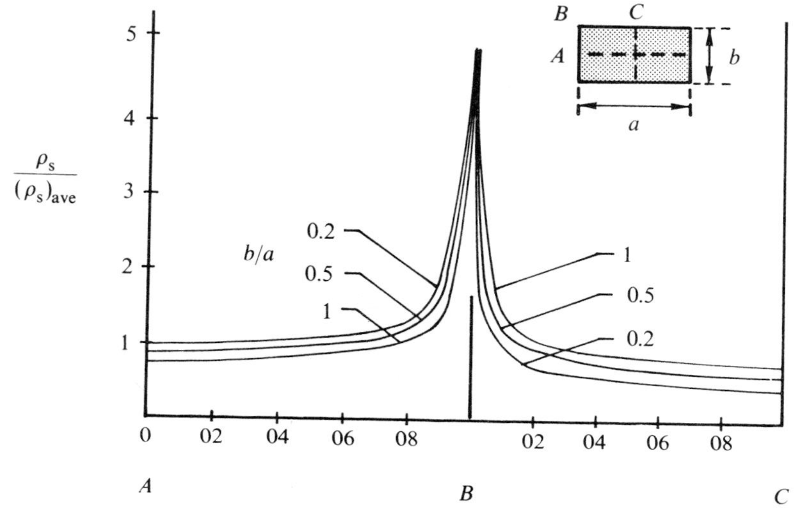

Fig. 4.14. Static charge density on a conducting rectangular cylinder. From Mei *et al.* (1963: p. 56) © IEEE.

Edges encountered in actual devices are never perfectly sharp. It is therefore interesting to investigate the influence of the curvature on, say, the distribution of the charge density on the metal. Data concerning a hyperbolic cylinder are shown in Fig. 4.17. They were obtained by introducing elliptic-cylindrical coordinates

$$x = a \cosh \eta \cos \psi, \qquad y = a \sinh \eta \sin \psi. \tag{4.28}$$

The hyperbola is defined by $\psi = \frac{1}{2}\alpha$ and $\psi = 2\pi - \frac{1}{2}\alpha$. The potential outside the cylinder is given by (R. De Smedt, private communication)

$$\phi = A \cosh \nu\eta \sin \nu(\psi - \tfrac{1}{2}\alpha) \quad (\tfrac{1}{2}\alpha \leqslant \psi \leqslant 2\pi - \tfrac{1}{2}\alpha). \tag{4.29}$$

Here ν is the singularity exponent defined in (4.10). The charge density follows as

$$\frac{\rho_s}{\varepsilon_0} \frac{a}{A\nu} = \frac{\cosh \nu\eta}{\sqrt{(\cosh^2\eta - \cos^2 \frac{1}{2}\alpha)}}. \tag{4.30}$$

At large distances from the tip, i.e. at large η, the charge density ρ_s becomes proportional to $r^{\nu-1}$, which is the variation associated with a sharp edge. The wedge may therefore be considered as 'sharp' as soon as the observer moves sufficiently far away from the tip. This rather trivial remark can be given quantitative content by plotting the actual variation of ρ_s along the sharp and rounded boundaries (Fig. 4.17b). In both cases the abscissa is the arc distance

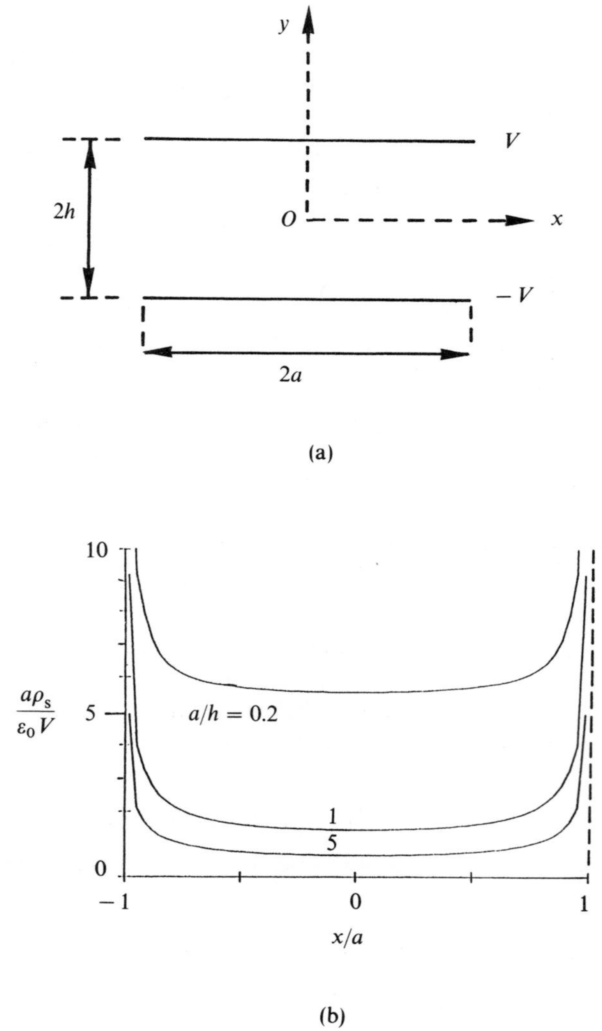

Fig. 4.15. (a) Geometry of a system of two perfectly conducting strips. (b) Charge density on conducting parallel strips. (Private communication from R. De Smedt.)

s, measured from the tip. The symbol R represents the radius of curvature of the hyperbola, equal to $a \cos \frac{1}{2}\alpha \tan^2 \frac{1}{2}\alpha$. It is found that the two laws differ by less than 2% as soon as the distance is of the order of $5R$. The transition from 'curved' to 'sharp' is further illustrated in Fig. 4.18, which shows a capped rectangular cylinder immersed in an incident E wave. In the limit of vanishing

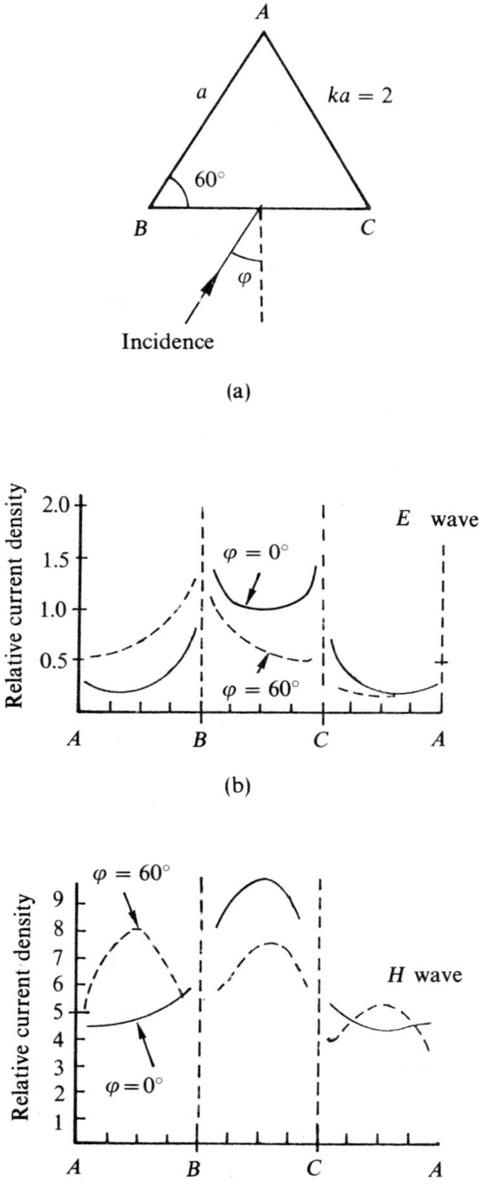

Fig. 4.16. (a) An equilateral triangular cylinder. (b) Current density in an E wave. (c) Current density in an H wave. From Iizuka et al. (1967: p. 798) © IEEE.

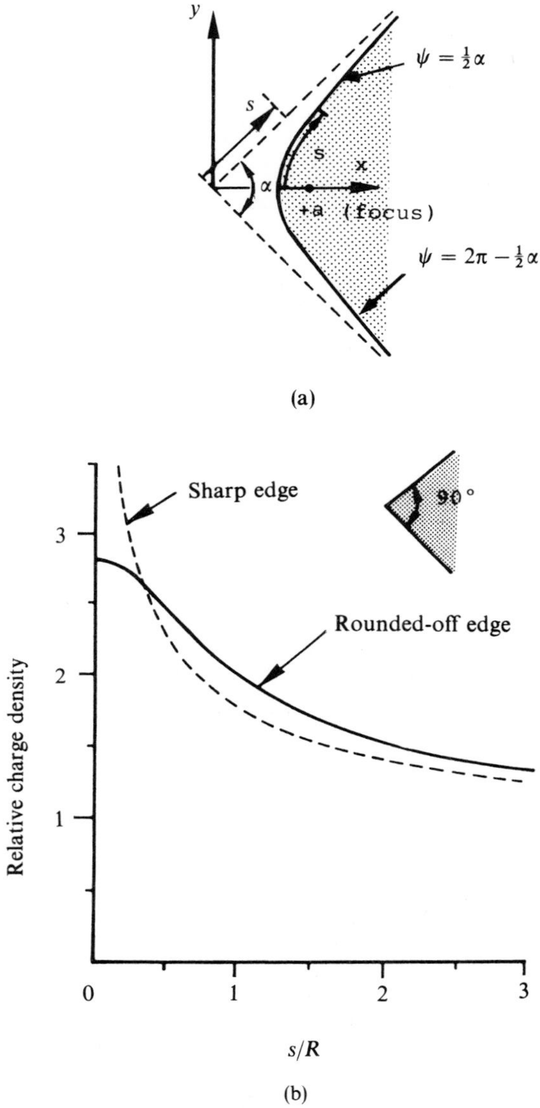

Fig. 4.17. (a) A hyperbolic conducting wedge. (b) Charge density on two wedges ($\alpha = 90°$).

thicknesses, J_{sz} is observed to behave as the current density on a strip. In particular, it tends to become infinite at the tip B.

The transition may also be followed—almost clinically—in situations where an analytical solution is available. One such example is the oblate

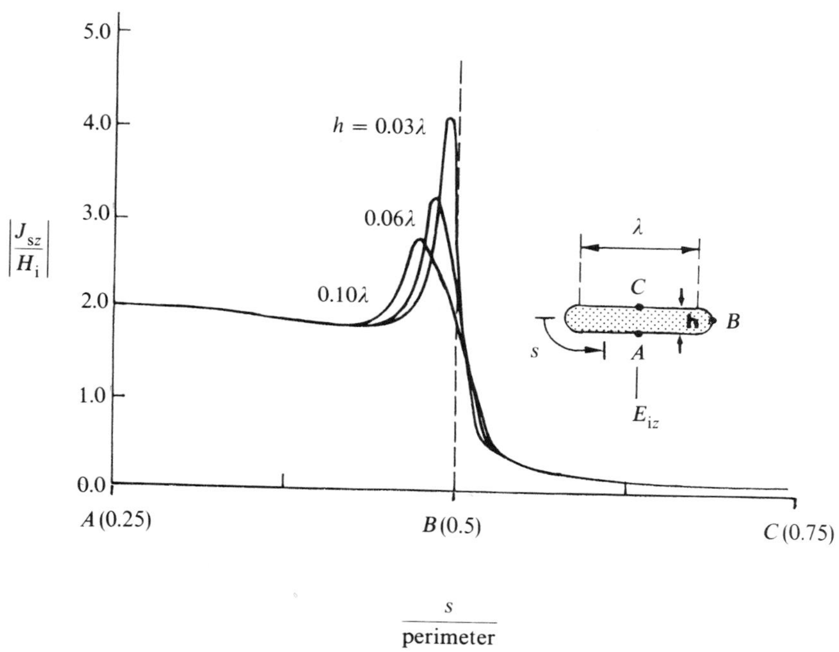

Fig. 4.18. Surface current density on a semi-cylinder-capped rectangular cylinder under E-illumination. (Personal communication from D. Wilton and C. Butler.)

spheroid, which becomes a flat circular disk in the limit of zero thickness. The solution shows that the expected edge behaviour comes about quite naturally in the course of the limit process (Kristensson *et al.* 1982; Kristensson 1985).

4.8 Uniqueness and the edge condition

The uniqueness problem may be conveniently formulated by asking the question: 'Is the solution of the potential problem of Fig. 4.19a determined uniquely by the requirement $\phi = 0$ on C, particularly when the boundary presents an edge?'

Before answering this question, let us go back to the hyperbola problem discussed in the previous section, and follow the transition from rounded tip to sharp edge obtained by letting R approach zero. During the transition process, the boundary condition $\phi = 0$, implicit in the potential value (4.29), remains satisfied on the contour, and in particular on the ever-shrinking tip region. The problem therefore remains 'well posed'. In the case of Fig. 4.19a, on the other hand, the requirement that ϕ vanish down to the edge does not

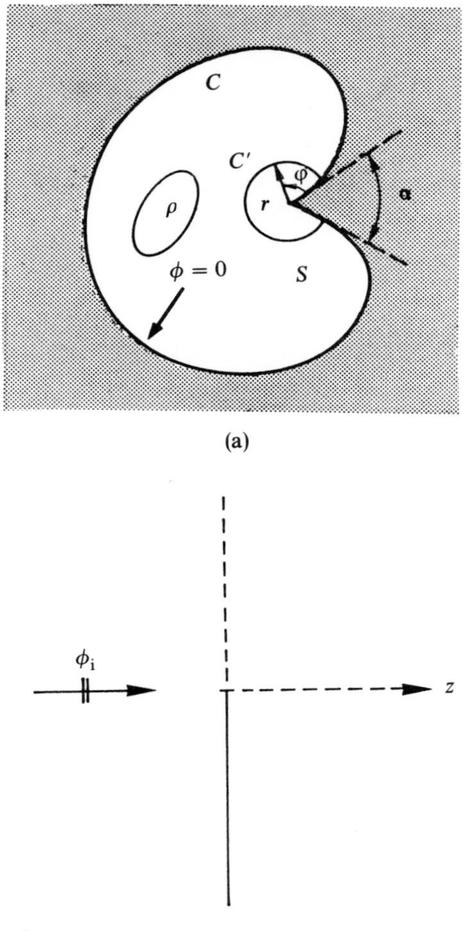

(b)

Fig. 4.19. (a) A cylindrical boundary presenting an edge. (b) A half infinite acoustic screen and incident wave.

explicitly take into account what happens *right at the tip*. Clearly, in a problem based on a differential equation, uniqueness might not obtain unless an additional requirement involving the tip region is included (Ziolkowski *et al.* 1987a,b; 1988, Volakis *et al.* 1989). We shall show that the edge condition fulfills this requirement.

The need for an additional condition was recognized early by Bouwkamp in his discussion of diffraction by a semi-infinite acoustic screen (Bouwkamp

1946, 1954). The pressure satisfies Helmholtz' equation (Fig. 4.19b)

$$\nabla^2\phi + k^2\phi = \rho \tag{4.31}$$

as well as the boundary conditions $\phi_1 = 0$ on a soft screen (the 'soft' solution) or $\partial\phi_2/\partial z = 0$ on a hard screen (the 'hard' solution). We shall assume that the sources ρ generate an incident plane wave e^{-jkz}. Bouwkamp remarked that $\partial\phi_2/\partial z$ is an acceptable solution for the 'soft' problem. The function obviously satisfies the 'soft' boundary condition. Furthermore, from

$$\nabla^2\phi_2 + k^2\phi_2 = \rho, \tag{4.32}$$

we deduce that

$$\nabla^2\frac{\partial\phi_2}{\partial z} + k^2\frac{\partial\phi_2}{\partial z} = \frac{\partial\rho}{\partial z}. \tag{4.33}$$

Substituting $\partial\rho/\partial z$ for ρ means that the original incident wave $\phi_{i2} = e^{-jkz}$ has been replaced by $-jk\,e^{-jkz}$. To within a trivial factor, therefore, $\partial\phi_2/\partial z$ and ϕ_1 are solutions of the same problem, and they should *coincide* if uniqueness were to hold under the stated boundary conditions. There is, in fact, no reason for this coincidence to occur. On the contrary, the analysis of Sommerfeld's problem in Section 4.6 clearly shows that the two solutions differ.

To prove that the edge condition removes the ambiguity, let us assume that two solutions, ϕ_a and ϕ_b, exist for the two-dimensional 'soft' problem illustrated in Fig. 4.19a. Both functions satisfy (4.31), hence their difference $\phi = \phi_a - \phi_b$ is a solution of

$$\nabla^2\phi = 0 \quad \text{in } S, \qquad \phi = 0 \quad \text{on } C. \tag{4.34}$$

Green's first theorem applied to ϕ gives

$$\iint_S (\phi\,\nabla^2\phi + |\operatorname{grad}\phi|^2)\,dS = \int_C \phi\frac{\partial\phi}{\partial n}\,dC, \tag{4.35}$$

from which it follows that grad $\phi = 0$, hence that $\phi = 0$ in S, which is turn implies that $\phi_a = \phi_b$. Green's theorem, however, can only be applied in a region where ϕ has no singularity. The immediate vicinity of the edge must therefore be excluded (Hönl *et al.* 1961). This can be achieved by closing C with a small circle C', centred on the edge, of radius r. Uniqueness now requires ϕ to behave in such a fashion that

$$\lim_{r\to 0}\int_{C'} \phi\frac{\partial\phi}{\partial r}\,r\,d\varphi = 0. \tag{4.36}$$

The edge condition requires ϕ to be proportional to $r^\nu \sin\nu\varphi$ (where $\nu \geqslant \frac{1}{2}$). Such a function satisfies (4.36); hence uniqueness is guaranteed.

The extension of the argument to the electromagnetic problem proceeds in an analogous fashion. Instead of (4.36), we now require the integral of Poynting's vector over C' to approach zero together with r (Jones 1952). The edge condition takes care of this requirement, fundamentally because it avoids radiation of energy from the edge (Maue 1949; Bouwkamp 1953; Hönl *et al.* 1961).

4.9 Enforcing the edge condition. One-dimensional series

Consider first an uncharged conducting strip, immersed in an incident E wave (Fig. 4.2a). At low frequencies, i.e. for $2a \ll \lambda$, the current density on the strip satisfies the integral equation (Butler 1985)

$$\int_{-a}^{a} J(x')\left(\log_e |x - x'| + \tilde{\gamma} + \log_e \frac{k}{2}\right) dx' = j \frac{2\pi}{kR_c} E_i(x) \tag{4.37}$$

where $\tilde{\gamma} = \gamma + \frac{1}{2} j\, \pi$, and γ is Euler's constant 0.5772. The known edge behaviour of J, given in (4.13), suggests expanding $J(x)$ in a series

$$J(x) = \frac{1}{\sqrt{[1 - (x/a)^2]}}\left(A_0 + \sum_{n=1}^{\infty} A_n \mathrm{T}_n(x/a)\right). \tag{4.38}$$

In this expression, the $\mathrm{T}_n(x)$ are the Chebyshev polynomials of the first kind. Their choice is justified by the interesting property

$$-\frac{1}{\pi}\int_{-a}^{a} \frac{1}{\sqrt{1 - (x'/a)^2}} \mathrm{T}_m(x'/a) \log_e |x - x'|\, dx' = \begin{cases} a \log_e [\frac{2}{a}] \mathrm{T}_0(x/a) & \text{for } m = 0, \\ (a/m) \mathrm{T}_m(x/a) & \text{for } m = 1, 2, \ldots \end{cases} \tag{4.39}$$

Thanks to this property, insertion of (4.38) into the left-hand member of (4.37) produces a series in $\mathrm{T}_n(x/a)$, which may then be equated to a similar series for $E_i(x)$. The coefficients of the latter are evaluated by means of the orthogonality property

$$\frac{1}{\pi}\int_{-a}^{a} \frac{1}{\sqrt{[1 - (x/a)^2]}} \mathrm{T}_m(x/a)\, \mathrm{T}_n(x/a)\, dx = \begin{cases} a & \text{for } m = n = 0, \\ \frac{1}{2} a & \text{for } m = n \neq 0, \\ 0 & \text{for } m \neq n. \end{cases} \tag{4.40}$$

Equating coefficients on both sides of the equation yields A_0, A_1, etc. The method may be extended to an H wave (Butler 1985).

A similar technique may be applied to 90° corners. In Fig. 4.20 the field in the gap—the main unknown—is conveniently expanded in terms of

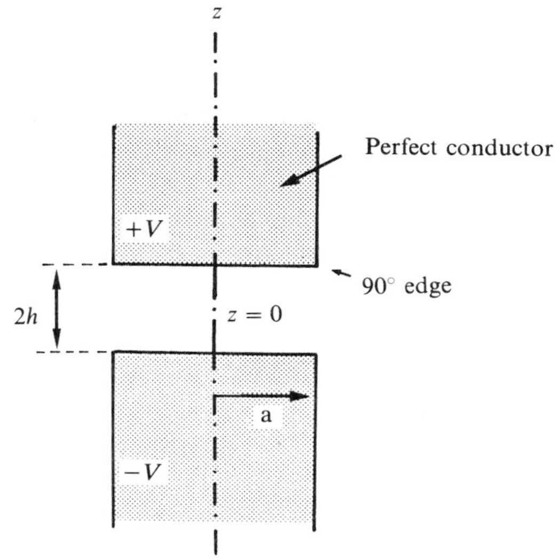

Fig. 4.20. A circular cylindrical antenna with gap.

Gegenbauer polynomials (Do Nhat *et al.* 1987). Thus,

$$E_z(a, z) = -\frac{V}{d}\frac{1}{[1-(z/h)^2]^{\frac{1}{3}}}\sum_{n=0}^{\infty} A_n C_{2n}^{\frac{1}{6}}(z/h). \tag{4.41}$$

These polynomials enjoy the orthogonality property

$$\int_{-1}^{1}\frac{1}{(1-x^2)^{\frac{1}{3}}} C_{2m}^{\frac{1}{6}}(x) C_{2n}^{\frac{1}{6}}(x)\,dx = \frac{\pi\Gamma(2m+\frac{1}{3})}{[\Gamma(\frac{1}{6})]^2 2^{\frac{1}{3}}(2m+\frac{1}{6})\,(2m)!}\,\delta_{mn}, \tag{4.42}$$

where $\Gamma(x)$ is the gamma function.

Another interesting example, featuring the use of a Fourier series, concerns a hollow circular cylinder centre-fed by a concentric ring of magnetic current of surface density $\boldsymbol{K}_s = \boldsymbol{u}_\varphi/r$, extending from $r = a$ to $r = b$ (Fig. 4.21a). The electric current density J_{sz} on the cylinder can be expanded in a series (Richmond 1980)

$$J_{sz}(z) = \sum_{m=1}^{\infty} I_m \cos\frac{m\pi z}{2h}. \tag{4.43}$$

Requiring the jump in H_φ at the tube surface to be J_{sz} leads to a matrix equation of the form

$$\sum_{n=1}^{\infty} Z_{mn} I_n = V_m, \tag{4.44}$$

where V_m can easily be evaluated from the knowledge of E_{iz}, the incident field generated by $\boldsymbol{K}_s$. A 'brute force' method of solution, based on letting m and n run from 1 to M only, gives the results shown in Fig. 4.21b. Much better convergence obtains upon splitting the current density in three terms, viz.

$$J_{sz}(z) = J_{sf}(z) + J_{se}(z) + \Delta J_s(z). \tag{4.45}$$

The term J_{sf} is the current on an infinite antenna (corresponding to $h = \infty$). Its Fourier series can be evaluated separately. The term J_{se}, which incorporates the edge behaviour, is conveniently written as

$$J_{se}(z) = A\sqrt{\left[1 - \left(\frac{z}{h}\right)^2\right]} = A \sum_{\substack{m=1 \\ m \text{ odd}}}^{\infty} \frac{2}{m} J_1\left(\frac{m\pi}{2}\right) \cos\frac{m\pi z}{2h}, \tag{4.46}$$

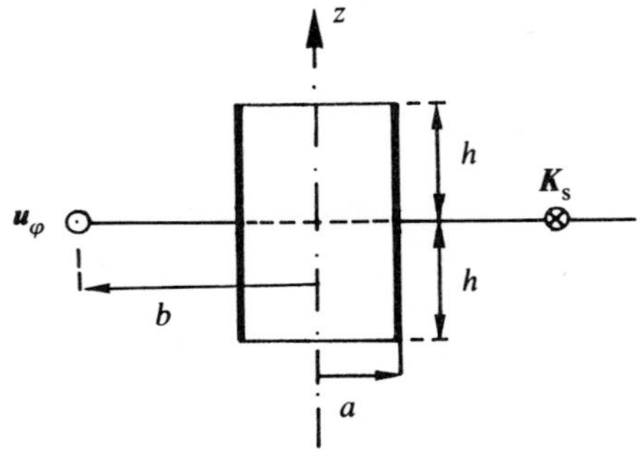

(a)

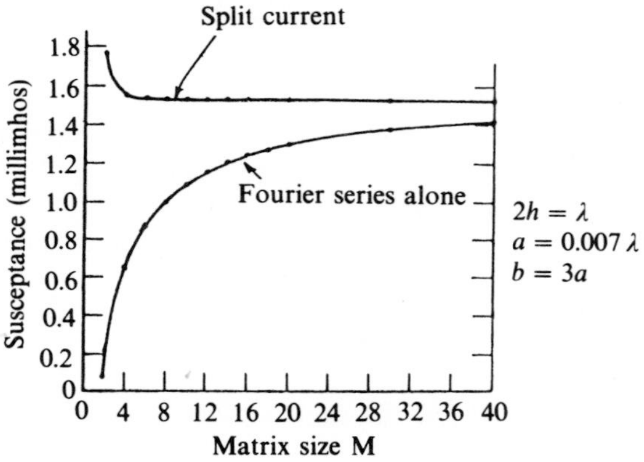

(b)

Fig. 4.21. (a) A hollow circular cylinder and ring excitation. (b) improved convergence obtained by incorporating the edge behaviour. From Richmond (1980: p. 919) © IEEE.

where A is a coefficient yet to be determined. The residual term ΔJ_s is a smooth slowly varying function, characterized by a fast-converging Fourier series. The improved numerical convergence resulting from the splitting is illustrated in Fig. 4.21b.

4.10 Enforcing the edge condition: one-dimensional subdomain functions

The use of subdomain functions to solve linear problems—a well-known technique—is discussed briefly in Appendix E. We apply the method to a bent strip immersed in an incident plane wave (Fig. 4.22). The induced current density satisfies the integral equation (Van Bladel 1985b)

$$E_z = E_{zi} - \frac{\omega\mu_0}{4}\int_{ABC} J_{sz}(S')\mathrm{H}_0^{(2)}(k|\boldsymbol{r}-\boldsymbol{r}'|)\,\mathrm{d}s' = 0 \quad (\boldsymbol{r}\text{ on } ABC), \tag{4.47}$$

where $\mathrm{H}_0^{(2)}$ is a Hankel function. To solve this equation with the help of 'pulse' basis functions, we subdivide ABC into N intervals, each of which carries a constant current density A_n. The integral in (4.47) now becomes a linear function of the constants A_n, as well as a continuous function of the strip coordinate s. This function, however, has an infinite slope at each pulse edge. Consequently, in a 'point-matching' solution, it is prudent to avoid placing matching points at these edges. Other points to be avoided are A, B, and C, where J_{sz} becomes infinite. Here again it pays to replace the pulse near to the edge (or the triangle if triangles are used as basis functions) by a function which intrinsically respects the edge condition (Glisson *et al.* 1980; Wilton *et al.* 1977; Umashankar 1988).

In Section 4.7, we have briefly discussed the problem of the charge distribution on a pair of parallel strips (Fig. 4.15). Numerical data clearly illustrate the improved convergence which obtains from enforcing the edge condition

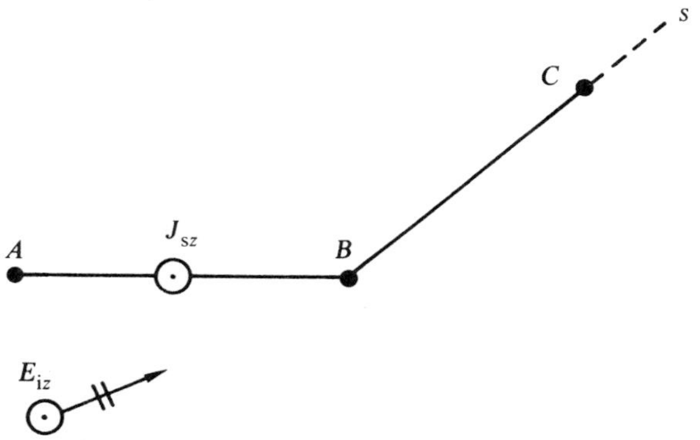

Fig. 4.22. A bent strip in an incident plane wave.

Table 4.1 Capacitance of parallel strips

N:	1	2	10	30	50
A	17.72	18.20	18.62	18.70	18.72
B	18.57	18.72	—	—	—
C	18.57	18.72	—	—	—

(McDonald *et al.* 1974). The charge density on the upper strip satisfies

$$\int_0^a \rho_s(x')G(x|x')\mathrm{d}x' = V, \tag{4.48}$$

where $2V$ is the voltage between the strips. For the aspect ratio $a/h = 1$, the kernel is explicitly

$$G(x|x') = \frac{1}{4\pi\varepsilon_0}\log_e \frac{[(x-x')^2 + 4a^2][(x+x')^2 + 4a^2]}{(x-x')^2(x+x')^2}. \tag{4.49}$$

From (F.2) and (F.4), the functional

$$J(\rho_s) = -\frac{2}{V^2}\int_0^a \rho_s(x)\int_0^a \rho_s(x')G(x|x')\mathrm{d}x'\,\mathrm{d}x + \frac{4}{V}\int_0^a \rho_s(x)\mathrm{d}x \tag{4.50}$$

is stationary in the vicinity of the sought density $\rho_s(x)$, and the stationary value of J is the linear capacitance of the stripline (in F m^{-1}). Table 4.1 shows the stationary values obtained by means of several trial-function methods:

A. The chosen trial function consists of N pulses of equal width. The heights of the pulses are the variational parameters.
B. The trial function is of the form $C_1/\sqrt{(a^2 - x^2)}$ plus $N - 1$ pulses. The total number of unknowns is again N.
C. The trial function has the form of a polynomial divided by $\sqrt{(a^2 - x^2)}$. The polynomial contains N variable coefficients.

The correct value of the linear capacitance is 18.72 pF m^{-1}. Table 4.1 confirms that the incorporation of the edge condition dramatically improves the convergence.

4.11 Enforcing the edge condition: two-dimensional entire-domain functions

The appropriate entire-domain functions for waveguides are the eigenfunctions of the modes. Figure 4.23 shows a few representative waveguide configurations, where the main unknowns are the aperture fields. When the waveguide consists of parallel plates, the mathematical problem becomes one-dimensional, and the techniques discussed in Section 4.9 are immediately

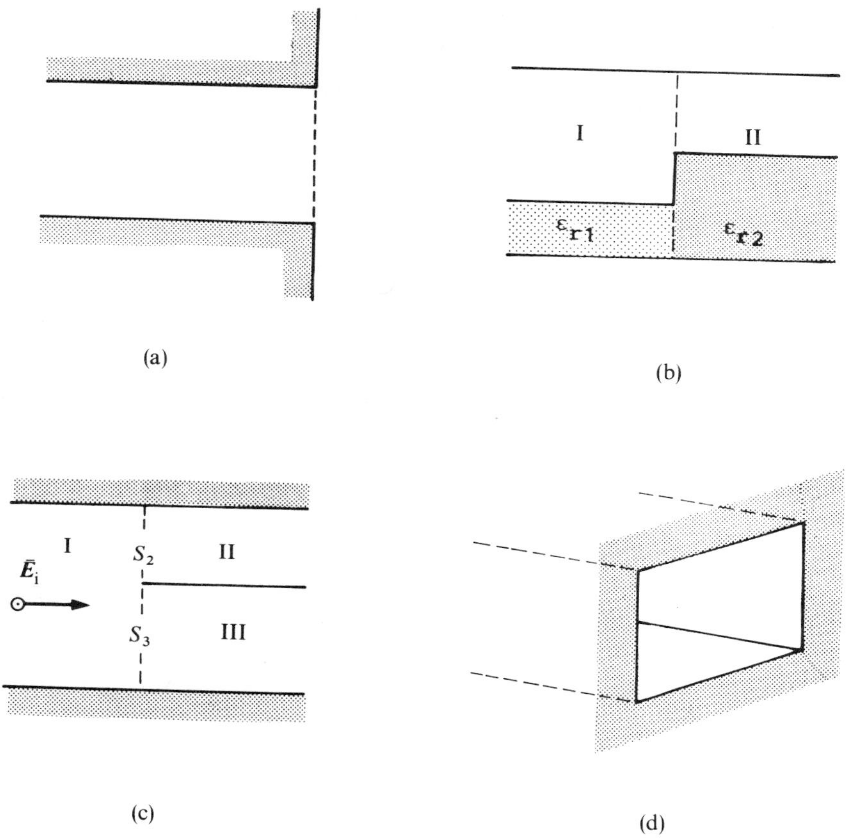

Fig. 4.23. (a) A flanged waveguide radiating into free space. (b) A waveguide with dielectric-step discontinuity. (c) A bifurcated waveguide. (d) A unit cell of a flanged waveguide antenna array.

applicable. In Fig. 4.23a, for example, the presence of a 90° corner suggests expanding the aperture field in Gegenbauer polynomials (Leong *et al.* 1988). In Fig. 4.23b, the modal expansions in guide I and II are conveniently augmented with terms which explicitly display the dielectric corner singularity (Shigesawa *et al.* 1986).

Truly two-dimensional problems, of the type shown in Figs 4.23c,d, add a degree of complexity to the analysis. In both figures the incident field is the lowest TE mode in the main guide. In Fig. 4.23c, the fields in the three regions of the truncated waveguide are expanded in normal modes, and the expansion coefficients are determined by matching fields at the interfaces S_2 and S_3. By use of the edge condition, it becomes possible to predict the asymptotic behaviour of the expansion coefficients at large mode indices (Mittra *et al.*

1971; Hockham 1975; Vassallo 1985). This information is highly advantageous, because it facilitates the inclusion of higher-order terms in the summations. Furthermore, if the asymptotic formula is not used, the accuracy of a numerical procedure can be tested by comparing the computed higher-order coefficients with their (known) asymptotic values.

4.12 Enforcing the edge condition: two-dimensional subdomain functions

A two-dimensional domain D is often subdivided, for numerical purposes, into elementary 'shells', normally of triangular shape, but sometimes rectangular or hexagonal. In the typical triangle A shown in Fig. 4.24a, a suitable subdomain basis function is

$$\rho(\mathbf{r}) = \rho_1 L_1(\mathbf{r}) + \rho_2 L_2(\mathbf{r}) + \rho_3 L_3(\mathbf{r}). \tag{4.51}$$

Here ρ_1, ρ_2, and ρ_3 are the (yet to be determined) values of ρ at points *1*, *2*, and *3*, and L_1, L_2, and L_3 are linear functions of x and y. In P, for example,

$$\begin{aligned} L_1(xy) &= \frac{\text{area of triangle P}\mathit{23}}{\text{area of triangle }\mathit{123}} \\ &= (1/2S_A)[x_2 y_3 - x_3 y_2 + x(y_2 - y_3) + y(x_3 - x_2)], \end{aligned} \tag{4.52}$$

where S_A is the area of the triangle. Circular permutations give the corresponding values of L_2 and L_3.

Functions $\rho(\mathbf{r})$ of the type appearing in (4.51) are linear in x and y. Their ensemble forms a continuous function in D, and is eminently suitable for use as a test function in a variational principle. For such use, the values of ρ_i become the variational parameters. Other forms of $\rho(\mathbf{r})$, involving polynomials of a higher order in x and y, are suitable as well, but will not be discussed here.

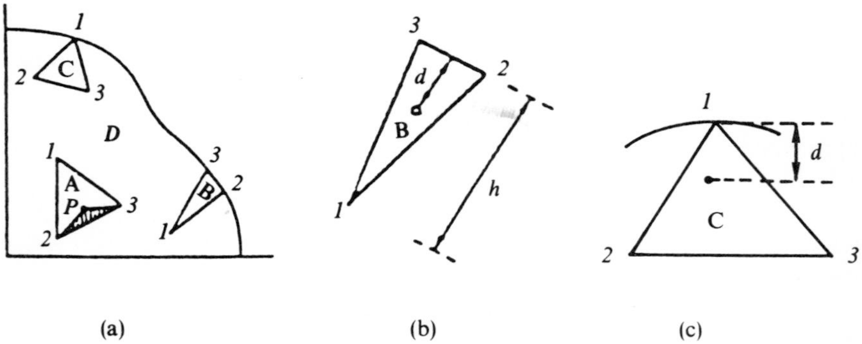

Fig. 4.24. (a) Subdivision of a domain into triangles. (b) A triangle of type B. (c) A triangle of type C. (From De Smedt *et al.* (1980: p. 705) © IEEE.

A form such as (4.51) has been successfully applied to the solution of integral equations. Consider, for example, the charge density $\rho_S(\boldsymbol{r})$ on a flat conductor immersed in an exterior field. It satisfies an integral equation of the form

$$\frac{1}{2\pi}\iint_S \rho_S(\boldsymbol{r}')\frac{1}{|\boldsymbol{r}-\boldsymbol{r}'|}\,\mathrm{d}S' = g(\boldsymbol{r}) \quad (\boldsymbol{r} \text{ in } S). \tag{4.53}$$

To determine the unknown $\rho_S(\boldsymbol{r})$ we insert (4.51) into (4.53), and obtain an equation of the form

$$\sum_m \frac{1}{2\pi}\iint_{S_m} \rho_m(\boldsymbol{r}')\frac{1}{|\boldsymbol{r}-\boldsymbol{r}'|}\,\mathrm{d}S' = g(\boldsymbol{r}) \quad (\boldsymbol{r} \text{ in } S), \tag{4.54}$$

where S_m stands for a partial triangle, and ρ_m is the value of $\rho(\boldsymbol{r})$ in S_m. For a triangle A, for example, the contribution to the integral is

$$\rho_1\frac{1}{2\pi}\iint_A \frac{L_1(\boldsymbol{r}')}{|\boldsymbol{r}-\boldsymbol{r}'|}\,\mathrm{d}S' + \rho_2\frac{1}{2\pi}\iint_A \frac{L_2(\boldsymbol{r}')}{|\boldsymbol{r}-\boldsymbol{r}'|}\,\mathrm{d}S' + \rho_3\frac{1}{2\pi}\iint_A \frac{L_3(\boldsymbol{r}')}{|\boldsymbol{r}-\boldsymbol{r}'|}\,\mathrm{d}S'$$
$$= \rho_1 F_1(\boldsymbol{r}) + \rho_2 F_2(\boldsymbol{r}) + \rho_3 F_3(\boldsymbol{r}). \tag{4.55}$$

The N unknowns ρ_i can be determined by point matching, i.e. by requiring (4.54) to be satisfied at the N vertices. The method requires evaluation of an integral such as $F_1(\boldsymbol{r})$ at two sets of points: at the vertices of the triangle itself, and at those of the other triangles of the net.

The subdivision into triangles is very flexible. It can be programmed to follow the contour with great accuracy, and also to concentrate triangles in regions where a rapid variation of the unknown is expected. To improve convergence, special forms must be used in triangles which make contact with an edge. In a triangle of type B, for example, the $1/\sqrt{d}$-dependence can be enforced by using the function (De Smedt *et al.*, 1980),

$$\rho(\boldsymbol{r}) = \rho_1\, 1/\sqrt{L_1} = \rho_1\sqrt{(h/d)}. \tag{4.56}$$

For a triangle such as C,

$$\rho(\boldsymbol{r}) = (\rho_2 L_2 + \rho_3 L_3)/(1 - L_1)^{\frac{3}{2}}. \tag{4.57}$$

This function behaves correctly at a point such as 1, since $1 - L_1$, L_2, and L_3 are proportional to d. In addition, (4.56) and (4.57) ensure continuity of ρ across the boundary of neighbouring triangles.

The subdomain functions (4.56) and (4.57) have been used successfully in problems involving acoustic and electromagnetic penetration through apertures and slots (De Smedt *et al.* 1980; De Smedt 1981). In some of these problems, e.g. in the penetration of microwave power through cracks around doors, or the leakage of electromagnetic pulses through imperfectly shielded windows, the slot is very thin. It may then be considered as one-dimensional, and a subdivision into rectangles becomes advisable. A suitable ρ is now (Fig. 4.25)

$$\rho = \frac{1}{\sqrt{[1-(n/d)^2]}}\left(\rho_1 + (\rho_2 - \rho_1)\frac{c}{L}\right). \tag{4.58}$$

This function, which is linear in c (the coordinate measured along the main fibre) incorporates the known singularity of ρ at the edges.

A suitable basis function for triangles touching the edge of a transmission line or a waveguide is (Tracey *et al.* 1977; Pantic *et al.* 1986)

$$\phi = \phi_1(1-\rho^\nu) + \phi_2\rho^\nu(1-\sigma) + \phi_3\rho^\nu\sigma. \tag{4.59}$$

Here ϕ is the potential function, ν is the singularity exponent, and ρ and σ are

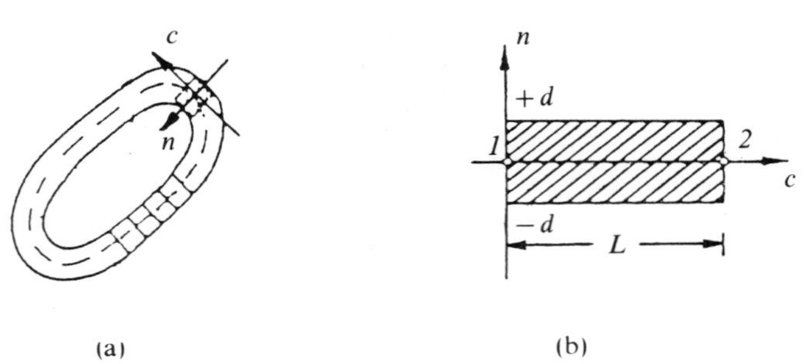

Fig. 4.25. (a) Subdivision of a slot into rectangular elements. (b) Details of a rectangular element. From De Smedt *et al.* (1980: p. 706) © IEEE.

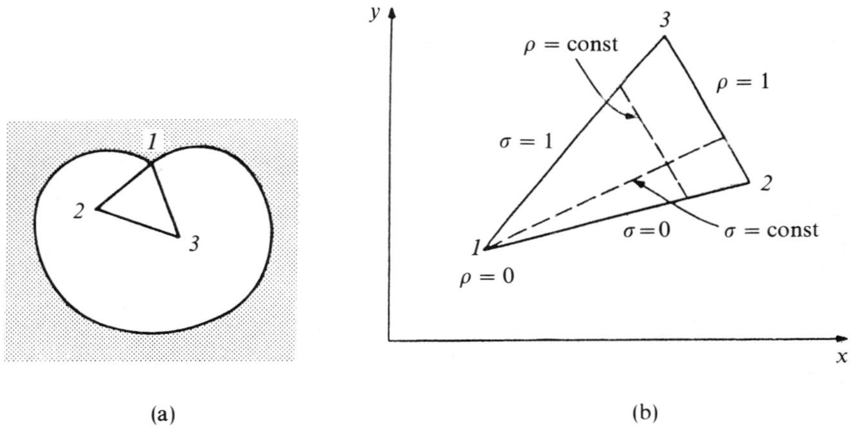

Fig. 4.26. (a) A cross-section of a waveguide. (b) Triangular polar coordinates. (From Pantic *et al.* (1986: p. 1098) © IEEE.

triangular polar coordinates. They are related to the cartesian coordinates by (Fig. 4.26)

$$x = x_1 + \rho[x_2 - x_1 + \sigma(x_3 - x_2)],$$
$$y = y_1 + \rho[y_2 - y_1 + \sigma(y_3 - y_2)]. \tag{4.60}$$

The ρ coordinate is small in the vicinity of 1, where it is proportional to d.

4.13 Singularities at a dielectric wedge. First symmetry

Examples of dielectric and magnetic wedges are shown in Figs 4.27b,c, where P is a typical edge point. To investigate the edge singularity we shall focus our attention on the quasi-static limit, valid in a region of dimensions small with respect to the wavelengths of concern in the problem. The details of the transition from the quasi-static region to the dynamic one will not be discussed here. It is useful to mention, however, that an expansion such as (4.4) must be complemented, for certain geometries, by terms of the form $r^m(\log_e r)^n$ (Meixner 1954a, b; Bach Andersen *et al.* 1978; Makarov *et al.* 1986, Marx 1990).

In an H wave, the electric field is transverse, and its static value in each dielectric region can be derived from a potential

$$\phi_k = A_k r^\nu \cos \nu(\varphi + \alpha_k) \quad (k = 1, 2). \tag{4.61}$$

The index k refers to regions 1 and 2 (Fig. 4.28a). The field itself is

$$\mathbf{E}_{tk} = \operatorname{grad} \phi_k = \frac{A_k \nu}{r^{1-\nu}} [\cos \nu(\varphi + \alpha_k)\mathbf{u}_r - \sin \nu(\varphi + \alpha_k)\mathbf{u}_\varphi]. \tag{4.62}$$

It can also be written as

$$\mathbf{E}_{tk} = \operatorname{grad} \Theta_k \times \mathbf{u}_z = \operatorname{grad}[A_k r^\nu \sin \nu(\varphi + \alpha_k)] \times \mathbf{u}_z, \tag{4.63}$$

where Θ_k is a flux function. The lines of constant Θ_k are the lines of force. The magnetic field corresponding to $\mathbf{E}_k$ is

$$H_{zk} = \mathrm{j}\omega\varepsilon_k A_k r^\nu \sin \nu(\varphi + \alpha_k). \tag{4.64}$$

In an E wave, singularities do not occur, since E_z is not perturbed by the presence of the dielectric. The reason is simple: the boundary conditions at the air–dielectric interface are automatically satisfied by the unperturbed fields. When the wedge is magnetic, however, $\mathbf{H}_t$ becomes discontinuous at the wedge surface, and the unperturbed fields must be supplemented by singular fields if the boundary conditions on $\mathbf{H}_t$ are to be satisfied. The point is further discussed in Section 4.15.

The value of the singularity exponent ν in an H wave is determined by the requirement that E_r and $\varepsilon_r E_\varphi$ be continuous at the air–dielectric interface. It turns out that two fundamental symmetries must be considered here (Bach

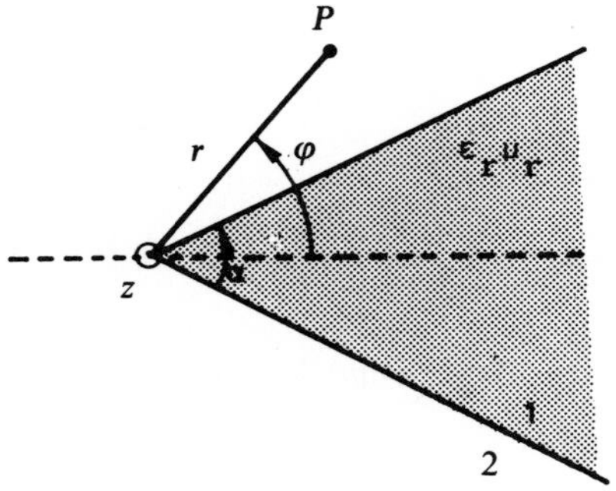

(a)

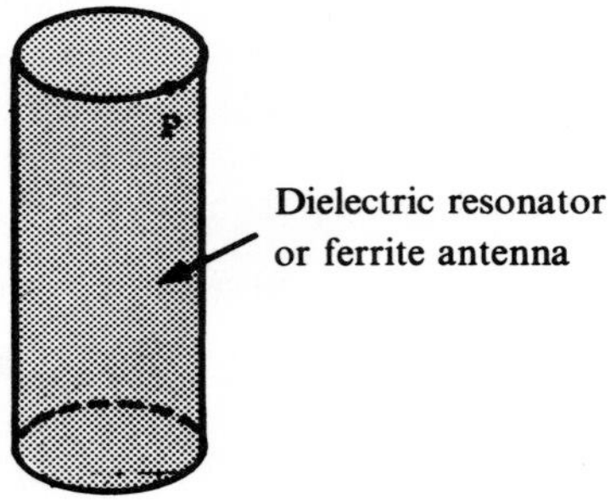

(b)

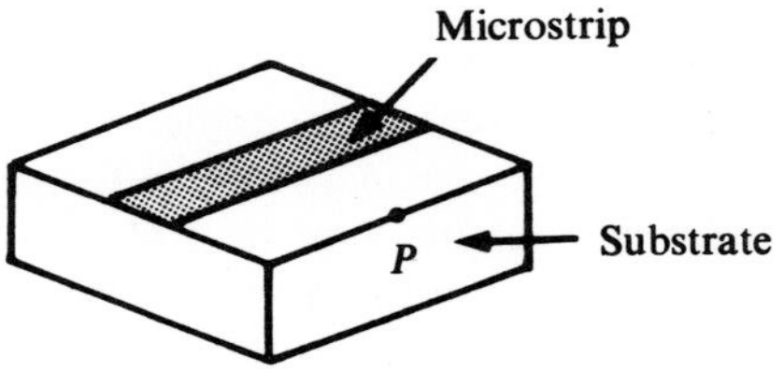

(c)

Fig. 4.27. (a) A dielectric wedge. (b, c) Examples of devices exhibiting dielectric wedges.

Andersen *et al.* 1978). In the first one, ϕ is *symmetric* with respect to the $(\varphi = 0)$ axis. Close to the edge,

$$\begin{aligned} \phi_1 &= Ar^\nu \cos \nu\varphi && \text{for } -\tfrac{1}{2}\alpha \leqslant \varphi \leqslant \tfrac{1}{2}\alpha, \\ \phi_2 &= A\,\frac{\cos \frac{1}{2}\nu\alpha}{\cos \nu(\pi - \frac{1}{2}\alpha)}\,r^\nu \cos \nu(\pi - \varphi) && \text{for } \tfrac{1}{2}\alpha \leqslant \varphi \leqslant 2\pi - \tfrac{1}{2}\alpha. \end{aligned} \tag{4.65}$$

Table 4.2. Lowest values of ν and τ

α (for ν)	ε_r: 1	2	5	10	38	50	100	∞	α (for τ)
0°	1.000	1.000	1.000	1.000	1.000	1.000	1.000	.5000	360°
20°	1.000	.9500	.8437	.7486	.6107	.5935	.5635	.5294	340°
40°	1.000	.9156	.7841	.6992	.6062	.5963	.5799	.5625	320°
60°	1.000	.8971	.7679	.6977	.6293	.6226	.6115	.6000	300°
80°	1.000	.8918	.7745	.7169	.6644	.6594	.6512	.6429	280°
100°	1.000	.8974	.7961	.7495	.7086	.7048	.6986	.6923	260°
120°	1.000	.9123	.8300	.7935	.7622	.7594	.7547	.7500	240°
140°	1.000	.9351	.8751	.8489	.8268	.8248	.8215	.8182	220°
160°	1.000	.9649	.9317	.9171	.9048	.9037	.9018	.9000	200°
180°	1.000	1.000	1.000	1.000	1.000	1.000	1.000	1.000	180°

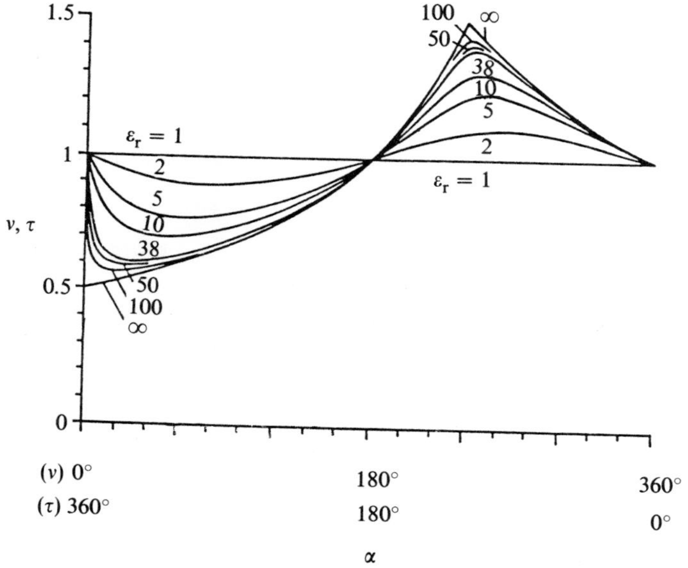

Fig. 4.28. The singularity exponent for a dielectric wedge. From Van Bladel (1985: p. 451) © IEEE.

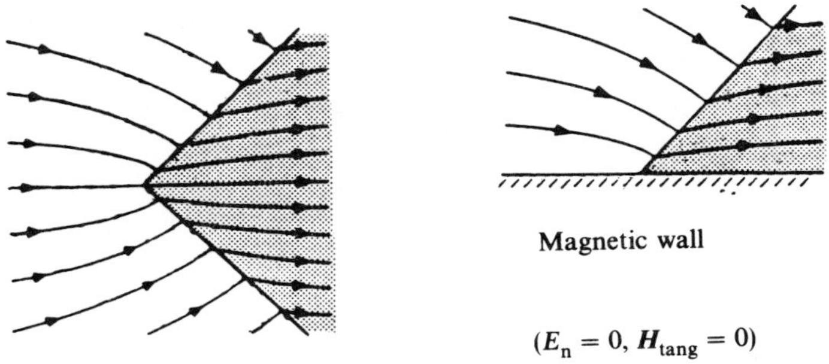

(a)

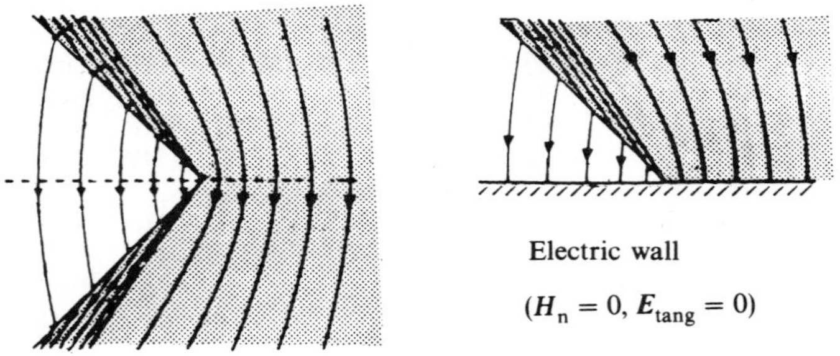

(b)

Fig. 4.29. (a) Lines of force of $\boldsymbol{E}$ near a sharp dielectric wedge ($\varepsilon_r = 10$). (b) Lines of force of $\boldsymbol{E}$ near a reentrant dielectric wedge ($\varepsilon_r = 10$). (From Van Bladel (1985: p. 451–3) © IEEE)

From the boundary conditions at $\varphi = \pm\frac{1}{2}\alpha$, it follows that

$$\varepsilon_r \tan\tfrac{1}{2}\nu\alpha = -\tan\nu(\pi - \tfrac{1}{2}\alpha). \qquad (4.66)$$

It is easy to check that this relationship ensures continuity of H_z at the air–dielectric interface. Numerical data for ν are given in Table 4.2 and Fig. 4.28 (Van Bladel 1985a). Clearly, singular fields occur only for sharp wedges, i.e. for $\alpha < \pi$. The lines of force of $\boldsymbol{E}$, shown in Fig. 4.29a, are left unperturbed when the symmetry plane is materialized into a magnetic wall.

4.14 Singularities at a dielectric wedge. Second symmetry

In a second symmetry the potentials are assumed to be *anti-symmetric* with respect to the $\varphi = 0$ plane. Thus,

$$\begin{aligned} \phi_1 &= Br^\tau \sin\tau\varphi && \text{for } -\tfrac{1}{2}\alpha \leqslant \varphi \leqslant \tfrac{1}{2}\alpha, \\ \phi_2 &= Br^\tau \frac{\sin\frac{1}{2}\tau\alpha}{\sin\tau(\pi - \frac{1}{2}\alpha)} \sin\tau(\pi - \varphi) && \text{for } \tfrac{1}{2}\alpha \leqslant \varphi \leqslant 2\pi - \tfrac{1}{2}\alpha. \end{aligned} \qquad (4.67)$$

The exponent τ is the solution of the transcendental equation

$$(1/\varepsilon_r)\tan(\tfrac{1}{2}\tau\alpha) = -\tan\tau(\pi - \tfrac{1}{2}\alpha). \tag{4.68}$$

Comparison with (4.66) shows that

$$\tfrac{1}{2}\tau\alpha = \nu(\pi - \tfrac{1}{2}\alpha). \tag{4.69}$$

We conclude that τ can again be determined from Table 4.2 and Fig. 4.28. The field singularities now occur for a *reentrant* wedge, and the lines of force of $\boldsymbol{E}$ are of the type shown in Fig. 4.29b. They are left undisturbed when the symmetry plane is metallized into an electric wall.

In the previous discussion, a dielectric medium of dielectric constant $\varepsilon = \varepsilon_r\varepsilon_0$ was in contact with a region of dielectric constant ε_0. The analysis remains valid when *two* dielectrics are in contact, but now ε_r must be interpreted as the ratio $\varepsilon_{r1}/\varepsilon_{r2}$ of the dielectric constants. The reentrant wedge ($\alpha > \pi$) now becomes a sharp dielectric wedge embedded in aa medium of higher dielectric constant.

4.15 Singularities at a magnetic wedge

We have already mentioned, in Section 4.13, that the singularity at a magnetic wedge is produced by the transverse component of the *magnetic* field. More precisely, the components of the singular field near a sharp wedge are (Fig. 4.27a)

$$\begin{aligned}
E_{z1} &= -A\, j\omega\mu_r\mu_0 r^{\nu}\sin\nu\varphi,\\
E_{z2} &= -A\, j\omega\mu_0\,\frac{\cos\frac{1}{2}\nu\alpha}{\cos\nu(\pi-\frac{1}{2}\alpha)}\,r^{\nu}\sin\nu(\pi-\varphi),\\
\boldsymbol{H}_{t1} &= A\,\mathrm{grad}(R^{\nu}\cos\nu\varphi) = \frac{A\nu}{r^{1-\nu}}[(\cos\nu\varphi)\boldsymbol{u}_r - (\sin\nu\varphi)\boldsymbol{u}_{\varphi}],\\
\boldsymbol{H}_{t2} &= A\frac{\cos\frac{1}{2}\nu\alpha}{\cos\nu(\pi-\frac{1}{2}\alpha)}\mathrm{grad}\,[r^{\nu}\cos\nu(\pi-\varphi)]\\
&= \frac{A\nu}{r^{1-\nu}}\frac{\cos\frac{1}{2}\nu\alpha}{\cos\nu(\pi-\frac{1}{2}\alpha)}\{[\cos\nu(\pi-\varphi)]\,\boldsymbol{u}_r + [\sin\nu(\pi-\varphi)]\,\boldsymbol{u}_{\varphi}]\}.
\end{aligned} \tag{4.70}$$

The singularity exponent ν has the value given in (4.66), but now μ_r must replace ε_r. Table 4.2 and Fig. 4.28 remain valid, but the lines of force shown in Fig. 4.29a become those of $\boldsymbol{H}$. The magnetic wall must be replaced by an electric one.

Similar considerations hold for the reentrant wedge, where the field components are of the form

$$E_{z1} = -B\,\mathrm{j}\omega\mu_r\mu_0\, r^{\tau} \cos\tau\varphi,$$

$$E_{z2} = -B\,\mathrm{j}\omega\mu_0 \frac{\cos\frac{1}{2}\tau\alpha}{\cos\tau(\pi - \frac{1}{2}\alpha)} r^{\tau} \cos\tau(\pi - \varphi),$$

$$\boldsymbol{H}_{t1} = B\,\mathrm{grad}(r^{\tau}\sin\tau\varphi) = \frac{B\tau}{r^{1-\tau}}[(\sin\tau\varphi)\boldsymbol{u}_r + (\cos\tau\varphi)\boldsymbol{u}_{\varphi})], \tag{4.71}$$

$$\boldsymbol{H}_{t2} = B\frac{\sin\frac{1}{2}\tau\alpha}{\sin\tau(\pi - \frac{1}{2}\alpha)}\mathrm{grad}[r^{\tau}\sin\tau(\pi - \varphi)]$$

$$= \frac{B\tau}{r^{1-\tau}}\frac{\sin\frac{1}{2}\tau\alpha}{\sin\tau(\pi - \frac{1}{2}\alpha)}\{[\sin\tau(\pi - \varphi)]\boldsymbol{u}_r - [\cos\tau(\pi - \varphi)]\boldsymbol{u}_{\varphi}\}.$$

The singularity exponent τ is determined by (4.68), but again μ_r must replace ε_r. Table 4.2 and Figs 4.28 and 4.29b remain valid, but a magnetic wall must replace the electric one.

Let us apply these considerations to the doubly salient rotor–stator configuration of Fig. 4.30a, which is often encountered in electric machines. If the material of the teeth is infinitely permeable, conditions (4.66) and (4.68) yield the values $\nu = \frac{2}{3}$ at a protruding corner (A or C) and $\tau = \frac{2}{3}$ at a reentrant one (B or D). The $\boldsymbol{H}$ field in the iron vanishes, but the magnetic induction does not. From (4.70) and (4.71) it is

$$\boldsymbol{B} = r^{-\frac{1}{3}}[(-\cos\tfrac{2}{3}\varphi)\boldsymbol{u}_r + (\sin\tfrac{2}{3}\varphi)\boldsymbol{u}_{\varphi}] \tag{4.72}$$

near A (Fig. 4.30b), and

$$\boldsymbol{B} = r^{-\frac{1}{3}}[(\sin\tfrac{2}{3}\varphi)\boldsymbol{u}_r + (\cos\tfrac{2}{3}\varphi)\boldsymbol{u}_{\varphi}], \tag{4.73}$$

near B (Fig. 4.30c). The lines of force shown in Fig. 4.30a were generated numerically without introducing the edge condition as a constraint in the algorithm. We observe that the profile of these lines is in general agreement with the behaviour predicted by (4.72) and (4.73), and with the patterns shown in Fig. 4.29 (Van Bladel 1981).

The two assumptions made above—namely infinite permeability and infinite sharpness of the corner—correspond to idealized conditions. The results obtained through this model are nevertheless useful. The infinite induction which exists at an edge when $\mu_r = \infty$ implies that saturation will probably occur in the edge region when μ_r is high (but finite). Further, the application of a law such as (4.72) can give a reasonable idea of the extent of the saturated region.

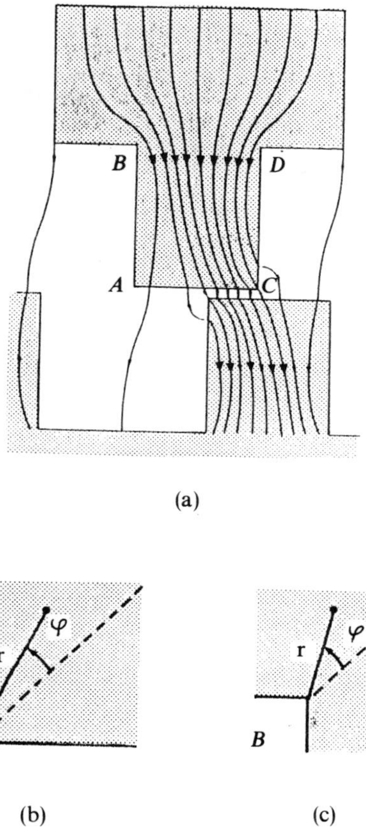

Fig. 4.30. (a) A doubly salient structure. (b) A protruding corner A. (c) A reentrant corner B. From Van Bladel (1981: 221) © IEE.

4.16 Composite wedges

The composite wedges discussed in this section are formed by adjacent metal and dielectric bodies. A first example is the 90° metallic wedge in contact with a dielectric (or magnetic) plane (Fig. 4.31a). A typical point on the edge is identified by the letter P. The singularity exponent σ for a *metal–dielectric* structure immersed in an H wave is given in Table 4.3 and Fig. 4.32. We observe that the exponent lies between the extreme values $\frac{2}{3}$, corresponding to a metallic corner in vacuum, and $\frac{1}{2}$, corresponding to the limit $\varepsilon_r \to \infty$. The exponent χ relative to a *metal–magnetic* wedge also appears in Table 4.3 and Fig. 4.32. A discussion of the analytical form of the fields, and of the general shape of the lines of force, is available elsewhere (Van Bladel 1985a).

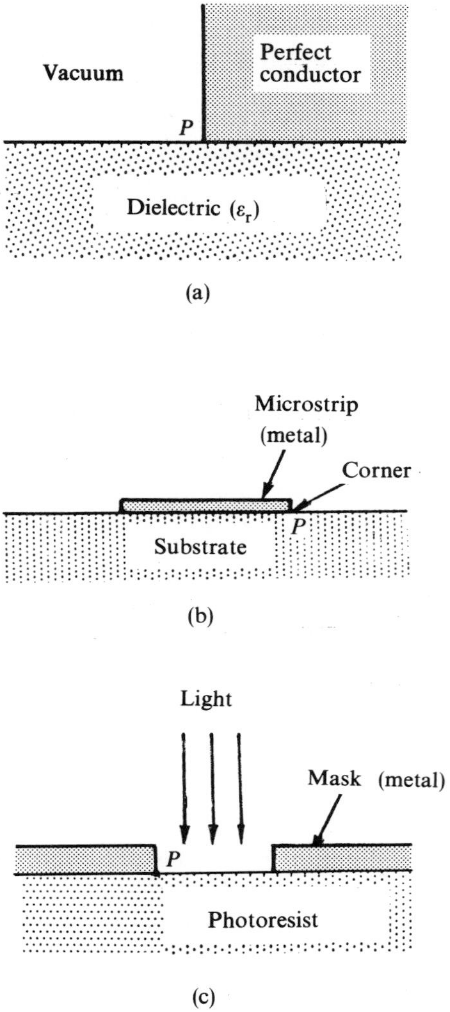

Fig. 4.31. (a) A conducting wedge in contact with a dielectric. (b, c) Examples of a conductor–dielectric contact.

A second interesting configuration consists of two adjacent 90° wedges, one of them metallic, the other dielectric (or magnetic) (Fig. 4.33). The singularity exponents are now α (for a dielectric wedge) and β (for a magnetic wedge). They are given in Table 4.3 and Fig. 4.32.

In a third configuration (Fig. 4.34a) the electric field in the absence of dielectric is tangent to the symmetry plane. The unperturbed field therefore

Table 4.3. Lowest values of α, β, σ, χ

ε_r or μ_r	σ, β	χ, α	ε_r or μ_r	σ, β	χ, α
1.0	.666667	.666667	10.0	.528977	.863222
1.5	.630990	.704833	15.0	.519907	.886866
2.0	.608173	.732280	20.0	.515163	.901373
2.5	.592231	.753248	25.0	.512246	.911431
3.0	.580431	.769946	38.0	.508163	.927762
4.0	.564094	.795167	50.0	.506242	.936862
5.0	.553300	.813570	100.0	.503152	.955171
7.5	.537535	.844042			

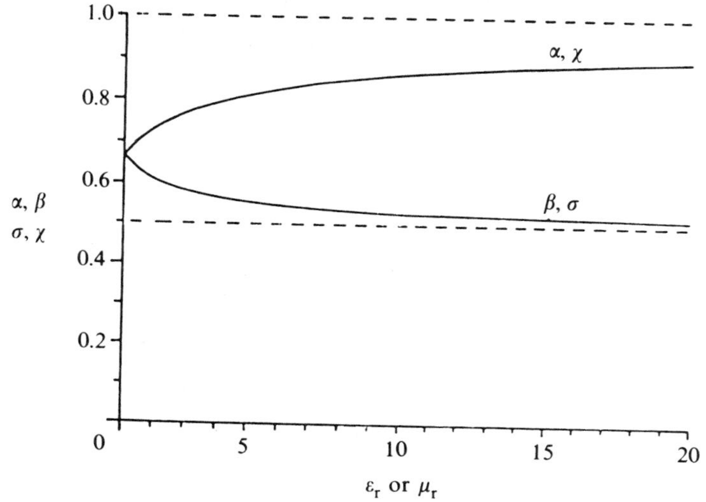

Fig. 4.32. Singularity exponents for a few composite wedges. From Van Bladel (1985: p. 453) © IEEE.

automatically satisfies the boundary condition 'D_n continuous' at the air–dielectric interface. We conclude that singularity exponent, fields, and lines of force are the same as when the metallic wedge is embedded in free space. An important particular case is that of the infinitely thin conducting sheet shown in Fig. 4.34b. The singularity exponent $\frac{1}{2}$ is the same as in vacuum, and the lines of force are the same as well. Consequently the 'edge-sensitive' basis functions discussed in previous Sections are pertinent for the determination of the current density in the sheet. They have been used successfully in the study of the microstrip structure shown in Fig. 4.34c

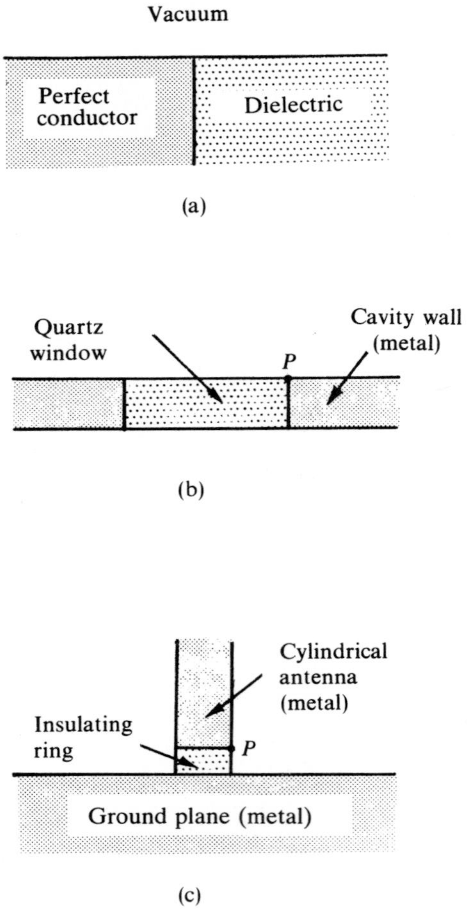

Fig. 4.33. (a) A 90° metallic wedge in contact with a 90° dielectric wedge. (b, c) Examples of 90°–90° contacts.

(Beyne *et al.* 1988; Faché *et al.* 1988, 1989). The strong edge singularity may cause avalanche breakdown when the substrate is a semiconductor (Lewis *et al.* 1970).

A few additional comments are in order:

(1) in a structure formed by two (or more) dielectrics, alone or in contact with a perfect conductor, a correct choice of geometry and material characteristics may result in the elimination of the singularity (Mittra *et al.* 1971; Hurd 1976; Brooke *et al.* 1977).

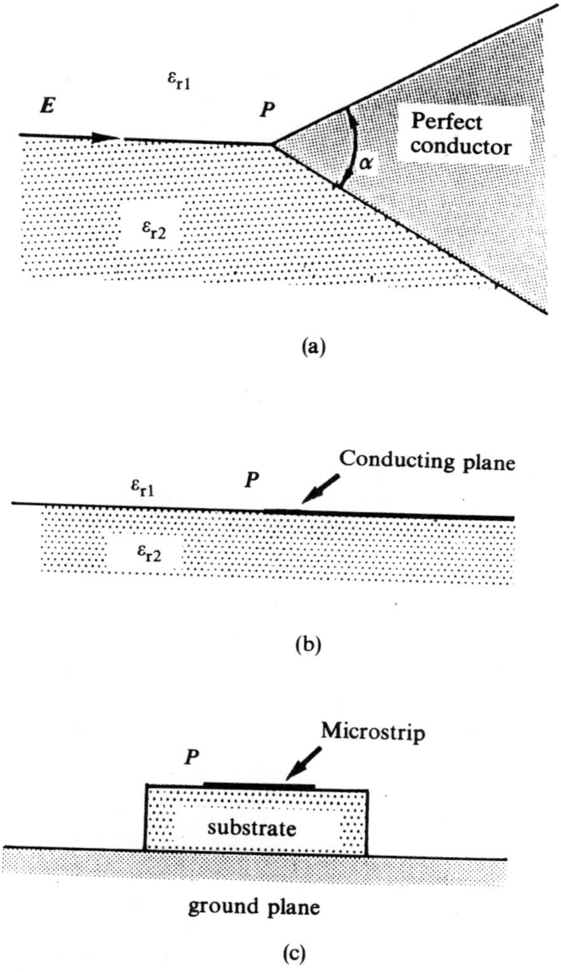

Fig. 4.34. (a) Dielectric interface coinciding with the bisectrix of the outer angle. (b) Metallic half plane resting on a dielectric. (c) Infinitely thin microstrip on a substrate.

(2) when the wedges are formed by anisotropic materials, the number of parameters becomes quite large. Because of this added complexity, only fragmentary numerical data are available in the literature (Lee *et al.* 1968; Brooke *et al.* 1977; Chan *et al.* 1988; Miyata *et al.* 1989).

(3) the assumption of perfect conductivity for the metal part of the wedge is not always appropriate for the metals encountered in practice. The presence of a finite conductivity inevitably increases the analytical complexity of the

problem. A detailed solution has nevertheless been given for the resistive half-plane, alone or in contact with a perfectly conducting coplanar half-plane (Braver *et al.* 1988). In the analysis, power expansion (4.4) must be replaced by

$$\psi = r^{\nu} \sum_{m=0}^{\infty} \sum_{n=0}^{m} C_{mn}(\varphi)\, r^{m} (\log_{e} r)^{n}. \tag{4.74}$$

References

BACH ANDERSEN, J., and SOLODUKHOV, V.V. (1978). Field behavior near a dielectric wedge. *IEEE Transactions on Antennas and Propagation*, **26**, 598–602.

BEYNE, L., and DE ZUTTER, D. (1988). Power deposition of a microstrip applicator radiating into a layered biological structure. *IEEE Transactions on Microwave Theory and Techniques*, **36**, 126–31.

BOLOMEY, J.C., and WIRGIN, A. (1971). Sur le comportement du champ électromagnétique aux arêtes, *Annales de l'Institut Henri Poincaré*, **14**, 97–112.

BOUWKAMP, C.J. (1946). A note on singularities occurring at sharp edges in electromagnetic diffraction theory. *Physica*, **12**, 467–74.

BOUWKAMP, C.J. (1953). *Diffraction theory*, pp. 59–66. Research Report EM-50, New York University.

BOUWKAMP, C.J. (1954). Diffraction theory. *Reports on Progress in Physics*, **17**, 35–100.

BRAVER, I.M., FRIDBERG, P.S., GARB, K.G. and JAKOVER, I.M. (1988). The behavior of the electromagnetic field near the edge of a resistive half-plane. *IEEE Transactions on Antennas and Propagation*, **36**, 1760–8.

BROOKE, G.H., and KHARADLY, M.M.Z. (1971). Field behavior near anisotropic and multidielectric edges. *IEEE Transactions on Antennas and Propagation*, **25**, 571–5.

BUTLER, C.M. (1980). Capacitance of a finite-length conducting cylindrical tube. *Journal of Applied Physics*, **51**, 5607–9.

BUTLER, C.M. (1985). General solutions of the narrow strip (and slot) integral equations. *IEEE Transactions on Antennas and Propagation*, **33**, 1085–90.

CHAN, C.H., PANTIC-TANNER, Z., and MITTRA, R. (1988). Field behaviour near a conducting edge embedded in an inhomogeneous anisotropic medium, *Electronics Letters*, **24**, 355–6.

DE SMEDT, R., and VAN BLADEL, J. (1980). Magnetic polarizability of some small apertures. *IEEE Transactions on Antennas and Propagation*, **28**, 703–7.

DE SMEDT, R. (1981). Low frequency scattering through an aperture in a rigid screen — Some numerical results. *Journal of Sound and Vibration*, **75**, 371–86.

DO-NHAT, T., and MACPHIE, R.H. (1987). The static electric field distribution between two semi-infinite circular cylinders: a model for the feed gap field of a dipole antenna. *IEEE Transactions on Antennas and Propagation*, **35**, 1273–80.

DURAND, E. (1964). *Electrostatique, I. Les distributions*, pp. 254, 292, 297. Masson, Paris.

FACHÉ, N., and DE ZUTTER, D. (1988). Rigorous full wave space domain solution for dispersive microstrip lines, *IEEE Transactions on Microwave Theory and Techniques*, **36**, 731–7.

FACHÉ, N., and DE ZUTTER, D. (1989). Circuit parameters for single and coupled microstrip lines by a rigorous full-wave space-domain analysis. *IEEE Transactions on Microwave Theory and Techniques*, **37**, 421–5.

Glisson, A.W., and Wilton, D.R. (1980). Simple and efficient numerical methods for problems of electromagnetic radiation and scattering from surfaces. *IEEE Transactions on Antennas and Propagation*, **28**, 593–603.

Hockham, G.A. (1975). Use of the edge condition in the numerical solution of waveguide antenna problems. *Electronics Letters*, **11**, 418–9.

Hönl, H., Maue, A.W., and Westpfahl, K. (1961). Theorie der Beugung. In *Crystal optics-diffraction* (ed. S. Flügge), pp. 252–66, Springer, Berlin.

Hurd, R.A. (1976). The edge condition in electromagnetics. *IEEE Transactions on Antennas and Propagation*, **24**, 70–73.

Iizuka, K. and Yen, J.L. (1967). Surface currents on triangular and square metal cylinders. *IEEE Transactions on Antennas and Propagation*, **15**, 795–801.

Jones, D.S. (1952). Diffraction by an edge and by a corner. *Quarterly Journal of Mechanics and Applied Mathematics*, **5**, 363–78.

Jones, D.S. (1964). *The theory of electromagnetism*, pp. 566–9. Pergamon Press, Oxford.

Jones, D.S (1986). *Acoustic and electromagnetic waves*, pp. 531–4. Clarendon Press, Oxford.

Kristensson, G., and Waterman, P.C. (1982). The T-matrix for acoustic and electromagnetic scattering by circular disks. *Journal of the Acoustical Society of America*, **72**, 1612–25.

Kristensson, G. (1985). The current distribution on a circular disc. *Canadian Journal of Physics*, **63**, 507–13.

Lee, S.W. and Mittra, R. (1968). Edge condition and intrinsic loss in uniaxial plasma. *Canadian Journal of Physics*, **46**, 111–20.

Leong, M.S., Kovi, P.S., and Chandra, (1988). Radiation from a flanged parallel-plate waveguide: solution by moment method with inclusion of edge condition. *IEE Proceedings*, **H135**, 249–55.

Lewis, J.A., and Wasserstrom, E. (1970). The field singularity at the edge of an electrode on a semiconductor surface. *Bell System Technical Journal*, **49**, 1183–94.

Makarov, G.I., and Osipov, A.V. (1986). Structure of Meixner's series. *Radio Physics and Quantum Electronics*, **29**, 544–9. Translated from *Izvestiya Vysshikh Nehebnykh Zavedenir, Radiofizika*, **29**, 714–20, June 1986.

Marx, E. (1990). Computed fields near the edge of a dielectric wedge. *IEEE Transactions on Antennas and Propagation*, **38**, 1438–42.

Maue, A.W. (1949). Zur Formulierung eines allgemeinen Beugungsproblems durch eine Integralgleichung. *Zeitschrift für Physik*, **126**, 601–18.

Mc Donald, B.H., Friedman, M., and Wexler, A. (1974). Variational solution of integral equations. *IEEE Transactions on Microwave Theory and Techniques*, **22**, 237–48.

Mei, K.K., and Van Bladel, J. (1963). Low-frequency scattering by rectangular cylinders. *IEEE Transactions on Antennas and Propagation*, **11**, 52–6.

Meixner, J. (1949). Die Kantenbedingung in der Theorie der Beugung electromagnetischer Wellen an vollkommen leitenden ebenen Schirmen, *Annalen der Physik*, **6**, 2–9.

Meixner, J. (1954a). *The behavior of electromagnetic fields at edges*. Research Report EM-72, New York University.

Meixner, J. (1954b). The behavior of electromagnetic fields at edges. *IEEE Transactions on Antennas and Propagation*, **20**, 442–6.

Mittra, R., and Lee, S.W. (1971). *Analytical techniques in the theory of guided waves*, pp. 4–11, 34–41. Macmillan, New York.

MIYATA, I., TAKEDA, T., and KUWAHARA, T. (1989). Electric fields on edges of anisotropic dielectric materials. *Electronics Letters*, **25**, 1185–6.

MORRIS, G. (1982). Currents induced on a conducting halfplane by a plane wave at arbitrary incidence. *Electronics Letters*, **18**, 610–3.

MOTZ, H. (1947). The treatment of singularities in relaxation methods. *Quarterly of Applied Mathematics*, **4**, 371–7.

PANTIC, Z., and MITTRA, R. (1986). Quasi-TEM analysis of microwave transmission lines by the finite-element method. *IEEE Transactions on Microwave Theory and Techniques*, **34**, 1096–1103.

POINCELOT, P. (1963). *Précis d'électromagnétisme théorique*, pp. 84–90, Dunod, Paris.

RICHMOND, J.H. (1980). On the edge mode in the theory of thick cylindrical monopole antennas. *IEEE Transactions on Antennas and Propagation*, **28**, 916–21.

RUCK, G.T., BARRICK, D.E., STUART, W.D., and KRICHBAUM, C.K. (1970). *Radar cross section handbook*, Vol. 1, pp. 117–8. Plenum Press, New York.

SHIGESAWA, H., and TSUJI, M. (1986). Mode propagation through a step discontinuity in dielectric planar waveguide. *IEEE Transactions on Microwave Theory and Techniques*, **34**, 205–12.

SOMMERFELD, A. (1964). *Optics*, pp. 245–65. Academic Press, New York.

SMYTHE, W.P. (1950). *Static and dynamic electricity*, p. 114. McGraw-Hill, New York.

TRACEY, D.M., and COOK, T.S. (1977). Analysis of power type singularities using finite elements. *International Journal for Numerical Methods in Engineering*, **11**, 1225–33.

UMASHANKAR, K.R. (1988). Numerical analysis of electromagnetic wave scattering and interaction based on frequency-domain integral equation and method of moments techniques. *Wave Motion*, **10**, 493–525.

VAN BLADEL, J. (1981). Induction at a corner in perfect iron. *IEE Proceedings*, **B128**, 219–21.

VAN BLADEL, J. (1985*a*). Field singularities at metal–dielectric wedges. *IEEE Transactions on Antennas and Propagation*, **33**, 450–5.

VAN BLADEL, J. (1985*b*). *Electromagnetic fields*. Hemisphere, Washington. Reprinted, with corrections, from a text published in 1964 by McGraw-Hill, New York, pp. 368–370, 382–7.

VASSALLO, C. (1985). *Théorie des guides d'ondes électromagnétiques*, Tome 2, pp. 524–6, 577–603, 645. Eyrolles, Paris.

VOLAKIS, J.L., and SENIOR, T.B.A. (1989). Application of a class of generalized boundary conditions to scattering by a metal-backed dielectric half plane. *Proceedings of the IEEE*, **77**, 796–805.

WILLIAMS, J.T., DUDLEY, D.G., and BUTLER, C.M. (1987). Surface currents induced on a circular strip. *Radio Science*, **22**, 216–226.

WILTON, D.R., and GOVIND, S. (1977). Incorporation of edge conditions in moment method solutions. *IEEE Transactions on Antennas and Propagation*, **25**, 845–50.

ZIOLKOWSKI, R.W., and JOHNSON, W.A. (1987*a*). Electromagnetic scattering of an arbitrary plane wave from a spherical shell with a circular aperture. *Journal of Mathematical Physics*, **28**, 1293–1314.

ZIOLKOWSKI, R.W., and JOHNSON, W.A. (1987*b*). Pseudocoupling ansatz for electromagnetic aperture coupling in three dimensions. *Radio Science*, **22**, 169–74.

ZIOLKOWSKI, R.W., MARSLAND, D.P., LIBELO, L.F., and PISANE, G.F. (1988). Scattering from an open spherical shell having a circular aperture and enclosing a concentric dielectric sphere. *IEEE Transactions on Antennas and Propagation*, **36**, 985–99.

5

SINGULARITIES AT THE TIP OF A CONE

Some conical structures found in practice are shown in Fig. 5.1. Figure 5.1a shows a finite cone and a cone–sphere combination, useful for the modelling of rockets. A biconical antenna is modelled in Fig. 5.1c. Figure 5.1d depicts a flat sector, which can serve to model the rudder of an airplane. Figure 5.1e represents a straight wire, conceived as the limit of a cone of a vanishing opening angle. Such wires are encountered in, for example, antennas and lightning arresters (Figs 5.1e, f). A series of cones, sometimes used as a model for a rough surface, are illustrated in Fig. 5.1g.

Cones considered as scatterers, and illuminated by various sources (plane waves, dipole waves, etc.), have been studied extensively in literature, especially with respect to their radar cross-section (Bowman *et al.* 1969; Ruck *et al.* 1970). It is not our purpose to discuss these aspects further. We shall instead focus our attention on the behaviour of the fields near the tip of the cone, at distances short with respect to the wavelength. A study based on separation of variables is available for circular cones; we shall discuss it in detail. Separation is also applicable to elliptic cones, but here we shall only quote a few numerical results, with the purpose of testing the accuracy of a more general method, valid for cones of arbitrary cross-section. This method, based on a variational principle, is our main pole of interest. We apply it to a pair of geometries of considerable practical importance: the flat sector and the pyramid, for which fairly extensive numerical results are given.

5.1 Perfectly conducting circular cone. Electric singularity

At short distances from the tip of the cone, the electric field can be derived from a potential ϕ, which is conveniently expanded in spherical harmonics. Thus (Fig. 5.2),

$$\phi(R, \theta, \varphi) = \sum_{m=0}^{\infty} \sum_{\nu} R^{\nu} \mathrm{P}_{\nu}^{m}(\cos\theta)(A_{m\nu} \sin m\varphi + B_{m\nu} \cos m\varphi), \tag{5.1}$$

where P_{ν}^{m} is an associated Legendre function. The subscript ν actually depends on two indices, and should be written as ν_{mn}. We write ν to lighten the notation. The value of ν is determined by requiring ϕ to vanish on the cone.

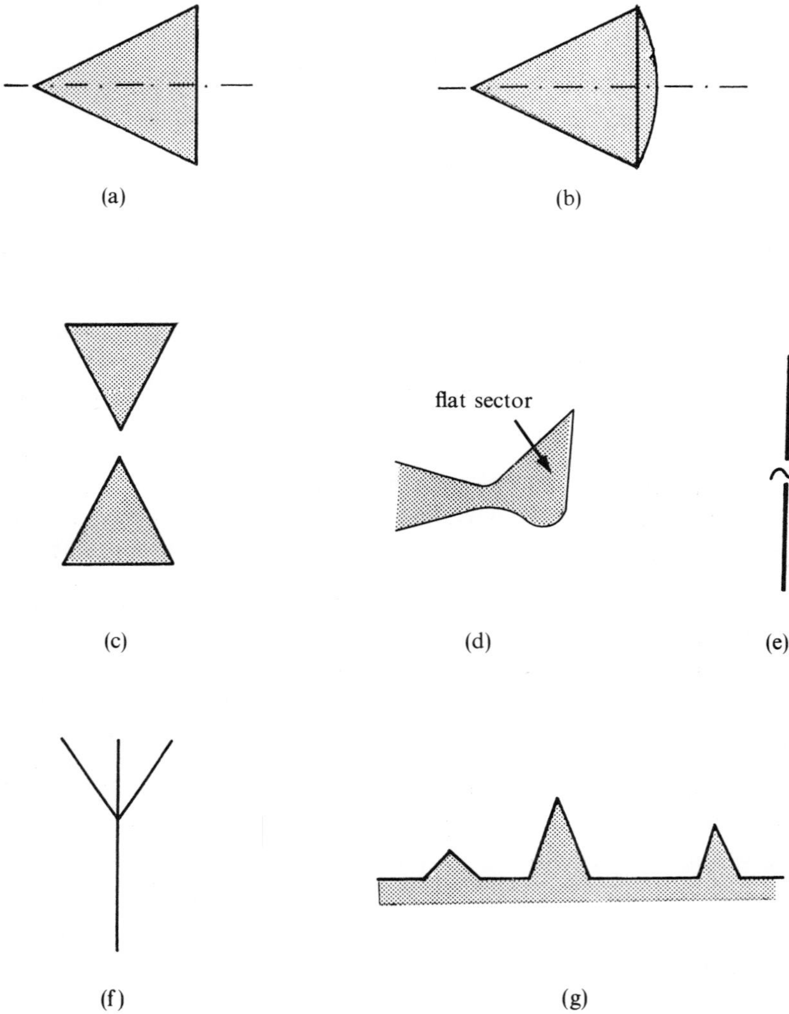

Fig. 5.1. Conical structures occurring in practice.

This leads to the condition

$$\mathrm{P}_\nu^m(\cos\theta_0) = 0. \tag{5.2}$$

This equation quantizes ν (Robin 1959; Bowman *et al.* 1969). Only values of ν less than 1 are of interest, since they alone give rise to infinite fields. Detailed calculations show that the $m = 0$ mode of revolution is the only one in which

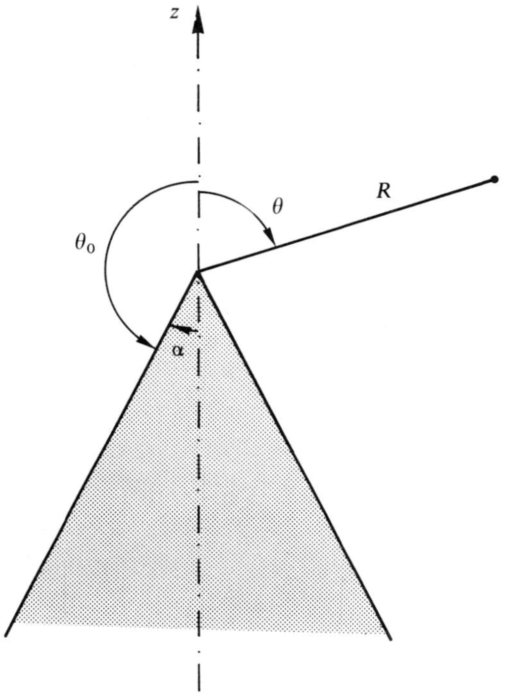

Fig. 5.2. Angles and coordinates for a circular cone.

infinities can occur. According to Table 5.1, this happens only for *sharp* cones, and for the *first* root (Van Bladel 1983a, b). The relevant data are displayed graphically in Fig. 5.3. The potential and fields associated with the singularity are, to within a trivial factor,

$$\phi = R^{\nu}\mathrm{P}_{\nu}(\cos\theta),$$

$$\boldsymbol{E} = \nu R^{\nu-1}\mathrm{P}_{\nu}(\cos\theta)\boldsymbol{u}_R + R^{\nu-1}\frac{\mathrm{dP}_{\nu}(\cos\theta)}{\mathrm{d}\theta}\boldsymbol{u}_{\theta}, \tag{5.3}$$

$$\boldsymbol{H} = -\,\mathrm{j}\omega\varepsilon_0\frac{R^{\nu}}{\nu+1}\frac{\mathrm{dP}_{\nu}(\cos\theta)}{\mathrm{d}\theta}\boldsymbol{u}_{\varphi},$$

where P_{ν} is a Legendre function. Typical lines of force for $\boldsymbol{E}$ are shown in Fig. 5.4, together with a few equipotentials. The electric field near the tip becomes infinite like $R^{\nu-1}$, and the same is true of the surface charge density

Table 5.1. Values of ν satisfying $P_\nu(\cos\theta_0) = 0$ (circular metallic cone)

θ_0	α	$\cos\theta_0$	Lowest ν	Second ν	Third ν
10°	170°	0.985	13.276	31.126	49.081
20°	160°	0.940	6.383	15.311	24.289
30°	150°	0.866	4.084	10.038	16.025
40°	140°	0.766	2.932	7.401	11.892
50°	130°	0.643	2.240	5.819	9.412
60°	120°	0.500	1.777	4.763	7.758
70°	110°	0.342	1.446	4.008	6.577
80°	100°	0.174	1.196	3.441	5.690
90°	90°	0.000	1.000	3.000	5.000
100°	80°	−0.174	0.8423	2.646	4.448
110°	70°	−0.342	0.7118	2.356	3.995
120°	60°	−0.500	0.6015	2.113	3.617
130°	50°	−0.643	0.0563	1.906	3.296
140°	40°	−0.766	0.4223	1.727	3.020
150°	30°	−0.866	0.3462	1.568	2.778
160°	20°	−0.940	0.2745	1.425	2.561
165°	15°	−0.966	0.2387	1.356	2.460
170°	10°	−0.985	0.2012	1.287	2.360
175°	5°	−0.996	0.1581	1.212	2.255
176°	4°	−0.998	0.1479	1.195	2.232
177°	3°	−0.999	0.1364	1.177	2.207
178°	2°	−0.999	0.1230	1.156	2.180
179°	1°	−0.999	0.1052	1.129	2.146

on the cone, which is given by

$$\rho_s = -\varepsilon_0[E_\theta]_{\theta=\theta_0} = -\varepsilon_0 R^{\nu-1}\left[\frac{dP_\nu(\cos\theta)}{d\theta}\right]_{\theta=\theta_0}$$
$$= \varepsilon_0 R^{\nu-1} P'_\nu(\cos\theta_0)\sin\theta_0, \tag{5.4}$$

where the prime denotes the derivative with respect to the argument (in this case, $\cos\theta$). The total charge between 0 and R is proportional to $R^{\nu+1}$, which means that no 'point charge' accumulates at the tip. We also observe that the magnetic field in (5.3) remains bounded, and that the same holds for the surface current.

The boundary condition (5.2) is also relevant for the acoustic pressure on a soft cone. At short distances, in the range $kR \ll 1$, the pressure P satisfies Laplace's equation, hence it should have the form given in (5.3) for ϕ. The actual value of P is known for both an incident plane wave and the incident field generated by a point source (Bowman *et al.* 1969). In both cases, the solution confirms the proportionality of P to $(kR)^\nu$.

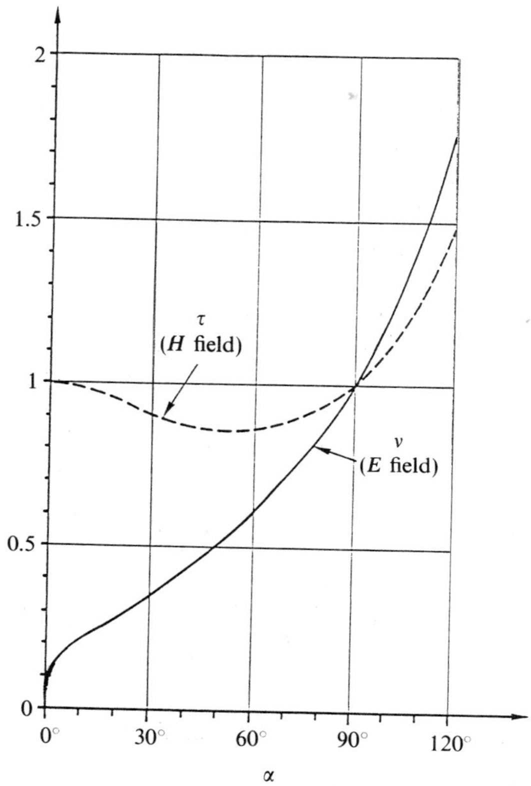

Fig. 5.3. Singularity exponents for a perfectly conducting circular cone. From Van Bladel (1983b: p. 902) © IEEE.

5.2 Perfectly conducting circular cone. Magnetic singularity

The magnetic field at short distances from the tip can be derived from a potential ψ of the general form (5.1). Because $\boldsymbol{H}$ must be tangent to the cone, the boundary condition at the surface becomes

$$\left[\frac{\mathrm{d}P_\tau^m(\cos\theta)}{\mathrm{d}\theta}\right]_{\theta=\theta_0} = 0. \tag{5.5}$$

This condition quantizes τ. It turns out that only one harmonic yields values of τ less than one, i.e. infinite fields. This harmonic corresponds to the lowest value of τ for $m = 1$ (Van Bladel 1983a). Numerical values are given in

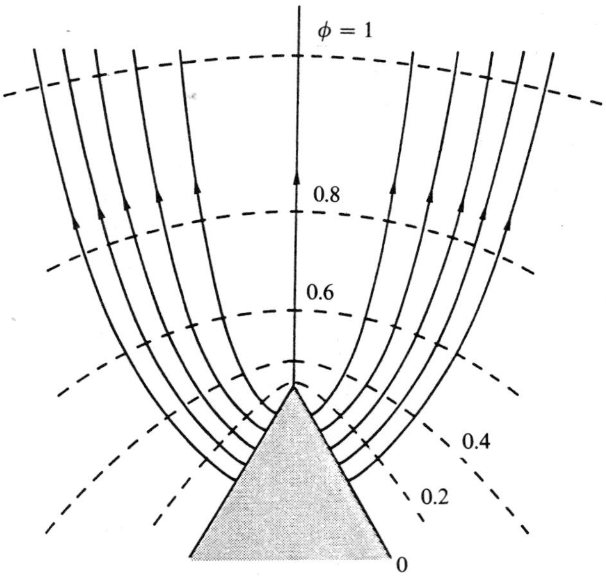

Fig. 5.4. Lines of force of the singular electric field near the tip of a perfectly conducting circular cone. From Van Bladel (1983b: p. 901) © IEEE.

Table 5.2 and Fig. 5.3. The relevant potential and fields are

$$\psi = R^{\tau} \mathrm{P}_{\tau}^{1}(\cos\theta)\cos(\varphi-\varphi_0),$$

$$\boldsymbol{H} = R^{\tau-1}\left\{[\tau \mathrm{P}_{\tau}^{1}(\cos\theta)\cos(\varphi-\varphi_0)]\boldsymbol{u}_R + \left[\left(\frac{\mathrm{d}}{\mathrm{d}\theta}\mathrm{P}_{\tau}^{1}(\cos\theta)\right)\cos(\varphi-\varphi_0)\right]\boldsymbol{u}_\theta - \left[\left(\frac{\mathrm{P}_{\tau}^{1}(\cos\theta)}{\sin\theta}\sin(\varphi-\varphi_0)\right)\boldsymbol{u}_\varphi\right]\right\}, \tag{5.6}$$

$$\boldsymbol{E} = \mathrm{j}\omega\mu_0\frac{R^{\tau}}{\tau+1}\left\{\left(\frac{\mathrm{P}_{\tau}^{1}(\cos\theta)}{\sin\theta}\sin(\varphi-\varphi_0)\right)\boldsymbol{u}_\theta + \left[\left(\frac{\mathrm{d}}{\mathrm{d}\theta}\mathrm{P}_{\tau}^{1}(\cos\theta)\right)\cos(\varphi-\varphi_0)\right]\boldsymbol{u}_\varphi\right\}.$$

These formulas show that $\boldsymbol{E}$, together with the charge density ρ_s at the surface, remains bounded near the tip. The magnetic field, on the other hand, becomes infinite when $\alpha < \frac{1}{2}\pi$. The lines of force of $\boldsymbol{H}$ in the meridian plane $\varphi = \varphi_0$ are shown in Fig. 5.5.

Table 5.2. Values of τ satisfying $\left[\frac{d}{d\theta} P_\tau^1(\cos\theta)\right]_{\theta_0} = 0$ (circular metallic cone)

θ_0	α	$\cos\theta_0$	Lowest τ	Second τ	Third τ
10°	170°	0.985	10.083	30.057	48.416
20°	160°	0.940	4.843	14.793	23.967
30°	150°	0.866	3.120	9.712	15.822
40°	140°	0.766	2.275	7.177	11.732
50°	130°	0.643	1.783	5.661	9.314
60°	120°	0.500	1.468	4.654	7.691
70°	110°	0.342	1.254	3.940	6.534
80°	100°	0.174	1.105	3.408	5.669
90°	90°	0.000	1.000	3.000	5.000
100°	80°	−0.174	0.9276	2.679	4.468
110°	70°	−0.342	0.8809	2.424	4.038
120°	60°	−0.500	0.8563	2.221	3.685
130°	50°	−0.643	0.8520	2.063	3.395
140°	40°	−0.766	0.8672	1.949	3.161
150°	30°	−0.866	0.9005	1.886	2.984
160°	20°	−0.940	0.9453	1.890	2.888
165°	15°	−0.966	0.9671	1.919	2.887
170°	10°	−0.985	0.9849	1.957	2.925
175°	5°	−0.996	0.9962	1.989	2.977
176°	4°	−0.998	0.9976	1.993	2.985
177°	3°	−0.999	0.9986	1.996	2.992
178°	2°	−0.999	0.9994	1.998	2.996
179°	1°	−0.999	0.9998	2.000	2.999

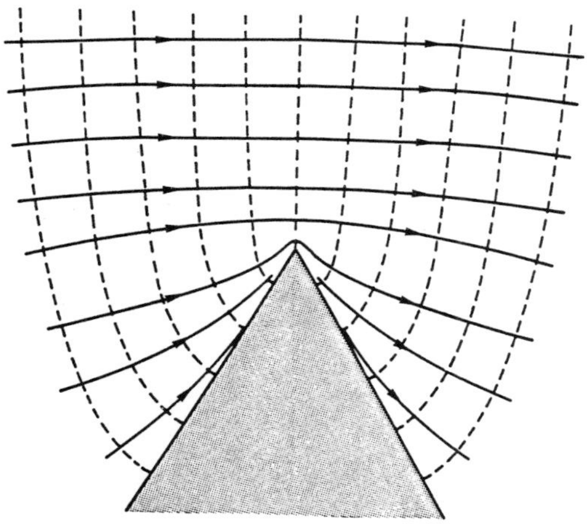

Fig. 5.5. Lines of force of the singular magnetic field near the tip of a perfectly conducting circular cone, in the $\varphi = \varphi_0$ plane. From Van Bladel (1983b: p. 901) © IEEE.

The surface current density on the cone is given by

$$J_s = [H \times u_\theta]_{\theta=\theta_0} = R^{\tau-1}\left[\left(\frac{P_\tau^1(\cos\theta_0)}{\sin\theta_0}\sin(\varphi-\varphi_0)\right)u_R + [\tau P_\tau^1(\cos\theta_0)\cos(\varphi-\varphi_0)]u_\varphi\right]. \tag{5.7}$$

This density becomes infinite at the tip, but the *total* current flowing to the tip is zero.

Boundary condition (5.5) is also relevant for the acoustic pressure on a hard cone. Results known for the cone immersed in either a plane wave or the field of a point source confirm that P is proportional to $(kR)^\tau$ at distances short with respect to λ. Data are also available for the electromagnetic field, particularly in connection with dipole sources (electric and magnetic) and plane waves (Bowman *et al.* 1969). For a linearly polarized incident plane wave

$$E_i = (u_{\theta_i}\sin\beta_i + u_{\varphi_i}\cos\beta_i) \times \exp\{jkR[\sin\theta\sin\theta_i\cos(\varphi-\varphi_i)+\cos\theta\cos\theta_i]\}, \tag{5.8}$$

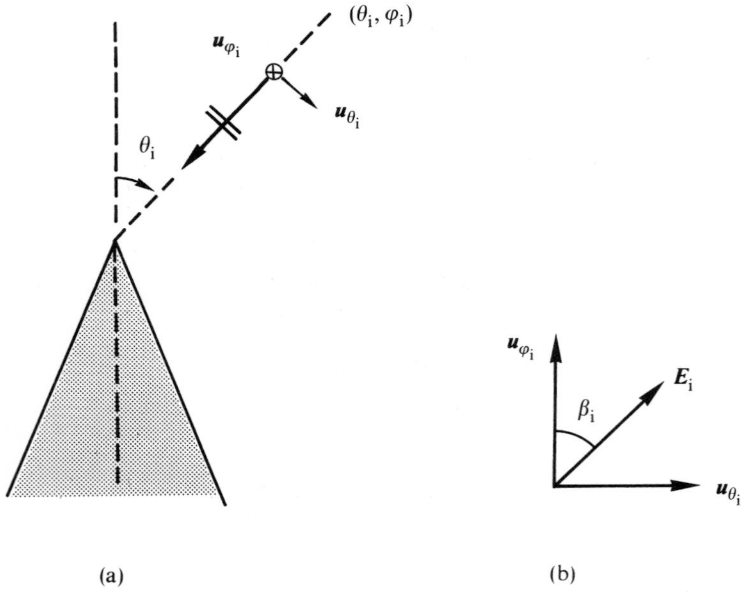

Fig. 5.6. A plane wave incident on a perfectly conducting circular cone.

the electric field at short distances is, to within a factor (Fig. 5.6),

$$\boldsymbol{E} \doteqdot [(kR)^{\nu-1} \sin \beta_i] \mathrm{P}_\nu^1(\cos \theta_i)\left(\boldsymbol{u}_R + \frac{1}{\nu}\boldsymbol{u}_\theta \frac{\partial}{\partial \theta}\right)\mathrm{P}_\nu(\cos \theta). \tag{5.9}$$

The electric singularity is apparent. The formula also shows that the singularity is not excited when $\beta_i = 0$, i.e. when $\boldsymbol{E}_i$ has no z component. Such a property could have been anticipated by the profile of the lines of force shown in Fig. 5.4. Finally, the magnetic field is given by

$$\boldsymbol{H} \doteqdot (kR)^{\tau-1}\left(\boldsymbol{u}_R + \frac{1}{\tau}\boldsymbol{u}_\theta \frac{\partial}{\partial \theta} + \frac{1}{\tau \sin \theta}\boldsymbol{u}_\varphi \frac{\partial}{\partial \varphi}\right)$$

$$\left(\frac{\sin \beta_i}{\sin \theta_i}\frac{\partial}{\partial \varphi_i} - \cos \beta_i \frac{\partial}{\partial \theta_i}\right)[\mathrm{P}_\tau^1(\cos \theta)\mathrm{P}_\tau^1(\cos \theta_i)\cos(\varphi - \varphi_i)]. \tag{5.10}$$

5.3 Dielectric cone of circular cross-section

Dielectric cones are often encountered in practice, for example at the tip of a radome, or as part of a high performance horn antenna (Fig. 5.7a). The electric field can be derived from a potential ϕ, which in turn can be expanded in spherical harmonics. The expansions are of the general form (Fig. 5.7b)

$$\phi_1(R, \theta, \varphi) = \sum_{m=0}^{\infty} \sum_{\nu} R^\nu \mathrm{P}_\nu^m(\cos \theta)(A_{m\nu} \sin m\varphi + B_{m\nu} \cos m\varphi) \tag{5.11}$$

in region 1, and

$$\phi_2(R, \theta, \varphi) = \sum_{m=0}^{\infty} \sum_{\nu} R^\nu \mathrm{P}_\nu^m(\cos \theta')(C_{m\nu} \sin m\varphi + D_{m\nu} \cos m\varphi) \tag{5.12}$$

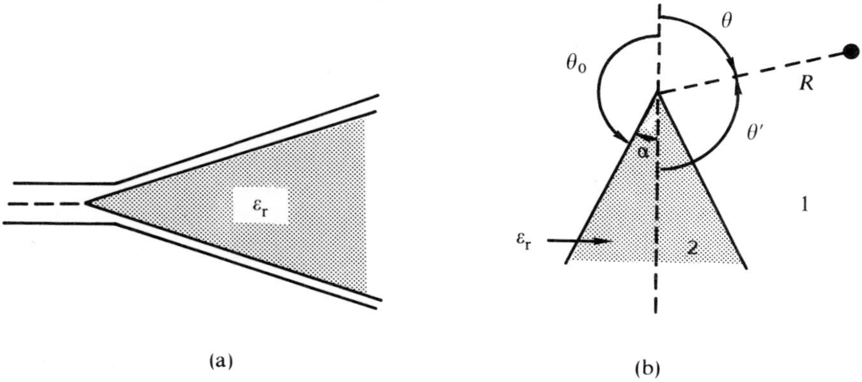

Fig. 5.7. Circular dielectric cones, and relevant coordinates.

in region 2. Notice tht $\theta' = \pi - \theta$. The boundary conditions at the air–dielectric interface are $\phi_1 = \phi_2$ and $\partial\phi_1/\partial\theta = \varepsilon_r \partial\phi_2/\partial\theta = -\varepsilon_r \partial\phi_2/\partial\theta'$. They quantize ν through the equation

$$[\mathrm{P}_\nu^m(\cos\theta)]_{\theta=\alpha}\left[\frac{\mathrm{dP}_\nu^m(\cos\theta)}{\mathrm{d}\theta}\right]_{\theta=\theta_0} + \varepsilon_r[\mathrm{P}_\nu^m(\cos\theta)]_{\theta=\theta_0}\left[\frac{\mathrm{dP}_\nu^m(\cos\theta)}{\mathrm{d}\theta}\right]_{\theta=\alpha} = 0. \tag{5.13}$$

The half-opening angle α is equal to $\pi - \theta_0$. In region 1, the electric field relative to the '$\cos m\varphi$' harmonic is of the form

$$\boldsymbol{E}_1 = \operatorname{grad}\phi_1 = A_{m\nu}R^{\nu-1}\left[[\nu\mathrm{P}_\nu^m(\cos\theta)\cos m\varphi]\boldsymbol{u}_R + \left(\frac{\mathrm{dP}_\nu^m(\cos\theta)}{\mathrm{d}\theta}\cos m\varphi\right)\boldsymbol{u}_\theta - \left(\frac{m\mathrm{P}_\nu^m(\cos\theta)}{\sin\theta}\sin m\varphi\right)\boldsymbol{u}_\varphi\right]. \tag{5.14}$$

The corresponding value in region 2 follows by replacing θ by θ', and $\boldsymbol{u}_\theta$ by $\boldsymbol{u}_{\theta'} = -\boldsymbol{u}_\theta$. It should be noted that the previous results remain valid when regions 1 and 2 are dielectrics with respective permittivities ε_{r1} and ε_{r2}. The symbol ε_r now stands for the ratio $\varepsilon_{r2}/\varepsilon_{r1}$.

Table 5.3. Singularity exponent ν for the $m = 0$ mode (circular dielectric cone)

α	$\varepsilon_r = 1$	$\varepsilon_r = 2$	$\varepsilon_r = 5$	$\varepsilon_r = 10$	$\varepsilon_r = 38$	$\varepsilon_r = 50$	$\varepsilon_r = 100$	$\varepsilon_r = \infty$
0°	1.0000	1.0000	1.0000	1.0000	1.0000	1.0000	1.0000	0.0000
2°	1.0000	0.99939	0.99754	0.99438	0.97505	0.96586	0.92184	0.12296
5°	1.0000	0.99618	0.98424	0.96281	0.82210	0.76320	0.58513	0.15814
10°	1.0000	0.98486	0.93672	0.85400	0.55957	0.49883	0.37559	0.20122
20°	1.0000	0.94561	0.80463	0.65536	0.41923	0.38886	0.33556	0.27450
30°	1.0000	0.90248	0.71945	0.58463	0.42501	0.40746	0.37797	0.34618
40°	1.0000	0.87317	0.69120	0.58292	0.47175	0.46044	0.44183	0.42228
50°	1.0000	0.86374	0.70262	0.61871	0.53950	0.53180	0.51927	0.50629
60°	1.0000	0.87403	0.74289	0.68025	0.62417	0.61887	0.61030	0.60151
70°	1.0000	0.90208	0.80699	0.76388	0.72655	0.72308	0.71750	0.71180
80°	1.0000	0.94538	0.89307	0.86983	0.85000	0.84817	0.84524	0.84225
90°	1.0000	1.0000	1.0000	1.0000	1.0000	1.0000	1.0000	1.0000
100°	1.0000	1.0582	1.1220	1.1538	1.8333	1.1862	1.1905	1.1956
110°	1.0000	1.1056	1.2299	1.3020	1.3884	1.3994	1.4194	1.4456
120°	1.0000	1.1259	1.2597	1.3275	1.3942	1.4010	1.4122	1.4241
130°	1.0000	1.1164	1.2154	1.2552	1.2869	1.2897	1.2942	1.2958
140°	1.0000	1.0888	1.1516	1.1739	1.1905	1.1922	1.1945	1.1958
150°	1.0000	1.0564	1.0918	1.1037	1.1125	1.1133	1.1145	1.1156
160°	1.0000	1,0274	1.0437	1.0491	1.0530	1.0534	1.0539	1.0544
170°	1.0000	1.0073	1,0117	1.0131	1.0142	1.0142	1.0144	1.0145
180°	1.0000	1.0000	1.0000	1.0000	1.0000	1.0000	1.0000	1.0000

The magnetic field near the apex of a circular *dielectric* cone remains bounded. In the case of a *magnetic* cone, however, the magnetic fields $\boldsymbol{H}_1$ and $\boldsymbol{H}_2$ can again be derived from scalar potentials of the form (5.11) and (5.12), and are therefore singular. The singularity exponent is given by (5.13), but ε_r must be replaced by μ_r.

As in the case of a metallic cone, field singularities only obtain for values of ν less than unity. Numerical solution of (5.13), for $\varepsilon_r > 1$, shows that singularities are found in two cases (Van Bladel 1984, 1985b):

(a) When the cone is sharp (i.e. for $0 < \alpha < 90°$), singularities arise for the lowest value of ν corresponding to the φ-independent mode. (Table 5.3 and Fig. 5.8). Typical lines of force are shown in Fig. 5.9, where equal fluxes of $\boldsymbol{d}$ flow between two successive surfaces generated by the rotation of the (solid) lines of force.

(b) When the cone is reentrant (i.e. for $90° < \alpha < 180°$), there are singularities for the lowest value of ν corresponding to the $m = 1$ mode. (Table 5.4 and Fig. 5.10). The lines of force in the meridian plane $\varphi = 0°$ are shown in Fig. 5.11. The shaded area is the region of higher ε_r. Consequently, the figure can be interpreted as showing a sharp cone of lower dielectric constant penetrating into a zone of higher ε_r.

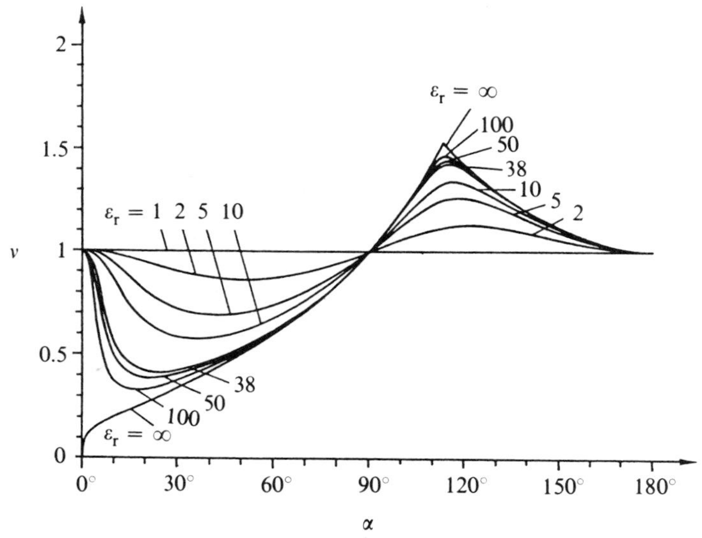

Fig. 5.8. The singularity exponent for a circular dielectric cone ($m = 0$ mode). From Van Bladel (1985b: p. 894) © IEEE.

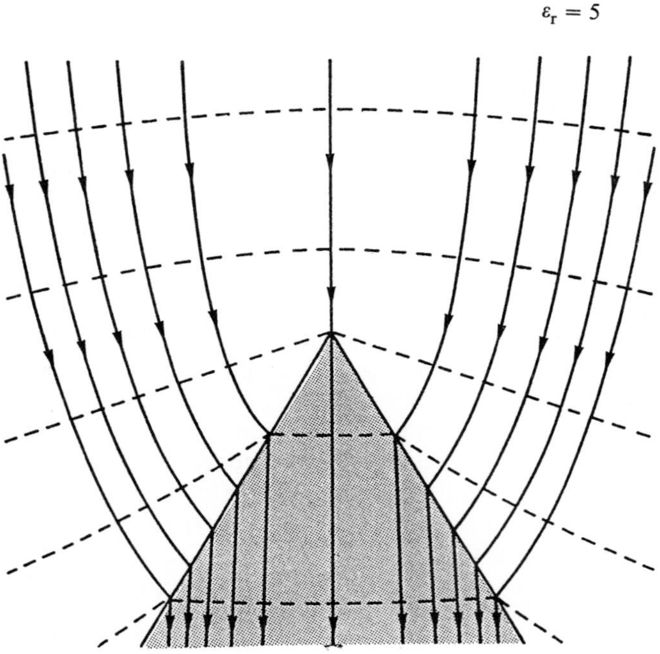

Fig. 5.9. Lines of force of the singular electric field near the tip of a sharp circular dielectric cone. From Van Bladel (1985b: p. 893) © IEEE.

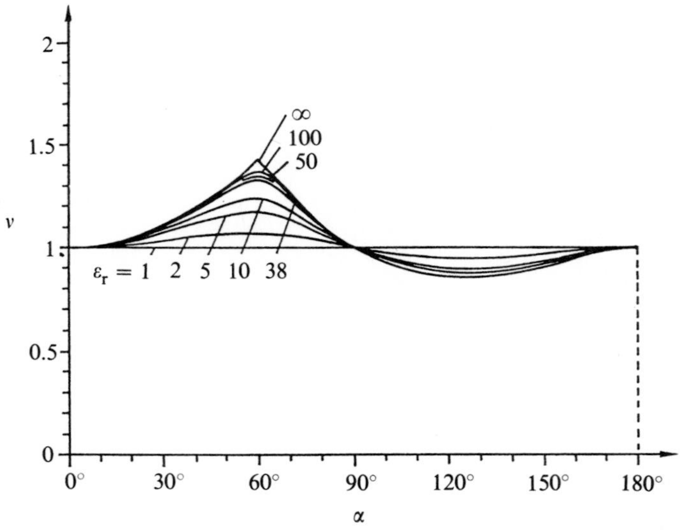

Fig. 5.10. The singularity exponent for a circular dielectric cone ($m = 1$ mode). From Van Bladel (1985b: p. 894) © IEEE.

$\varepsilon_r = 5$

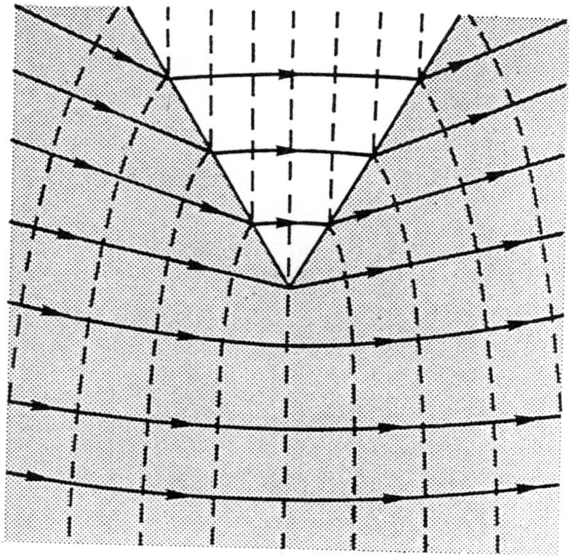

Fig. 5.11. Lines of force of the singular electric field near the tip of a reentrant circular dielectric cone. Van Bladel (1985b: p. 893) © IEEE.

5.4 Metallic cone of arbitrary cross-section

Let C be the intersection of the cone boundary with a sphere of unit radius, centred at the tip of the cone (Fig. 5.12). The surface S' is the intersection of the cone interior with the unit sphere, and S is the part of the unit sphere complementary to S'. The three-dimensional curve C, which defines the shape of the cone, can be represented parametrically by the functions $\theta(t)$ and $\varphi(t)$. In the vicinity of the apex, the electric field is derivable from a potential ϕ, which can be expanded in a series

$$\phi(R, \theta, \varphi) = \sum_{m,n=0}^{\infty} A_{mn} R^{\nu_{mn}} Y_{mn}(\theta, \varphi). \tag{5.15}$$

This series, in which Y_{mn} is a generalized spherical harmonic, is an extension of the expansion (5.1) used for the circular cone. Because ϕ satisfies Laplace's equation, and vanishes on the metal, Y_{mn} is an eigenfunction of the problem (De Smedt *et al.* 1986)

$$\nabla^2_{\theta\varphi} Y_{mn} + \nu_{mn}(\nu_{mn} + 1)\, Y_{mn} = 0 \quad \text{in } S, \qquad Y_{mn} = 0 \quad \text{on } C. \tag{5.16}$$

Table 5.4. Singularity exponent τ for the $m = 1$ mode (circular dielectric cone)

α	$\varepsilon_r = 1$	$\varepsilon_r = 2$	$\varepsilon_r = 5$	$\varepsilon_r = 10$	$\varepsilon_r = 38$	$\varepsilon_r = 50$	$\varepsilon_r = 100$	$\varepsilon_r = \infty$
0°	1.0000	1.0000	1.0000	1.0000	1.0000	1.0000	1.0000	1.0000
10°	1.0000	1.0049	1.0098	1.0119	1.0138	1.0140	1.0143	1.0145
20°	1.0000	1.0183	1.0365	1.0447	1.0517	1.0523	1.0534	1.0544
30°	1.0000	1.0370	1.0757	1.0937	1.1094	1.1109	1.1132	1.1156
40°	1.0000	1.0564	1.1214	1.1542	1.1844	1.1873	1.1520	1.1965
50°	1.0000	1.0702	1.1614	1.2149	1.2720	1.2780	1.2850	1.2968
60°	1.0000	1.0712	1.1710	1.2380	1.3311	1.3446	1.3719	1.4241
70°	1.0000	1.0563	1.1309	1.1764	1.2280	1.2338	1.2435	1.2543
80°	1.0000	1.0299	1.0642	1.0818	1.0982	1.0995	1.1024	1.1050
90°	1.0000	1.0000	1.0000	1.0000	1.0000	1.0000	1.0000	1.0000
100°	1.0000	0.97344	0.94946	0.93928	0.93082	0.93005	0.92882	0.92757
110°	1.0000	0.95406	0.91500	0.89598	0.88590	0.88472	0.88254	0.88093
120°	1.0000	0.94363	0.89667	0.87762	0.86217	0.86078	0.85856	0.85631
130°	1.0000	0.94264	0.89407	0.87423	0.85809	0.85663	0.85431	0.85196
140°	1.0000	0.95045	0.90650	0.85811	0.87297	0.87160	0.86941	0.86719
150°	1.0000	0.96488	0.93172	0.91730	0.90519	0.90408	0.90231	0.90051
160°	1.0000	0.98168	0.96343	0.95518	0.94810	0.94745	0.94640	0.54533
170°	1.0000	0.99502	0.98997	0.98765	0.98565	0.98546	0.98516	0.98486
175°	1.0000	0.99873	0.99746	0.99688	0.99637	0.99633	0.99625	0.99618
180°	1.0000	1.0000	1.0000	1.0000	1.0000	1.0000	1.0000	1.0000

The Laplacian operator is here

$$\nabla^2_{\theta\varphi} F = \frac{1}{\sin\theta}\frac{\partial}{\partial\theta}\left(\sin\theta\,\frac{\partial F}{\partial\theta}\right) + \frac{1}{\sin^2\theta}\frac{\partial^2 F}{\partial\varphi^2}. \tag{5.17}$$

The Y_{mn} form a complete set, and are orthogonal in the sense that

$$[\nu_{mn}(\nu_{mn}+1) - \nu_{pq}(\nu_{pq}+1)] \iint_S Y_{mn}(\theta, \varphi)\, Y_{pq}(\theta, \varphi) \sin\theta \,\mathrm{d}\theta\,\mathrm{d}\varphi = 0. \tag{5.18}$$

Each partial harmonic generates fields of the form

$$\begin{aligned} \boldsymbol{E} &= \operatorname{grad}\phi = R^{\nu_{mn}-1}(\nu_{mn} Y_{mn} \boldsymbol{u}_R + \operatorname{grad}_{\theta\varphi} Y_{mn}), \\ \boldsymbol{H} &= -\,\mathrm{j}\omega\varepsilon_0 \frac{R^{\nu_{mn}}}{\nu_{mn}+1}\, \boldsymbol{u}_R \times \operatorname{grad}_{\theta\varphi} Y_{mn}, \end{aligned} \tag{5.19}$$

where

$$\operatorname{grad}_{\theta\varphi} F = \frac{\partial F}{\partial\theta}\boldsymbol{u}_\theta + \frac{1}{\sin\theta}\frac{\partial F}{\partial\varphi}\boldsymbol{u}_\varphi. \tag{5.20}$$

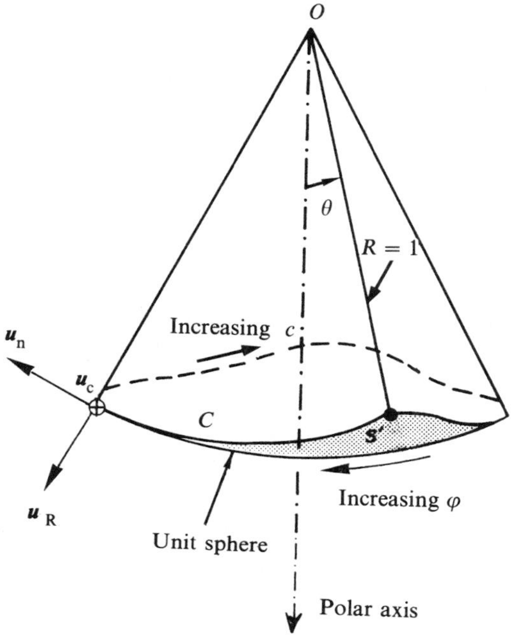

Fig. 5.12. A cone of arbitrary cross-section.

Clearly, values of ν less than 1 result in a singular electric field. The problem therefore narrows down to a search for the values of ν less than one. A method to solve this problem is described in Section 5.5.

A similar analysis holds for the magnetic field. This field can again be derived from a scalar potential of type (5.15), but the eigenfunctions (denoted by Z_{mn} instead of Y_{mn}) must now satisfy

$$\nabla^2_{\theta\varphi} Z_{mn} + \tau_{mn}(\tau_{mn} + 1) Z_{mn} = 0 \quad \text{in } S, \qquad \boldsymbol{u}_n \cdot \operatorname{grad}_{\theta\varphi} Z_{mn} = 0 \quad \text{on } C. \tag{5.21}$$

Here $\boldsymbol{u}_n$, unit vector in the tangent plane of the sphere, is perpendicular to both $\boldsymbol{u}_c$ and the conical surface. Thus,

$$\boldsymbol{u}_n = \boldsymbol{u}_c \times \boldsymbol{u}_R = \left(\sin\theta \frac{d}{dt}\, \boldsymbol{u}_\theta - \frac{d\theta}{dt}\, \boldsymbol{u}_\varphi\right)\left[\left(\frac{d\theta}{dt}\right)^2 + (\sin^2\theta)\left(\frac{d\varphi}{dt}\right)^2\right]^{-\frac{1}{2}}. \tag{5.22}$$

The Z_{mn} satisfy the orthogonality condition (5.18), and the fields they generate are given by

$$E = j\omega\mu_0 \frac{R^{\tau_{mn}}}{\tau_{mn}+1} u_R \times \operatorname{grad}_{\theta\varphi} Z_{mn},$$
$$H = \operatorname{grad}\psi = R^{\tau_{mn}-1}(\tau_{mn} Z_{mn} u_R + \operatorname{grad}_{\theta\varphi} Z_{mn}). \tag{5.23}$$

The problem, once again, consists in finding the values of τ less than one.

Charge and current densities at the surface of the cone are, for the harmonics $R^{\nu_{mn}} Y_{mn}$ and $R^{\tau_{mn}} Z_{mn}$,

$$\rho_s = \varepsilon_0 u_n \cdot E = \varepsilon_0 R^{\nu_{mn}-1} u_n \cdot \operatorname{grad}_{\theta\varphi} Y_{mn},$$
$$J_s = u_n \times H = R^{\tau_{mn}-1}\left(-\tau_{mn} Z_{mn} u_c + \frac{\partial Z_{mn}}{\partial c} u_R\right), \tag{5.24}$$

where c is the arc length along C, counted positive in the direction of u_c.

5.5 Numerical procedure

Eigenvalue problems (5.16) and (5.21) can be solved elegantly by means of variational methods. Consider first the functional

$$J_1(\phi) = \iint_S [|\operatorname{grad}_s \phi|^2 - \mu(\mu+1)\phi^2] \, dS, \tag{5.25}$$

where ϕ is a trial function, and grad_s is the operator defined in (2.109). On the unit sphere, using (5.20),

$$J_1(\phi) = \iint_S \left[\left(\frac{\partial\phi}{\partial\theta}\right)^2 + \frac{1}{\sin^2\theta}\left(\frac{\partial\theta}{\partial\varphi}\right)^2 - \mu(\mu+1)\phi^2\right] \sin\theta \, d\theta \, d\varphi. \tag{5.26}$$

It is proved in Appendix E that $J_1(\phi)$ is stationary with respect to Z_{mn}. The analysis shows that (5.21) is the appropriate Euler equation, that μ coincides with τ_{mn}, and that the natural boundary condition is $u_n \cdot \operatorname{grad}_s \phi_0 = 0$ on C (the condition which characterizes the magnetic problem). The Z_{mn} and τ_{mn} can therefore be obtained by inserting trial functions in (5.26), no restrictions being put on their behaviour on boundary curve C. If, on the other hand, the 'basket' of trial functions is restricted to functions which vanish on C, the natural boundary condition disappears, and $J_1(\phi)$ becomes stationary with respect to the electric eigenfunctions Y_{mn}. The parameter μ is now equal to ν_{mn}.

Instead of $J_1(\phi)$ one can equally well use the functional

$$J_2(\phi) = \iint_S |\operatorname{grad}_s \phi|^2 \, dS \Big/ \iint_S \phi^2 \, dS. \tag{5.27}$$

On the unit sphere:

$$J_2(\phi) = \iint_S \left[\left(\frac{\partial \phi}{\partial \theta} \right)^2 + \frac{1}{\sin^2 \theta} \left(\frac{\partial \phi}{\partial \varphi} \right)^2 \right] \sin \theta \, d\theta \, d\varphi \Big/ \iint_S \phi^2 \sin \theta \, d\theta \, d\varphi. \quad (5.28)$$

From Appendix F, $J_2(\phi)$ is stationary with respect to either Y_{mn} or Z_{mn}, and the stationary values are $\nu_{mn}(\nu_{mn} + 1)$ and $\tau_{mn}(\tau_{mn} + 1)$, respectively. In the magnetic problem the smallest eigenvalue is zero, and the corresponding eigenfunction is Z = constant. Because of the orthogonality property (5.18), the other eigenfunctions must have zero average value on S. This property can serve as a guide to select the test functions efficiently.

It would be too space-consuming to describe detailed numerical procedures based on either $J_1(\phi)$ or $J_2(\phi)$. In a version developed by De Smedt (De Smedt *et al.* 1985; De Smedt 1986) the following points were taken into account:

(1) The trial functions are formed by the method of finite elements. These functions are normally chosen to be polynomials of the third degree, although linear functions are used in a few cases. The procedure leads to matrix problems of the form

$$\mathbf{S} \cdot \boldsymbol{\phi} = \mu(\mu + 1) \mathbf{T} \cdot \boldsymbol{\phi} \quad (\text{for } J_1), \quad (5.29)$$

$$(\boldsymbol{\phi} \cdot \mathbf{S} \cdot \boldsymbol{\phi})/(\boldsymbol{\phi} \cdot \mathbf{T} \cdot \boldsymbol{\phi}) \quad \text{stationary} \quad (\text{for } J_2). \quad (5.30)$$

(2) The matrix elements must be evaluated with care. This holds in particular when the elements are close to the $\theta = 0$ or $\theta = \pi$ boundaries, in which case quasi-singularities occur through the $1/\sin \theta$ term in (5.26) or (5.28).

(3) Attention must be paid to the singularities which occur in the (θ, φ) plane at angular points of the $(\theta(t), \varphi(t))$ curve. Such points are found, for example, at the edge of a sector, or at the intersection of two planes in a pyramid. The solution consists in introducing barycentric coordinates of the type defined in (4.60).

5.6 The elliptic cone

The accuracy of the variational procedure can be tested on the example of a cone of elliptic cross-section (Fig. 5.13a). Such a cone may serve to model a hill, and its geometry is therefore of interest for propagation studies. Taking φ as the parameter t, the equation of the relevant C curve in the (θ, φ) plane becomes (De Smedt *et al.* 1986),

$$\theta = \theta(\varphi) = \arctan \frac{\varepsilon \tan \theta_{\mathrm{m}}}{\sqrt{(\sin^2 \varphi + \varepsilon^2 \cos^2 \varphi)}}, \quad (5.31)$$

where $\varepsilon = b/a$. Instead of θ_m, half-opening angle of the major axis, we could use the half-opening angle θ_0 of the minor axis. The two are related by

$$\tan\theta_0 = \varepsilon \tan\theta_m. \tag{5.32}$$

Curve C has two axes of symmetry. Consequently, the fundamental symmetries of the electric potential ϕ must be of the kind shown in Fig. 5.13b. Extensive calculations, based on the methods of Section 5.5, show that the

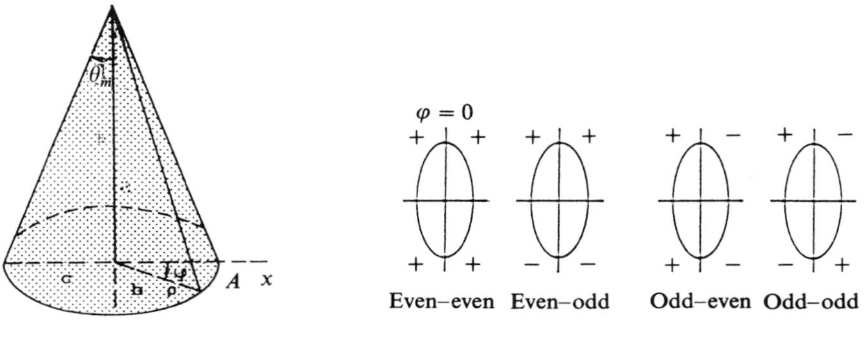

(a) (b)

Fig. 5.13. An elliptic cone, with basic symmetries for ϕ. From De Smedt *et al.* (1986: pp. 866, 867) © IEEE.

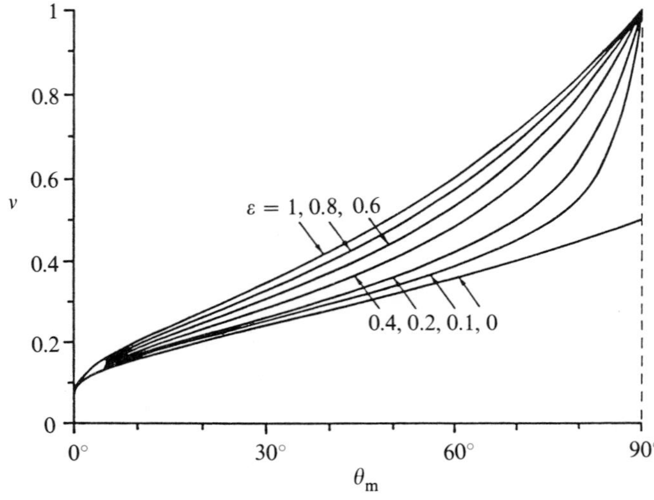

Fig. 5.14. The electric singularity exponent ν for a perfectly conducting elliptic cone (even–even symmetry of ϕ). From De Smedt *et al.* (1986: p. 867) © IEEE.

only value of ν less than unity, for all ε and θ_m, is the lowest one relative to the even–even symmetry (De Smedt *et al.* 1985). Relevant results are shown in Table 5.5 and Fig. 5.14. To check the accuracy of these results we have added values obtained by an alternate method: separation of variables. In the case of the elliptic cone, the method requires the use of spheroconal coordinates (see e.g. Kraus *et al.* 1961; Blume *et al.* 1985a, b; Vafiadis *et al.* 1984, 1985). These coordinates are connected to Cartesian coordinates by the relationships

$$x = R \sin\theta \cos\varphi,$$
$$y = R\sqrt{(1 - k^2 \cos^2\theta)} \sin\varphi, \qquad z = (R\cos\theta)\sqrt{(1 - k'^2 \sin^2\varphi)}, \tag{5.33}$$

where $k, k' > 0$, $k^2 + k'^2 = 1$, $0 \leqslant R < \infty$, $0 \leqslant \theta \leqslant \pi$, $0 \leqslant \varphi \leqslant 2\pi$. The coordinate surfaces ($\theta = \theta_0 =$ constant) are cones of elliptic cross-section. Very accurate values of the singularity exponents have been obtained at the

Table 5.5a, b. Singularity exponent ν for even–even symmetry (elliptic metallic cone)

(a) From variational principle

θ_m	$\varepsilon = 0$	$\varepsilon = 0.1$	$\varepsilon = 0.2$	$\varepsilon = 0.4$	$\varepsilon = 0.6$	$\varepsilon = 0.8$	$\varepsilon = 1$
5°	0.13150	0.13365	0.13621	0.14193	0.14774	0.15312	0.15814
10°	0.15878	0.16305	0.16700	0.17594	0.18495	0.19334	0.20122
20°	0.20213	0.20948	0.21604	0.23129	0.24670	0.26104	0.27450
30°	0.24124	0.25046	0.26094	0.28386	0.30646	0.32718	0.34618
40°	0.27850	0.29213	0.30744	0.34043	0.37080	0.39823	0.42228
50°	0.31667	0.33625	0.35859	0.40381	0.44458	0.47847	0.50629
60°	0.35749	0.38691	0.41963	0.48348	0.53980	0.57239	0.60151
70°	0.40034	0.44960	0.50182	0.58868	0.64703	0.68548	0.71180
80°	0.44781	0.55091	0.64244	0.74637	0.79686	0.82488	0.84225
90°	0.50000	1.00000	1.00000	1.00000	1.00000	1.00000	1.00000

(b) By separation of variables (Eindhoven)

θ_m	$\varepsilon = 0$	$\varepsilon = 0.1$	$\varepsilon = 0.2$	$\varepsilon = 0.4$	$\varepsilon = 0.6$	$\varepsilon = 0.8$	$\varepsilon = 1$
5°	.13004	.13331	.13644	.14236	.14790	.15315	.15814
10°	.15822	.16308	.16777	.17674	.18525	.19339	.20122
20°	.20168	.20969	.21752	.23266	.24719	.26114	.27450
30°	.24010	.25181	.26333	.28573	.30712	.32732	.34618
40°	.27759	.29410	.31042	.34203	.37155	.39838	.42228
45°	.29658	.31615	.33553	.37283	.40695	.43709	.46310
50°	.31595	.33924	.36232	.40628	.44535	.47862	.50629
60°	.35635	.39035	.42392	.48504	.53447	.57250	.60151
70°	.39982	.45393	.50602	.59043	.64747	.68555	.71180
80°	.44736	.55612	.64486	.74709	.79697	.82489	.84225
90°	.50000	1.00000	1.00000	1.00000	1.00000	1.00000	1.00000

Table 5.6a, b. Singularity exponent τ_{oe} for odd–even symmetry (elliptic metallic cone)

(a) From variational principle

θ_m	$\varepsilon = 0$	$\varepsilon = 0.1$	$\varepsilon = 0.2$	$\varepsilon = 0.4$	$\varepsilon = 0.6$	$\varepsilon = 0.8$	$\varepsilon = 1$
0°	1.00000	1.00000	1.00000	1.00000	1.00000	1.00000	1.00000
5°	0.99825	0.99810	0.99794	0.99741	0.99696	0.99654	0.99618
10°	0.99234	0.99212	0.99115	0.98953	0.98780	0.98624	0.98486
20°	0.96640	0.96613	0.96297	0.95724	0.95195	0.94804	0.94533
30°	0.91977	0.92024	0.91517	0.90653	0.90113	0.89953	0.90051
40°	0.85349	0.85695	0.85200	0.84706	0.85083	0.85791	0.86719
45°	0.81560	0.81894	0.81898	0.82157	0.83012	0.84313	0.85715
50°	0.77640	0.78656	0.78676	0.79748	0.81400	0.83336	0.85196
60°	0.69812	0.71708	0.73025	0.76242	0.79958	0.83080	0.85636
70°	0.62526	0.66249	0.69640	0.76449	0.81658	0.85403	0.88093
80°	0.55937	0.64951	0.72544	0.82577	0.87709	0.90782	0.92757
90°	0.50000	1.00000	1.00000	1.00000	1.00000	1.00000	1.00000

(b) By separation of variables (Eindhoven)

θ_m	$\varepsilon = 0$	$\varepsilon = 0.1$	$\varepsilon = 0.2$	$\varepsilon = 0.4$	$\varepsilon = 0.6$	$\varepsilon = 0.8$	$\varepsilon = 1$
0°	1.00000	1.00000	1.00000	1.00000	1.00000	1.00000	1.00000
5°	0.99807	0.99788	0.99769	0.99730	0.99692	0.99655	0.99618
10°	0.99209	0.99131	0.99053	0.98902	0.98756	0.98617	0.98486
20°	0.96615	0.96308	0.96020	0.95509	0.95091	0.94767	0.94533
30°	0.91904	0.91339	0.90872	0.90222	0.89913	0.89879	0.90051
40°	0.85257	0.84680	0.84345	0.84287	0.84817	0.85690	0.86719
45°	0.81466	0.81047	0.80955	0.81526	0.82729	0.84202	0.85717
50°	0.77543	0.77409	0.77686	0.79105	0.81109	0.83220	0.85196
60°	0.69748	0.70604	0.72062	0.75847	0.79676	0.82964	0.85631
70°	0.62461	0.65243	0.68844	0.75956	0.81416	0.85304	0.88093
80°	0.55880	0.64037	0.72058	0.82258	0.87608	0.90741	0.92757
90°	0.50000	1.00000	1.00000	1.00000	1.00000	1.00000	1.00000

Technical University of Eindhoven (Boersma *et al.* 1990). These values are quoted in Tables 5.5–8, together with those generated by the variational method. The agreement between the two methods is excellent.

The magnetic singularity relative to a circular cone was shown, in Section 5.2, to be associated with a 'cos φ' type of potential. Such φ-dependence corresponds to both even–odd and odd–even symmetries of the magnetic potential. Extensive calculations indicate that, for the elliptic cone, the only values of τ less than 1 are precisely those associated with these two symmetries. The results are shown in Tables 5.6–7 and Figs. 5.15–6. We observe that the values appearing in the column $\varepsilon = 1$ (which corresponds to the circular

Table 5.7a, b. Singularity exponent τ_{eo} for even–odd symmetry (elliptic metallic cone)

(a) From variational principle

θ_m	$\varepsilon = 0$	$\varepsilon = 0.1$	$\varepsilon = 0.2$	$\varepsilon = 0.4$	$\varepsilon = 0.6$	$\varepsilon = 0.8$	$\varepsilon = 1$
0°	1.00000	1.00000	1.00000	1.00000	1.00000	1.00000	1.00000
5°	1.00000	0.99971	0.99950	0.99888	0.99811	0.99719	0.99618
10°	1.00000	0.99901	0.99815	0.99569	0.99263	0.98900	0.98486
20°	1.00000	0.99655	0.99302	0.98365	0.97235	0.95946	0.94533
30°	1.00000	0.99310	0.98538	0.96658	0.94540	0.92298	0.90051
40°	1.00000	0.98899	0.97674	0.94927	0.92000	0.89235	0.86719
45°	1.00000	0.98689	0.97257	0.94085	0.90999	0.88173	0.85715
50°	1.00000	0.98496	0.96882	0.93431	0.90258	0.87496	0.85196
60°	1.00000	0.98236	0.96347	0.92821	0.89742	0.87420	0.85636
70°	1.00000	0.98152	0.96268	0.93047	0.90828	0.89244	0.88093
80°	1.00000	0.98311	0.97023	0.95102	0.94064	0.93298	0.92757
90°	1.00000	1.00000	1.00000	1.00000	1.00000	1.00000	1.00000

(b) By separation of variables (Eindhoven)

θ_m	$\varepsilon = 0$	$\varepsilon = 0.1$	$\varepsilon = 0.2$	$\varepsilon = 0.4$	$\varepsilon = 0.6$	$\varepsilon = 0.8$	$\varepsilon = 1$
0°	1.00000	1.00000	1.00000	1.00000	1.00000	1.00000	1.00000
5°	1.00000	0.99979	0.99954	0.99893	0.99817	0.99725	0.99618
10°	1.00000	0.99917	0.99819	0.99577	0.99274	0.98910	0.98486
20°	1.00000	0.99683	0.99309	0.98393	0.97272	0.95973	0.94533
30°	1.00000	0.99340	0.98570	0.96740	0.94624	0.92353	0.90051
40°	1.00000	0.98955	0.97757	0.95041	0.92144	0.89318	0.86719
45°	1.00000	0.98771	0.97378	0.94303	0.91172	0.88267	0.85717
50°	1.00000	0.98606	0.97045	0.93698	0.90457	0.87599	0.85196
60°	1.00000	0.98379	0.96602	0.93048	0.89974	0.87530	0.85631
70°	1.00000	0.98361	0.96607	0.93434	0.91055	0.89343	0.88093
80°	1.00000	0.98683	0.97365	0.95406	0.94167	0.93340	0.92757
90°	1.00000	1.00000	1.00000	1.00000	1.00000	1.00000	1.00000

cone) reproduce, with excellent accuracy, those given in Table 5.2. The same observation holds for Table 5.1 and the $\varepsilon = 1$ column of Table 5.5.

5.7 The flat sector. Singularity of the electric field

Flat plates exhibiting corners (salient or reentrant) occur in numerous structures, for example in aircraft wings, and in some types of antennas. In an early search for the value of the singularity exponent, Braunbek (1956) remarked that the integrability of the electric energy near the corner requires the

Table 5.8a, b. Singularity exponents for a sector of opening angle δ

(a) From variational principles

Salient			Reentrant		
δ	ν	τ	δ	ν	τ
10°	0.131	0.998	180°	0.500	0.500
20°	0.159	0.992	200°	0.560	0.448
40°	0.202	0.966	220°	0.625	0.400
60°	0.241	0.920	240°	0.698	0.357
80°	0.279	0.853	260°	0.776	0.317
90°	0.297	0.816	270°	0.816	0.297
100°	0.317	0.776	280°	0.853	0.279
120°	0.357	0.698	300°	0.920	0.241
140°	0.400	0.625	320°	0.966	0.202
160°	0.448	0.560	340°	0.992	0.159
180°	0.500	0.500	350°	0.998	0.131

(b) By separation of variables (Eindhoven)

Salient			Reentrant		
δ	ν	τ	δ	ν	τ
10°	0.13004	0.99807	180°	0.50000	0.50000
20°	0.15822	0.99209	200°	0.55880	0.44736
40°	0.20168	0.96615	220°	0.62461	0.39982
60°	0.24010	0.91904	240°	0.69748	0.35635
80°	0.27759	0.85257	260°	0.77543	0.31595
90°	0.29658	0.81466	270°	0.81466	0.29658
100°	0.31595	0.77543	280°	0.85257	0.27759
120°	0.35635	0.69748	300°	0.91904	0.24010
140°	0.39982	0.62461	320°	0.96615	0.20168
160°	0.44736	0.55880	340°	0.99209	0.15822
180°	0.50000	0.50000	350°	0.99807	0.13004

integral

$$E = \int_0^R |\mathrm{grad}\,\phi|^2 R^2 \,\mathrm{d}R \tag{5.34}$$

to be finite. For an electric potential ϕ of the form $R^\nu Y(\theta, \varphi)$, this requirement leads to the condition $\nu > -\frac{1}{2}$. By means of some elementary considerations, Braunbek further concluded that ν should be lower than $\frac{1}{2}$ for a salient corner, and higher than $\frac{1}{2}$ for a reentrant one. The transition value $\frac{1}{2}$ corresponds to the semi-infinite plane, a geometry discussed in detail in Section 4.6.

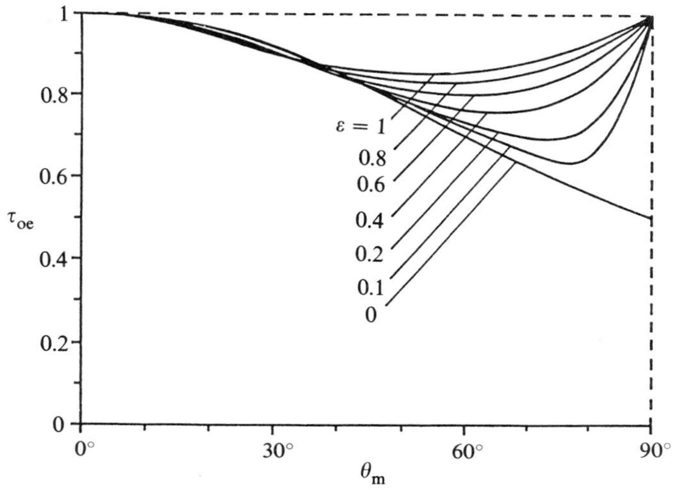

Fig. 5.15. The magnetic singularity exponent τ_{oe} for a perfectly conducting elliptic cone (odd–even symmetry of ψ). From De Smedt *et al.* (1986: p. 868) © IEEE.

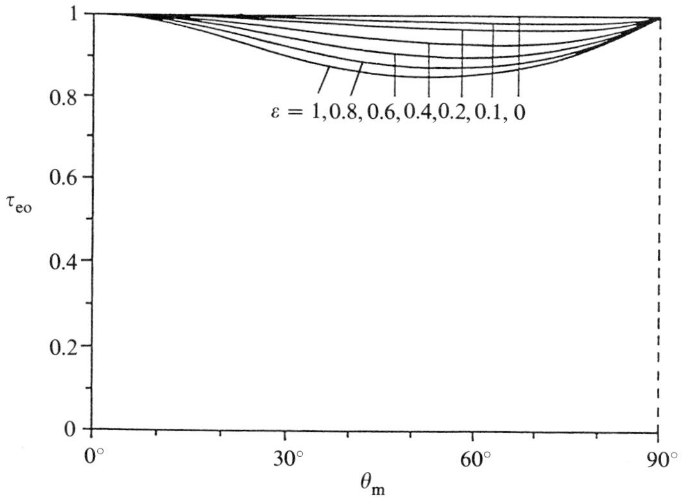

Fig. 5.16. The magnetic singularity exponent τ_{eo} for a perfectly conducting elliptic cone (even-odd symmetry of ψ). From De Smedt *et al.* (1986: p. 869) © IEEE.

The predictions of Braunbek are confirmed by specific values of ν, obtained by assimilating the sector to an elliptic cone of ε equal to zero. The relevant values are found in the $\varepsilon = 0$ column of Table 5.5. They are reproduced in Table 5.8 and Fig. 5.17 in terms of the total opening angle $\delta = 2\theta_m$ of the sector (De Smedt *et al.* 1987). Separation of variables in spheroconal coordinates is again applicable to the sector (see e.g. Blume 1967; Blume *et al.* 1969; Satterwhite *et al.* 1970; Satterwhite 1974; Morrison *et al.* 1976; Sahalos *et al.* 1983a, b). Results obtained by this method are included in Table 5.8.

The electric singularity is associated with a potential ϕ endowed with even–even symmetry (Fig. 5.18a). In that particular symmetry, the electric field is perpendicular to the sector, but lies *in* the sector plane at points of the plane complementary to the sector. The intersection C of cone and unit sphere reduces, in the (θ, φ) plane, to the segments $(\varphi = 0, 0 \leqslant \theta \leqslant \theta_m)$ and $(\varphi = \pi, 0 \leqslant \theta \leqslant \theta_m)$, which therefore define the position of the sector (Fig. 5.18b). The figure displays equipotetial (or equi-Y) curves, where $Y(\theta, \varphi)$ has been normalized such that its maximum is unity.

It is interesting to obtain a qualitative feeling for the way the electric field behaves near the tip of a sector. In the case of an aperture endowed with corners, for example, the component of $\boldsymbol{E}$ in the aperture plane is the main

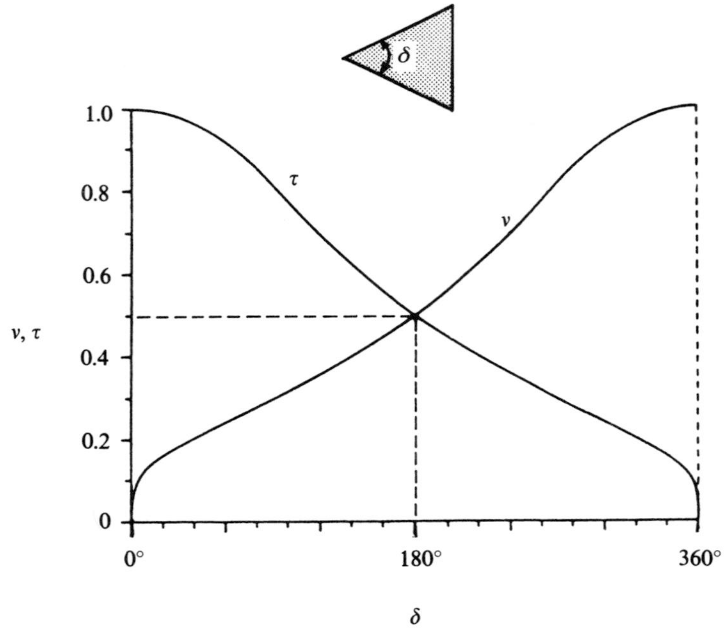

Fig. 5.17. Singularity exponents for a flat metallic sector. De Smedt *et al.* (1987: p. 695) © IEE.

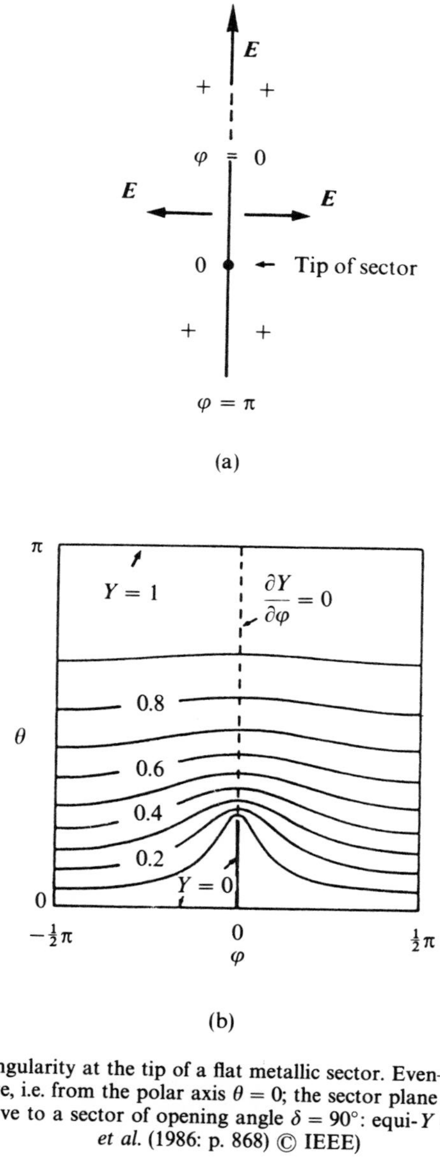

(b)

Fig. 5.18. Electric singularity at the tip of a flat metallic sector. Even–even symmetry. (a) The sector seen from above, i.e. from the polar axis $\theta = 0$; the sector plane is $\varphi = 0$, $\varphi = \pi$. (b) The function $Y(\theta, \varphi)$ relative to a sector of opening angle $\delta = 90°$: equi-Y curves. From De Smedt *et al.* (1986: p. 868) © IEEE)

unknown, and any information on its behaviour facilitates the solution of the field problem (Fig. 5.19a). Bearing this in mind we have plotted, in Fig. 5.19b, c, lines of force of $\boldsymbol{E}$ near the tip of the corner for two important opening angles, $\delta = 90°$ and $\delta = 270°$. Similar patterns are obtained for

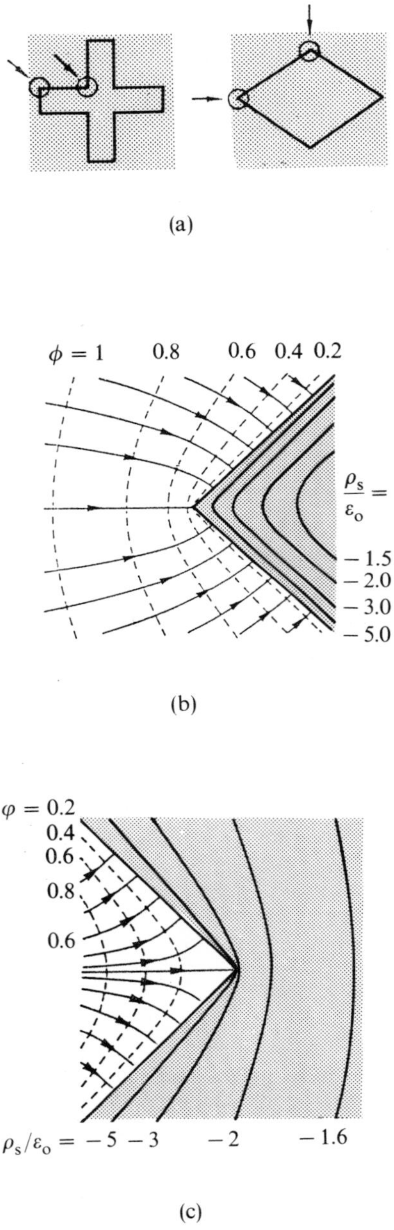

Fig. 5.19. (a) Two typical apertures with corners. (b, c) Lines of force of $\boldsymbol{E}$ and equipotentials, together with lines of constant ρ_s, for a sector with $\delta = 90°$ and $\delta = 270°$, respectively. From De Smedt *et al.* (1987: pp. 694–6) © IEE.

$\delta = 60°, 120°, 240°, 300°$ (De Smedt *et al.* 1987). The correctness of the general profile of the equipotentials is confirmed by the data appearing in Fig. 5.20, which show the variation of ϕ over the *total* area of two small apertures: the diamond and the rectangle (De Meulenaere *et al.* 1977).

The surface charge density on the sector, in the presence of a singular field grad $[R^{\nu} Y(\theta, \varphi)]$, is given by

$$\rho_s/\varepsilon_0 = 2R^{\nu-1}\mathbf{u}_{\varphi}\cdot \mathrm{grad}_{\theta\varphi} Y = 2R^{\nu-1}(1/\sin\theta)\frac{\partial Y}{\partial \varphi}. \tag{5.35}$$

The factor 2 in the second member results from adding the charges on both sides of the sector. Lines of constant ρ_s are shown in Fig. 5.19b, c. In the case of Fig. 5.19b, ρ_s near the tip is proportional to $R^{-0.703}$. Sufficiently far away from the tip, however, the singularity goes progressively over to the 'edge' type, characterized by the $1/\sqrt{d}$ law given in (4.11) and (4.24). This behaviour

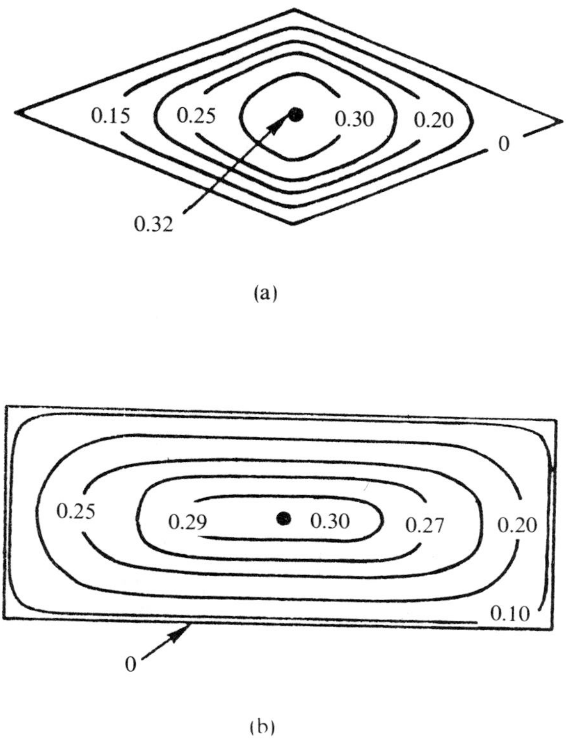

Fig. 5.20. Equipotentials in two apertures of dimensions small with respect to λ. From De Meulenaere *et al.* (1987: p. 199) © IEEE.

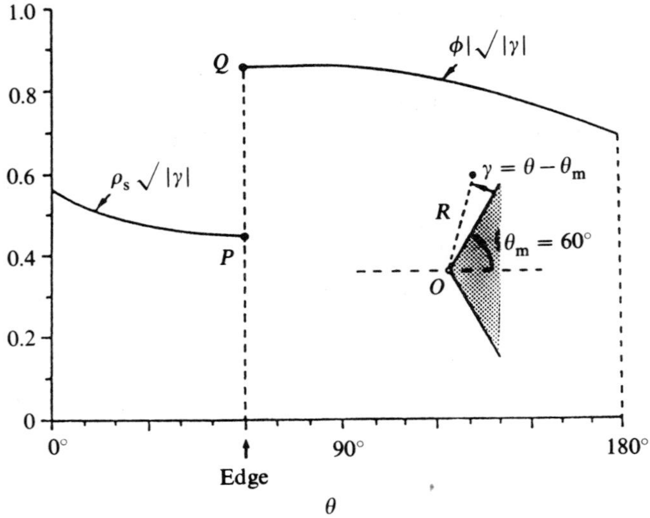

Fig. 5.21. The variation of ρ_s and ϕ along a circle of constant radius R in the sector plane. The opening angle δ is 120°. From De Smedt *et al.* (1987: p. 696) © IEE.

is confirmed by the numerical results displayed in Fig. 5.21. Close to the edge of the sector, $|\gamma|$ is proportional to d, the distance to the edge. Because the surface charge density ρ_s on the sector is proportional to $1/\sqrt{d}$, we expect the $\rho_s\sqrt{|\gamma|}$ curve to approach a finite, non-zero value at 'edge' point P. The figure confirms this behaviour. The potential ϕ outside the sector, on the other hand, should be proportional to $\sqrt{d}$, i.e. to $\sqrt{|\gamma|}$. It is now $\phi/\sqrt{|\gamma|}$ which should approach a finite nonzero value at the edge point Q. The figure again confirms this prediction.

5.8 The flat sector. Singularity of the magnetic field

The data of Tables 5.6 and 5.7 show that the even–odd singularity disappears in the case of the sector, but that the odd–even one remains. The general behaviour of $\boldsymbol{H}$ in the surviving symmetry is shown in Fig. 5.22a. Data concerning the singularity exponent τ are presented in Table 5.8 and Fig. 5.17. We observe that

$$\tau(\delta) = \nu(2\pi - \delta). \tag{5.36}$$

This property is not accidental. It can be predicted by comparing the variation of $Z(\theta, \varphi)$, depicted in Fig. 5.22b for $\delta = 270°$, with that of $Y(\theta, \varphi)$ shown in Fig. 5.18b for $\delta = 90°$. By turning one of the pictures upside down,

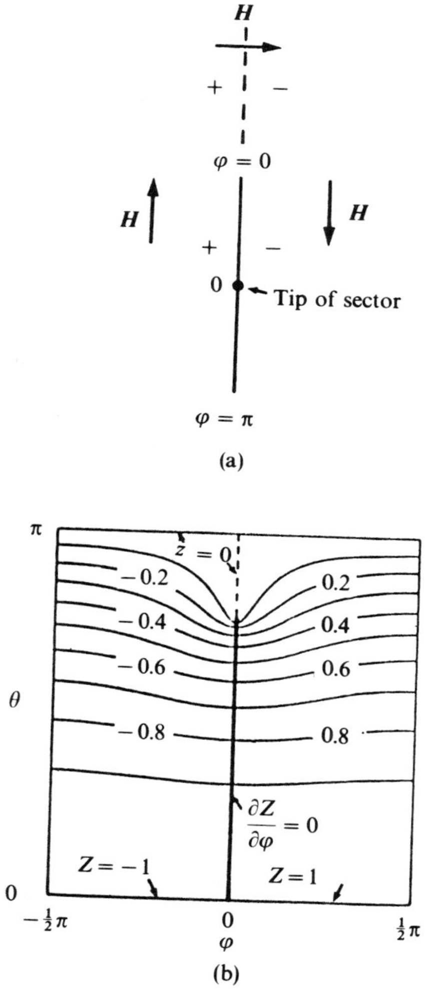

Fig. 5.22. Magnetic singularity at the tip of a flat metallic sector. Odd–even symmetry. (a) The sector seen from above, i.e. from the polar axis $\theta = 0$. The sector plane is $\varphi = 0$, $\varphi = \pi$. (b) The function $Z(\theta, \varphi)$ relative to a sector of opening angle $\delta = 270°$. Equi-Z curves. From De Smedt *et al.* (1987: p. 695) © IEE.

we observe that

$$Z(\theta, \varphi, \delta) = \begin{cases} Y(\pi - \theta, \varphi, 2\pi - \delta) & (0 < \varphi \leqslant \pi), \\ -Y(\pi - \theta, \varphi, 2\pi - \delta) & (-\pi < \varphi \leqslant 0). \end{cases} \tag{5.37}$$

A less geometrical proof is based on Babinet's principle (Boersma 1990). This principle establishes a parallel between the problem of a flat disk of surface S, immersed in a field $\mathbf{E}_i = \mathbf{f}$, with $\mathbf{H}_i = \mathbf{g}$, and that of an aperture S in a metallic screen, immersed in a field $\mathbf{E}_i = -R_c\mathbf{g}$, with $\mathbf{H}_i = \pm(\mathbf{f}/R_c)$ (see e.g. Van Bladel 1985a). Property (5.36) can also be proven by solving the problem in spheroconical coordinates (Blume *et al.* 1990).

The surface current density associated with a potential $\psi = R^\tau Z(\theta, \varphi)$, is given by

$$\mathbf{J}_s = 2\mathbf{u}_\varphi \times \operatorname{grad}\psi = 2\mathbf{u}_\varphi \times \operatorname{grad}(R^\tau Z) = 2R^{\tau-1}\left(-\frac{\partial Z}{\partial\theta}\mathbf{u}_R + \tau Z\mathbf{u}_\theta\right). \quad (5.38)$$

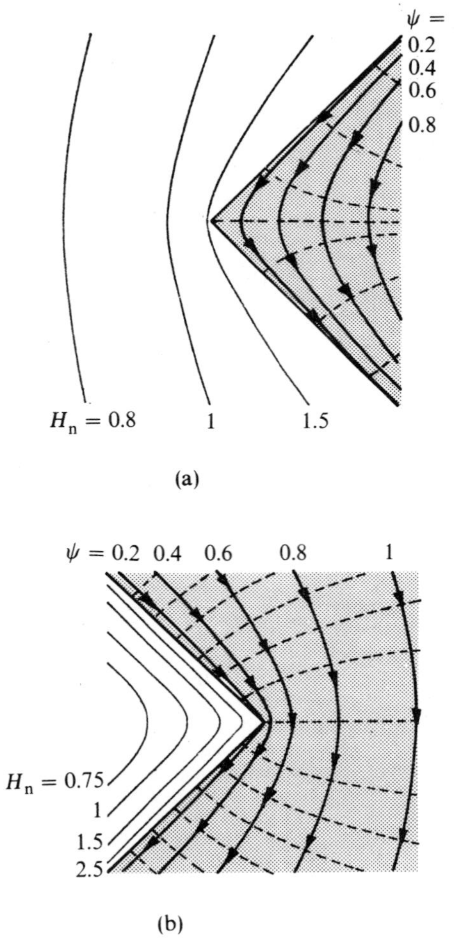

Fig. 5.23. Lines of current on a sector, together with values of the normal magnetic component H_n in the sector plane, for a sector with $\delta = 90°$ and $\delta = 270°$, respectively. From De Smedt *et al.* (1987: p. 697) © IEE.

This expression shows that the curves ψ = constant, drawn as solid lines in Fig. 5.23, are also lines of current. Furthermore, ψ is a flux function for the current. These general results are confirmed by data obtained by other methods (Ko *et al.* 1977; Glisson 1978; Butler 1980; Glisson *et al.* 1980; Sahalos *et al.* 1983a).

Because of the importance of knowing the profile of the lines of current, we give an expanded view of their general shape in Fig. 5.24. The figure contains information on, for example, the variation of a component such as J_{sy} along the line ABC. Around and beyond A, the edge-singularity dominates, and J_{sy} must be proportional to $1/\sqrt{x}$. Then follows a region around B where the corner singularity should dominate. Finally, in a third region close to C, the edge singularity requires J_{sy} to vanish like $\sqrt{y}$. This varying behaviour is confirmed by the data appearing in Fig. 5.24–5.

The singularities relative to the sector are not automatically excited by the incident fields. To test for the existence of the electric singularity, for example, we split the incident potential ϕ_i into four components, corresponding to the four symmetries shown in Fig. 5.13b. The singularities in $\boldsymbol{E}$ and ρ_s only occur when ϕ_i has an even–even component. It is interesting to observe that, because $\boldsymbol{E}$ lies in the plane of the sector (see Fig. 5.18a), the singularity is not influenced by the presence of a dielectric substrate in one of the half-spaces.

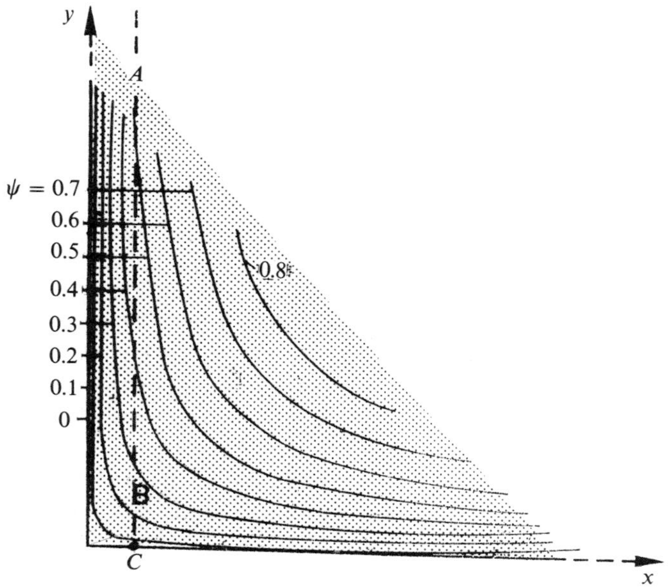

Fig. 5.24. Lines of current near the tip of a sector of opening angle $\delta = 90°$ $(1 - \tau = 0.185)$. From De Smedt *et al.* (1986: p. 869) © IEEE.

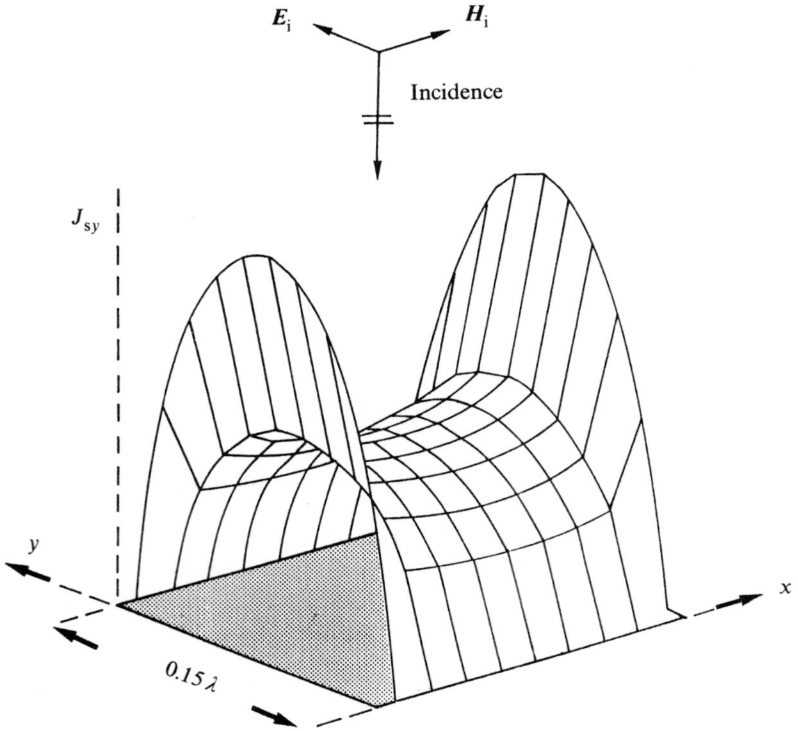

Fig. 5.25. Current induced on a 0.15λ square plate by a normally incident plane wave. From Glisson *et al.* (1980: p. 599) © 1980, IEEE.

This is the case, for example, near the corner P of a rectangular patch antenna (Fig. 5.26).

In the case of the magnetic singularity, the incident potential ψ_i must have an odd–even component if the singularity is to be excited. There, again, the presence of a nonmagnetic dielectric in one of the half-spaces does not affect the singularity and, in particular, the profile of the lines of current on the metallic corner.

The data presented in this section concern potential problems of a significance which transcends their application to electromagnetism. They are relevant, for example, for the study of the aerodynamics of a delta-wing placed at zero incidence in a uniform air stream (Brown *et al.* 1969; Taylor 1971).

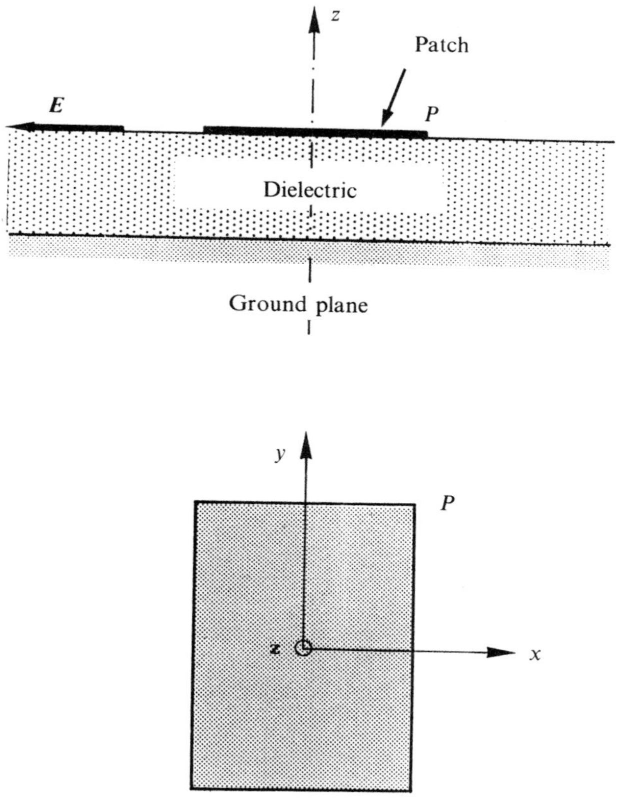

Fig. 5.26. A rectangular patch antenna lying on a dielectric substrate. The $\boldsymbol{E}$ singularity is the same as in free space.

5.9 The metallic pyramid

Given the multiplicity of angles involved, we limit our analysis to pyramids with equal faces. For such bodies, the three values of α are equal, and the cross-section perpendicular to the z axis is an equilateral triangle (Fig. 5.27a). The angles α and θ_1 are connected by the relationship

$$\cos\alpha = 1 - \tfrac{3}{2}\sin^2\theta_1 . \tag{5.39}$$

The $\alpha = 90°$ corner, in particular, is often encountered in practice, for example, at the apex of a cube.

The pyramidal structure is endowed with six symmetries: (1) three rotational symmetries, with respect to a rotation over 120°; (2) three reflectional symmetries, with respect to the three planes containing an edge and the z axis.

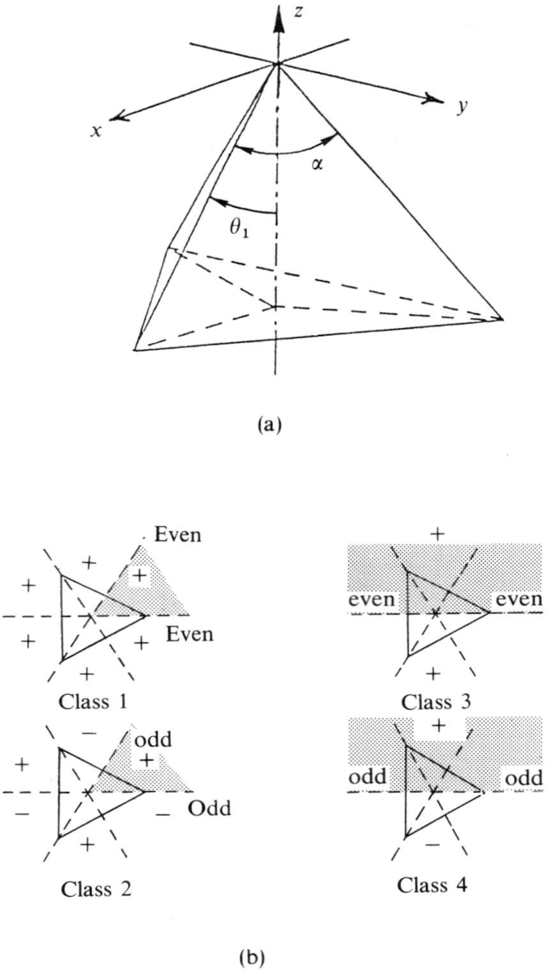

Fig. 5.27. A pyramid with equal faces, together with fundamental symmetries. From De Smedt *et al.* (1986: p. 869) © IEEE.

Four basic symmetries must be considered for the potentials. They are shown in Fig. 5.27b, together with the shaded areas in which the fundamental field problem must be solved. Numerical results obtained by variational methods (De Smedt *et al.* 1986) show that the only electric singularity is associated with the lowest value of ν in the 'class 1' symmetry. The magnetic singularities occur in classes 3 and 4, which are degenerate in the sense that they yield identical values for τ, while the corresponding eigenfunctions are different. The relevant values of ν and τ are given in Table 5.9 and Fig. 5.28.

Table 5.9 Singularity exponent for a metallic pyramid

θ_1	α	electric type class 1	magnetic type classes 3 and 4
0.000°	0.000°	ν = 0.00000	τ = 1.00000
5.000°	8.657°	0.14955	0.99828
10.000°	17.298°	0.18106	0.99239
20.000°	34.459°	0.23728	0.96813
30.000°	51.318°	0.29210	0.93102
40.000°	67.652°	0.35071	0.88901
45.000°	75.522°	0.38278	0.86980
50.000°	83.122°	0.41766	0.85332
54.736°	90.000°	0.45410	0.84108
60.000°	97.181°	0.49975	0.83351
70.000°	108.937°	0.60839	0.84075
80.000°	117.050°	0.76421	0.89048
85.000°	119.249°	0.86970	0.93672
90.000°	120.000°	1.00000	1.00000

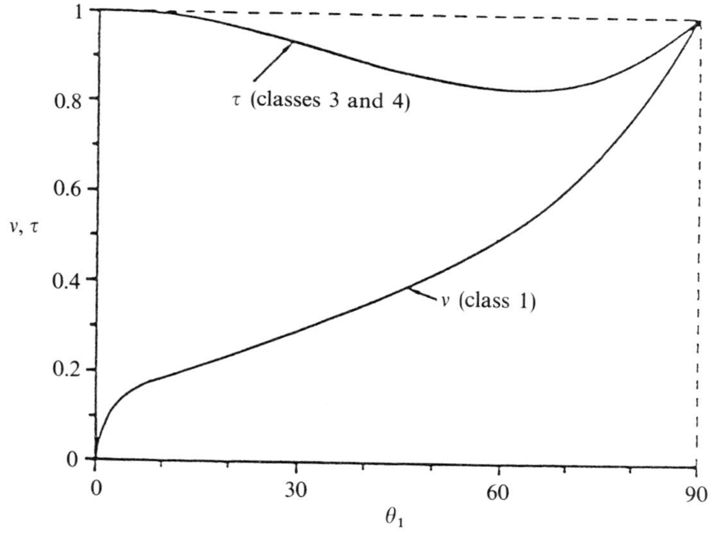

Fig. 5.28. Singularity exponents for a metallic pyramid, as a function of the opening angle θ_1. Exponent ν concerns the electric field, exponent τ the magnetic field. From De Smedt *et al.* (1986: p. 870) © IEEE.

Possible infinities of ρ_s and $\boldsymbol{J}_s$ are associated with the electric and magnetic singularities, respectively. This point has been discussed in Section 5.4. Relevant data on the variation of ρ_s and $\boldsymbol{J}_s$ are shown in Fig. 5.29. It is particularly interesting to follow the profile of the lines of current. The figure indicates that J_s may be parallel to an edge (Fig. 5.29c), in which case it tends to infinity in the vicinity of that edge. In other cases (Fig. 5.29b), the line of current crosses an edge, and is tangent to the latter at the crossing point. Details on this behaviour, and on other aspects—such as the variation of $Y(\theta, \varphi)$ and $Z(\theta, \varphi)$ in the (θ, φ) plane—are available elsewhere (De Smedt 1986).

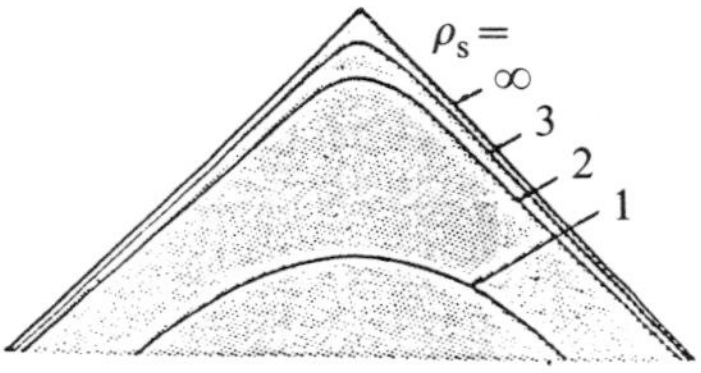

(a)

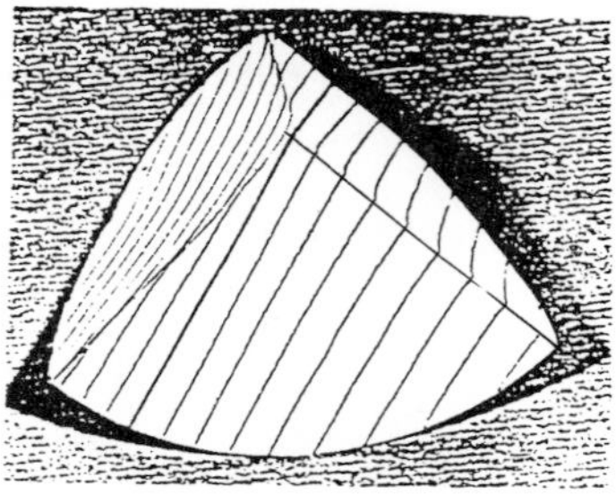

(b)

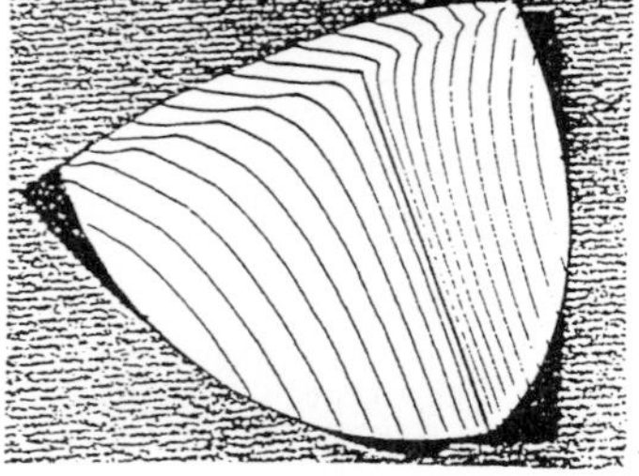

(c)

Fig. 5.29. (a) Lines of constant charge density on one of the faces of an $\alpha = 90°$ pyramid. (b) Lines of current on an $\alpha = 90°$ pyramid, class 3 symmetry. (c) Lines of current on an $\alpha = 90°$ pyramid, class 4 symmetry. From De Smedt *et al.* (1986: p. 870) © IEEE).

Knowledge of the singular behaviour is of obvious interest for the evaluation of the near fields. In applications where the *far field* is the main unknown, methods which do not specifically incorporate the corner behaviour may be satisfactory. Fundamentally this is because the effect of the singularity remains local. As a result, the error resulting from a small deviation from the correct corner behaviour does not propagate over large distances (Mei *et al.* 1987).

5.10 The sharp metallic cone

The curves in Fig. 5.28 show that ν approaches zero (and τ approaches 1) when θ_1 becomes very small, i.e. when the pyramid degenerates into a sharp needle. Consequently $\mathbf{E}$ becomes infinite like $1/R$ near the tip, whereas the singularity of $\mathbf{H}$ disappears. The actual value of $\mathbf{E}$ may be obtained from the electrostatic problem of the charged prolate spheroid, the 'sharp' limit of which is the needle. The solution is (Durand 1964)

$$\mathbf{E} = \frac{1}{R}\mathbf{u}_R - \frac{1}{R}(\tan\tfrac{1}{2}\theta)\mathbf{u}_\theta. \tag{5.40}$$

The approach of ν to zero (and of τ to one) may also be monitored on Fig. 5.3 (the circular cone), Figs. 5.14–6 (the elliptic cone), and Fig. 5.17 (the sector). In the case of the *circular* cone the following approximation becomes useful (Gray 1953):

$$\mathrm{P}_\nu(\cos\theta) \sim 1 + 2\nu\log_e\cos\tfrac{1}{2}\theta \quad (\nu \text{ small}). \tag{5.41}$$

Inserting this value into (5.2) gives, with $\dfrac{\theta_0}{2} = \dfrac{\pi}{2} - \dfrac{\alpha}{2}$ (Robin 1959),

$$\nu = 1/2\log_e(2/\alpha) = -1/2\log_e\tfrac{1}{2}\alpha \quad (\alpha \text{ small}). \tag{5.42}$$

For $\alpha = 1^\circ$, for example, (5.42) predicts $\nu = 0.1055$, in good agreement with the value 0.1052 given in Table 5.1.

In the case of a cone of *arbitrary* cross-section, the nature of the singularity can be investigated by means of perturbation techniques (Morrison 1977). The value of ν turns out to be (De Smedt 1988)

$$\nu = 1/2\log_e(2A/\alpha_m). \tag{5.43}$$

In this expression, α_m is the maximum half-opening angle of the cone (so chosen that the circle $\theta = \pi - \alpha_m$ on the unit sphere just encloses C), and A is a factor which depends only on the shape of the cross-section of the cone, but not on α_m. For a circular cone, A is unity; whereas, for an elliptic cone,

$$A = 2/(1+\varepsilon) = 2\alpha_{max}/(\alpha_{max} + \alpha_{min}). \tag{5.44}$$

The particular case of the sector is obtained by setting $\varepsilon = 0$, which yields $A = 2$. For a polyhedron with N equal faces, A can be approximated by

$$A = 1 + \pi^2/3N^2 \quad (N > 3). \tag{5.45}$$

5.11 Dielectric cone of arbitrary cross-section

The dielectric cone is shown in Fig. 5.12, a figure which was already used to define the geometry of a metallic cone. The surface S' is now the intersection of the *dielectric* body with the unit sphere. The conical volume is filled with a material of dielectric constant $\varepsilon_{r1}(\theta, \varphi)$, and is further embedded in a medium of dielectric constant $\varepsilon_{r2}(\theta, \varphi)$. In this section, we shall denote the surfaces S and S' by S_1 and S_2 respectively, for convenience. The curve C remains the boundary between S_1 and S_2.

The analysis of Section 5.3 is still valid in the present case. The electric field can be derived from a potential ϕ, which in turn is expanded in spherical harmonics of the general form (De Smedt 1987)

$$\begin{aligned} \phi_1(R, \theta, \varphi) &= R^\nu Y_1(\theta, \varphi) \quad \text{in region 1}, \\ \phi_2(R, \theta, \varphi) &= R^\nu Y_2(\theta, \varphi) \quad \text{in region 2}. \end{aligned} \tag{5.46}$$

The eigenvalue problem becomes (De Smedt 1985, 1987):

$$\begin{aligned} &\operatorname{div}_{\theta\varphi}(\varepsilon_{ri}\operatorname{grad}_{\theta\varphi} Y_i) + \nu(\nu+1)\varepsilon_{ri} Y_i = 0 \quad \text{in } S_i \quad (i = 1, 2), \\ &Y_i \text{ and } \varepsilon_{ri}\boldsymbol{u}_n \cdot \operatorname{grad}_{\theta\varphi} Y_i \text{ are continuous on } C \quad (i = 1, 2), \end{aligned} \tag{5.47}$$

where

$$\operatorname{div}_{\theta\varphi} \boldsymbol{f} = \frac{1}{\sin\theta}\frac{\partial}{\partial\theta}(\sin\theta\, f_\theta) + \frac{1}{\sin\theta}\frac{\partial f_\varphi}{\partial\varphi}. \tag{5.48}$$

The problem is again amenable to a variational solution. The functionals are now

$$\begin{aligned} J_1(\Gamma) &= \sum_{i=1,2} \iint_{S_i} [\varepsilon_{ri}|\operatorname{grad}_{\theta\varphi}\Gamma_i|^2 - \nu(\nu+1)\varepsilon_{ri}\Gamma_i^2]\sin\theta\, \mathrm{d}\theta\, \mathrm{d}\varphi, \\ J_2(\Gamma) &= \frac{\displaystyle\sum_{i=1,2} \iint_{S_i} \varepsilon_{ri}|\operatorname{grad}_{\theta\varphi}\Gamma_i|^2 \sin\theta\, \mathrm{d}\theta\, \mathrm{d}\varphi}{\displaystyle\sum_{i=1,2} \iint_{S_i} \varepsilon_{ri}\Gamma_i^2 \sin\theta\, \mathrm{d}\theta\, \mathrm{d}\varphi}. \end{aligned} \tag{5.49}$$

The symbol Γ stands for the ensemble of the trial functions Γ_1 and Γ_2. Both functionals are stationary with respect to $\Gamma = (Y_1, Y_2)$. The stationary value of J_2, in particular, is $\nu(\nu+1)$. In both cases, the Euler equation is (5.47), and the natural boundary condition coincides with the condition on $\partial Y_i/\partial n$

appearing in (5.47). Consequently, the only requirement for the trial functions Γ_1 and Γ_2 is to be equal on C.

The numerical results presented in this section were obtained by variational means. They only concern pyramids filled by (and embedded in) homogeneous media (Fig. 5.27a). The important electromagnetic parameter of the structure is the ratio $\varepsilon_r = \varepsilon_{r1}/\varepsilon_{r2}$. When the pyramid has a dielectric

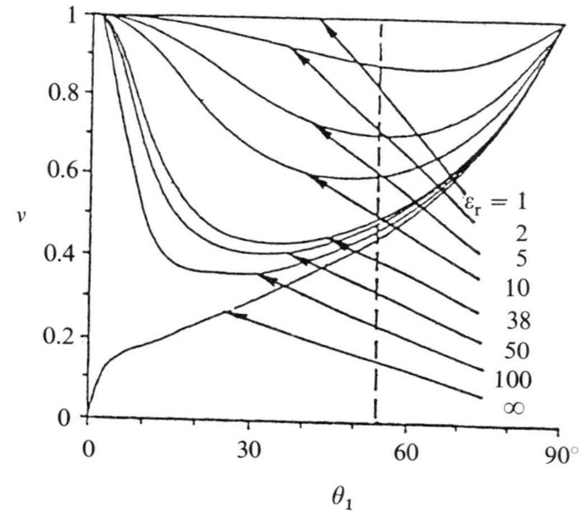

(a)

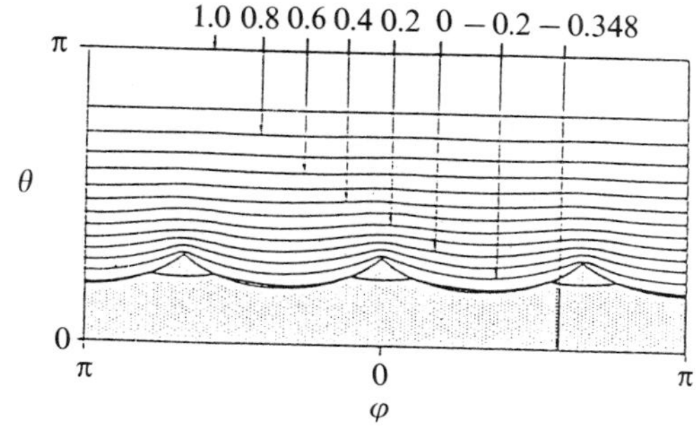

(b)

Fig. 5.30. A pyramid having dielectric constant higher than that of the surroundings ($\varepsilon_r > 1$). (a) Electric singularity exponent as a function of the opening angle θ_1. The vertical line refers to the orthogonal corner ($\alpha = 90°$). (b) The function $Y(\theta, \varphi)$ for $\varepsilon_r = 10$ and $\alpha = 90°$. From De Smedt (1987: pp. 1193–4) © Amer. Geoph. Union.

constant higher than the surroundings (i.e. when $\varepsilon_r > 1$), the electric singularity occurs in the class 1 symmetry (Fig. 5.27b). The relevant values of ν are shown in Fig. 5.30a, where we notice that the $\varepsilon_r \to \infty$ limit reproduces the values obtained in Fig. 5.28 for a perfectly conducting pyramid. A typical variation of $Y(\theta, \varphi)$ is given in Fig. 5.30b for the $\alpha = 90°$ pyramid. This type of pyramid is found, for example, at the corners of a rectangular dielectric resonator.

When the pyramid has a lower dielectric constant than the surroundings, which would be the case for a pyramidal hole in a dielectric slab, the electric singularity occurs in the class 3 and 4 symmetries, with a common value of ν (Fig. 5.31).

It is noteworthy that the data in Figs. 5.30–1 remain valid when ε_r is replaced by μ_r, i.e. when the regions 1 and 2 are filled with media of different magnetic permeabilities. For such a case, the singular field is the *magnetic field.*

To conclude this section, a few data concerning two additional corner structures are presented (De Smedt 1987). The first one is a semi-infinite 90° metal corner, in contact with a semi-infinite dielectric corner (Fig. 5.32a). The singular electric potential is symmetric with respect to the symmetry plane $x = y$. The relevant exponent ν is plotted in Fig. 5.32c, as a function of the contrast parameter $(\varepsilon_{r1} - \varepsilon_{r2})/(\varepsilon_{r1} + \varepsilon_{r2})$. For $\varepsilon_{r1} = \varepsilon_{r2}$, we obtain (as expected) the value 0.454 found in Table 5.9 for the isolated 90° metallic corner.

The second type of structure consists of a semi-infinite 90° metal corner resting on a dielectric base (Fig. 5.32b). Here, again, the singular electric potential is symmetric with respect to the bisecting plane $x = y$. The singularity exponent is plotted in Fig. 5.32c.

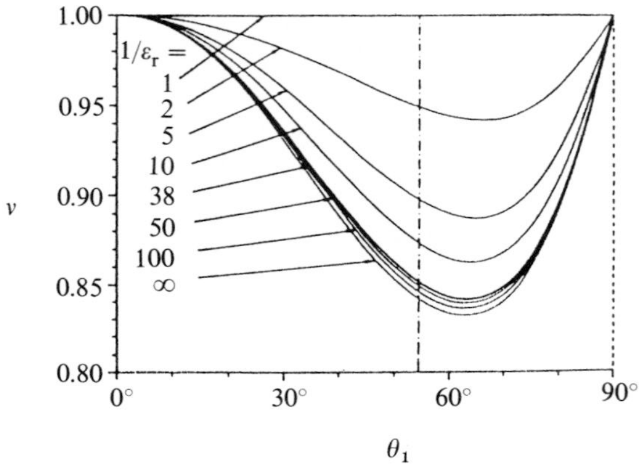

Fig. 5.31. A pyramid of lower dielectric constant than the surroundings ($\varepsilon_r < 1$). Electric singularity exponent as a function of the opening angle θ_1. The vertical line refers to the orthogonal corner ($\alpha = 90°$). From De Smedt (1987: p. 1193) © Amer. Geoph. Union.

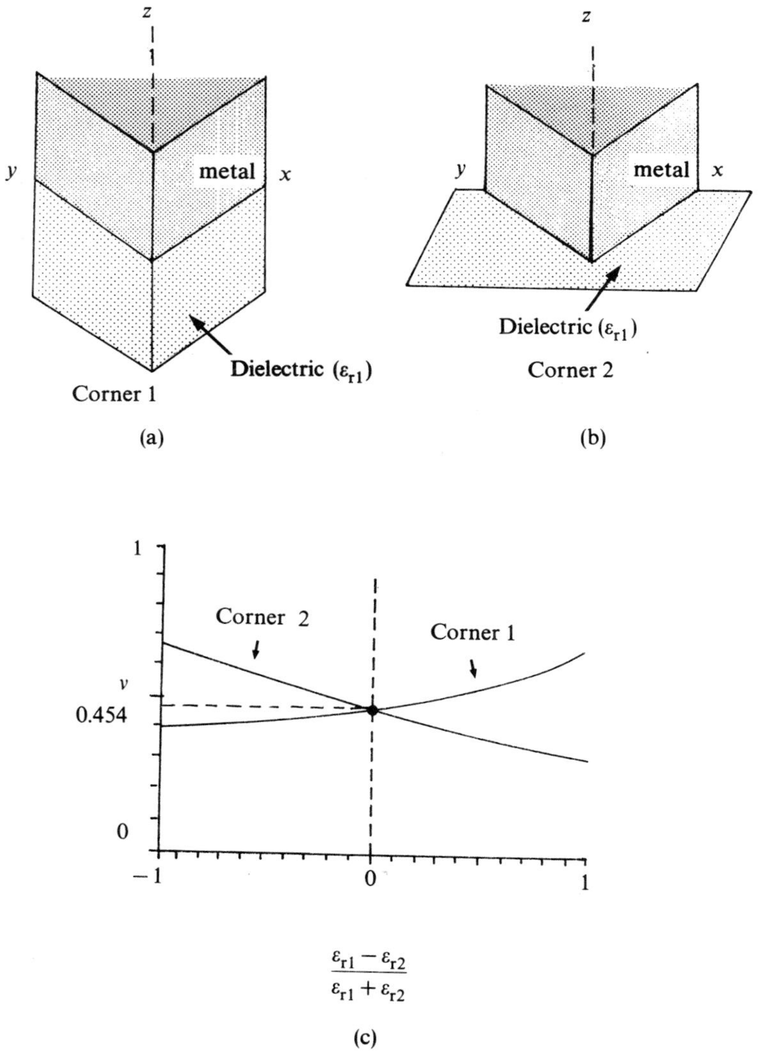

Fig. 5.32. Metallic and dielectric corners: ε_{r1} refers to the dielectric corner, ε_{r2} to the outside medium. From De Smedt p. 1194, © 1987 Amer. Geop. Union.

References

Blume, S. (1967). Theorie der Beugung an einem sektorförmigen Schlitz. *Optik*, **26**, 150–68, 192–210.

Blume, S., and Kirchner, M. (1969). Das Beugungsnahfeld an der Spitze eines sektorförmigen Schlitzes in einem ebenen Schirm. *Optik*, **29**, 185–203.

BLUME, S. (1971). Theorie der Beugung elektromagnetischer Wellen am Kegel elliptischen Querschnitts. Annalen der Physik, **26**, 302–8.

BLUME, S., and KAHL, G. (1985*a*). Field singularities at the tip of a cone with elliptic cross section. *Optik*, **70**, 170–5.

BLUME, S., and KAHL, G. (1985*b*). Comments on 'Fields at the tip of an elliptic cone'. *Proceedings of the IEEE*, **73**, 1857–8.

BLUME, S., and KLINKENBUSCH, L. (1990). Eigenmoden und Eigenwerte komplementärer sektorförmiger Strukturen: Eine sphärische Multipolanalyse des elektromagnetischen Randwertproblems. Accepted for publication in *Archiv für Elektrotechnik*.

BOERSMA, J., and JANSEN, J.K.M. (1990). *Electromagnetic field singularities at the tip of an elliptic cone*. EUT Report 90-01. Department of Mathematics and Computing Science. Eindhoven University of Technology.

BOERSMA, J. (1990). Note on the singularity exponents for complementary sectors. Submitted for publication.

BOWMAN, J.J., SENIOR, T.B.A., and USLENGHI, P.L.E. (1969). *Electromagnetic and Acoustic Scattering by Simple Shapes*, pp. 637–701. North-Holland, Amsterdam.

BRAUNBEK, W. (1956). On the diffraction field near a plane-screen corner. *IRE Transactions on Antennas and Propagation*, **4**, 219–23.

BROWN, S.N., and STEWARTSON, K. (1969). Flow near the apex of a plane delta wing. *Journal of the Institute of Mathematics and its Applications*, **5**, 206–16.

BUTLER, C.M. (1980). Investigation of a scatterer coupled to an aperture in a conducting screen. *Proceedings of the IEE*, **H127**, 161–9.

DE MEULENAERE, F., and VAN BLADEL, J. (1977). Polarizability of some small apertures. *IEEE Transactions on Antennas and Propagation*, **25**, 198–205.

DE SMEDT, R. (1985). Singular fields at the tip of a dielectric pyramid. *Actes du Colloque Optique Hertzienne et Diélectriques*, **54**, 1–4. Grenoble, France.

DE SMEDT, R., and VAN BLADEL, J. (1985). *Field singularities at the tip of a metallic cone of arbitrary cross section*, pp. 1–82. Internal Report 85–3, Laboratory of Electromagnetism and Acoustics, University of Ghent.

DE SMEDT, R., and VAN BLADEL, J. (1986). Field singularities at the tip of a metallic cone of arbitrary cross section. *IEEE Transactions on Antennas and Propagation*, **34**, 865–70.

DE SMEDT, R. (1986). *Numerieke Studie van enkele Elektromagnetische Verstrooiingsen Resonantieproblemen*, pp. 241–88. Ph.D. thesis. Laboratory of Electromagnetism and Acoustics, University of Ghent.

DE SMEDT, R., and VAN BLADEL, J. (1987). Field singularities near aperture corners. *Proceedings of the IEE*, **A134**, 694–8.

DE SMEDT, R. (1987). Singular field behavior near the tip of a dielectric or dielectric–metallic corner. *Radio Science*, **22**, 1190–6.

DE SMEDT, R. (1988). Electric singularity near the tip of a sharp cone. *IEEE Transactions on Antennas and Propagation*, **36**, 152–5.

DURAND, E. (1964). *Electrostatique, Tome I: les distributions*, p. 161. Masson, Paris.

GLISSON, A.W. (1978). On the development of numerical techniques for treating arbitrarily-shaped surfaces. Ph.D. thesis. University of Mississippi.

GLISSON, A.W., and WILTON, D.R. (1980). Simple and efficient numerical methods for problems of electromagnetic radiation and scattering from surfaces. *IEEE Transactions on Antennas and Propagation*, **28**, 593–603.

GRAY, M.C. (1953). Legendre functions of fractional order. *Quarterly Journal of Applied Mathematics*, **11**, 311–8.

KO, W.L., and MITTRA, R. (1977). A new approach based on a combination of integral equation and asymptotic techniques for solving electromagnetic scattering problems. *IEEE Transactions on Antennas and Propagation*, **25**, 187–97.

KRAUS, L., and LEVINE, L. M. (1961). Diffraction by an elliptic cone. *Communications on Pure and Applied Mathematics*, **XIV**, 49–68.

MEI, K.K., and CANGELLARIS, A.C. (1987). Application of field singularities at wedges and corners to time domain finite difference or finite element methods of field computations. *Radio Science*, **22**, 1239–46.

MORRISON, J.A. and LEWIS, J.A. (1976). Charge singularity at the corner of a flat plate. *SIAM Journal on Applied Mathematics*, **31**, 233–50.

MORRISON, J.A. (1977). Charge singularity at the vertex of a slender cone of general cross-section. *SIAM Journal on Applied Mathematics*, **33**, 127–32.

ROBIN, L. (1959). *Fonctions sphériques de Legendre et fonctions sphéroïdales*, Tome III, pp. 61–75. Gauthier-Villars, Paris.

RUCK, G.T., BARRICK, D.E., STUART, W.D., and KRICHBAUM, C.K. (1970). *Radar Cross Section Handbook*, pp. 118–121, 378–442. Plenum Press, New York.

SAHALOS, J.N., and THIELE, G.A. (1983*a*). The eigenfunction solution for scattered fields and surface currents of a vertex. *IEEE Transactions on Antennas and Propagation*, **31**, 206–11.

SAHALOS, J.N., and THIELE, G.A. (1983*b*). Periodic Lamé functions and applications to the diffraction by a plane angular sector. *Canadian Journal of Physics*, **61**, 1583–91.

SATTERWHITE, R. and KOUYOUMJIAN, R.G. (1970). *Electromagnetic diffraction by a perfectly-conducting plane angular sector*, pp. 41, 57. Technical Report 2183–2, ElectroScience Laboratory, Ohio State University.

SATTERWHITE, R. (1974). Diffraction by a quarterplane, the exact solution, and some numerical results. *IEEE Transactions on Antennas and Propagation*, **22**, 500–3.

TAYLOR, R.S. (1971). A new approach to the delta wing problem. *Journal of the Institute of Mathematics and its Applications*, **7**, 337–47.

VAFIADIS, E. and SAHALOS, J.N. (1984). Fields at the tip of an elliptic cone. *Proceedings of the IEEE*, **72**, 1089–91.

VAFIADIS, E. and SAHALOS, J.N. (1985). Analysis of semi-infinite elliptic conical scatterer. *Journal of Physics (Applied Physics)*, **D18**, 1475–93.

VAN BLADEL, J. (1983*a*). *Field singularities near an edge and the tip of a circular cone*, pp. 1–47. Internal Report 83–1. Laboratory of Electromagnetism and Acoustics, University of Ghent.

VAN BLADEL, J. (1983*b*). Field singularities at the tip of a cone. *Proceedings of the IEEE*, **71**, 901–2.

VAN BLADEL, J. (1984). *Field singularities at the tip of a dielectric cone*. Internal Report 84–3. Laboratory of Electromagnetism and Acoustics, University of Ghent.

VAN BLADEL, J. (1985*a*). *Electromagnetic fields*, p. 410. Hemisphere, Washington. Reprinted, with corrections, from a text published in 1964 by McGraw-Hill, New York.

VAN BLADEL, J. (1985*b*). Field singularities at the tip of a dielectric cone. *IEEE Transactions on Antennas and Propagation*, **33**, 893–5.

APPENDIX A
Double layers

A.1 Double layer of charge

A dipole layer generates a potential

$$\phi(\boldsymbol{r}_0) = \frac{1}{4\pi\varepsilon_0} \iint_S \tau(\boldsymbol{r}) \frac{\partial}{\partial n} \frac{1}{|\boldsymbol{r} - \boldsymbol{r}_0|} \, \mathrm{d}S. \tag{A.1}$$

The function $\tau(\boldsymbol{r})$ is the dipole density. Expression (A.1) is obtained by replacing the layer by a series of elementary dipoles of moment $\tau \boldsymbol{u}_\mathrm{n} \, \mathrm{d}S$, and adding the partial potentials. The dipole layer is traditionally generated by starting from two single layers separated by a distance h, and letting h approach zero while $\rho_\mathrm{s} h$ keeps a finite nonzero value τ (Fig. A.1). To generate dipoles, the charge $\rho_\mathrm{s} \, \mathrm{d}S$ on $\mathrm{d}S$ must be equal to minus the charge on $\mathrm{d}S'$, i.e. to $-\rho'_\mathrm{s} \, \mathrm{d}S'$. In general, the required ρ'_s is not equal to ρ_s, because $\mathrm{d}S'$ differs from $\mathrm{d}S$ when the surface is curved. To clarify this statement, let us investigate two methods of layer formation:

(a) We equate ρ'_s to ρ_s, which creates a double layer termed the 'equi-density' layer.
(b) We equate $\rho'_\mathrm{s} \, \mathrm{d}S'$ to $\rho_\mathrm{s} \, \mathrm{d}S$, which creates a 'dipole' layer. The main problem here is to determine ρ'_s.

The key element in our analysis is the relationship between $\mathrm{d}S'$ and $\mathrm{d}S$. The element of surface of S is given by

$$\mathrm{d}S = h_1 h_2 \, \mathrm{d}v_1 \, \mathrm{d}v_2. \tag{A.2}$$

The principal radii of curvature are related to the h_i by (Fig. 1.2)

$$\frac{1}{R_i} = -\frac{1}{h_i} \frac{\partial h_i}{\partial n} \quad (i = 1, 2). \tag{A.3}$$

The h_1 and h_2 are functions of the normal coordinate n. More precisely, at a short distance n from S,

$$h_i(n) = h_i(0) + n \left[\frac{\partial h_i}{\partial n} \right]_0 + \cdots = h_i(0) \left(1 - \frac{n}{R_i(0)} \right) + \cdots. \tag{A.4}$$

Consequently,

$$\mathrm{d}S(n) = \mathrm{d}S(0) \left[1 - n \left(\frac{1}{R_1} + \frac{1}{R_2} \right)_0 \right] + \cdots = \mathrm{d}S(0) \, [1 - J(0) n] + \cdots. \tag{A.5}$$

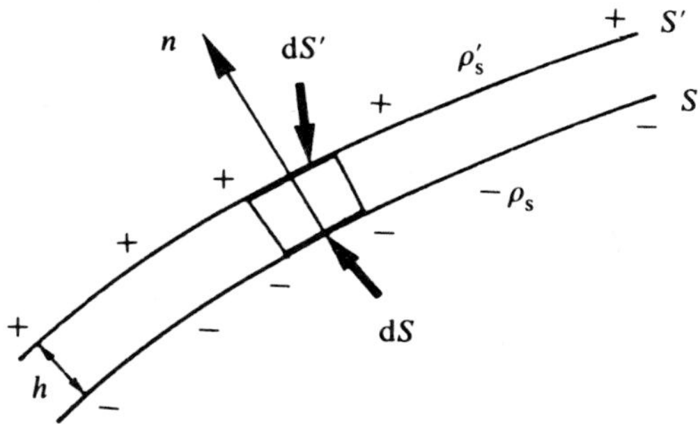

Fig. A.1. A double layer of charge.

where the symbol J denotes the first curvature

$$J = -\frac{1}{h_1}\frac{\partial h_1}{\partial n} - \frac{1}{h_2}\frac{\partial h_2}{\partial n} = \frac{1}{R_1} + \frac{1}{R_2}. \tag{A.6}$$

From (A.5), the value of ρ'_s appropriate for a dipole layer must be

$$\rho'_s = \rho_s(1 + nJ) = \rho_s + J\tau. \tag{A.7}$$

This value ensures overall charge neutrality of the layer and, as a result, continuity of the normal component of $\bar{e}$. We observe that the equidensity layer can be formed by adding a layer of (residual) density $\rho_s = -J\tau$ to the dipole layer. In distributional terms, from (1.104),

$$\rho_{\text{equi}} = -\tau \frac{\partial \delta_s}{\partial n} - J\tau\delta_s. \tag{A.8}$$

The nonzero charge density of the equidensity layer leads to the discontinuity condition

$$e_n^+ - e_n^- = -J\tau/\varepsilon_0. \tag{A.9}$$

The previous results can also be obtained on the basis of the $\delta(n)$ representation. In this approach the volume density of the *equidensity* layer is (Fig. A.1)

$$\begin{aligned} \rho_{\text{equi}} &= \rho_s(v_1, v_2)\,\delta(n - h) - \rho_s(v_1, v_2)\delta(n) \\ &= -\rho_s h\,\delta'(n) = -\tau(v_1, v_2)\,\delta'(n). \end{aligned} \tag{A.10}$$

From (1.104) and (A.8), the charge density of the dipole layer takes the form

$$\rho_{\text{dip}} = -\tau(v_1, v_2)\,\delta'(n) + J(v_1, v_2)\tau(v_1, v_2)\,\delta(n). \tag{A.11}$$

A comparison of (A.8) and (A.10) shows that $\partial\delta_s/\partial n$ and $\delta'(n)$ are *not* interchangeable when the surface has a nonvanishing total curvature J. The correct relationship is (Van Bladel 1989c)

$$\frac{\partial \delta_s}{\partial n} \to \delta'(n) - J\,\delta(n), \qquad \delta'(n) \to \frac{\partial \delta_s}{\partial n} + J\delta_s. \tag{A.12}$$

A.2 Double layer of current

Similar considerations hold for double layers of *surface current* (Fig. A.2a). Here, three ways of forming the layer are particularly interesting:

(1) the 'equidensity' layer, characterized by $\boldsymbol{j}'_s = \boldsymbol{j}_s$;
(2) the 'zero' layer, across which the magnetic field is continuous;
(3) the 'doublet' layer, consisting of a layer of electric dipole doublets.

A doublet consists of two antiparallel dipoles separated by a short distance h. The distributional representation of the current of such a doublet is given in (2.71). The double layer will be equivalent to an ensemble of doublets if the total tangential currents on dS and dS' are equal in magnitude, but antiparallel (Fig. A.2a). This will happen if

$$\boldsymbol{j}'_s\,\mathrm{d}S' = \boldsymbol{j}_s\,\mathrm{d}S. \tag{A.13}$$

From (A.5), and in the limit of small h,

$$(\boldsymbol{j}'_s)_{\mathrm{do}} = \boldsymbol{j}_s + J\boldsymbol{j}_s h. \tag{A.14}$$

If the product $(\boldsymbol{j}_s h)$ keeps a constant value $\boldsymbol{c}_s$ when h approaches zero:

$$(\boldsymbol{j}'_s)_{\mathrm{do}} = \boldsymbol{j}_s + J\boldsymbol{c}_s. \tag{A.15}$$

The tangential vector $\boldsymbol{c}_s$ is the 'strength' of the layer.

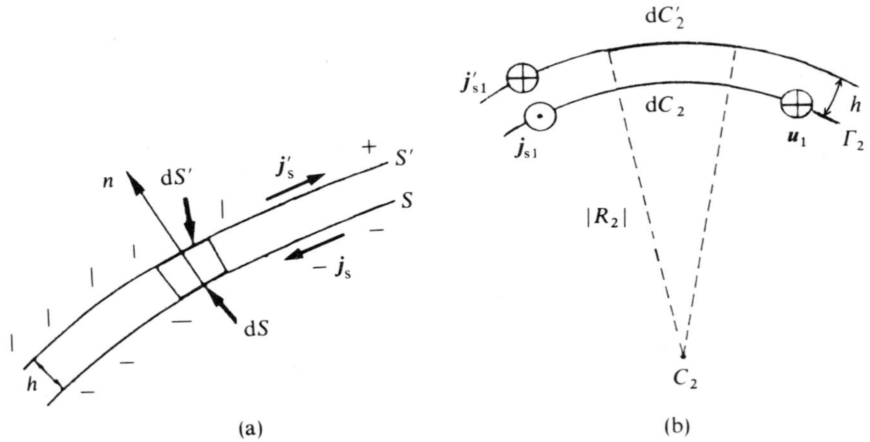

Fig. A.2. (a) A double layer of current. (b) Currents flowing in the $\boldsymbol{u}_1$ direction.

The vector potential of a doublet layer can be obtained by inserting, into (2.46), the appropriate value of $\boldsymbol{J}$, i.e. (2.71). Remembering that x is the coordinate in the direction perpendicular to the doublet, we write

$$A(\boldsymbol{r}) = \frac{\mu_0}{4\pi} \iint_S (J_s h) \frac{\partial}{\partial n} \frac{e^{-jk|\boldsymbol{r}-\boldsymbol{r}'|}}{|\boldsymbol{r}-\boldsymbol{r}'|} \, dS'. \tag{A.16}$$

Comparison with (1.102) shows that the volume current density of the doublet layer can be written as

$$(\boldsymbol{j})_{do} = -\boldsymbol{c}_s(v_1, v_2) \frac{\partial \delta_s}{\partial n}. \tag{A.17}$$

The next problem is to check whether the net tangential current of a doublet layer vanishes. The answer to this question requires a formula for the residual current $\boldsymbol{j}_{st}$. We start with the component of that current in the $\boldsymbol{u}_1$ direction. From Fig. A.2b,

$$j_{st1} \, dc_2 = j'_{s1} \, dc'_2 - j_{s1} \, dc_2. \tag{A.18}$$

But

$$\frac{dc'_2}{dc_2} = \frac{|R_2| + h}{|R_2|} = 1 - \frac{h}{R_2}. \tag{A.19}$$

Consequently,

$$\boldsymbol{j}_{st} \cdot \boldsymbol{u}_1 = \boldsymbol{j}'_s \cdot \boldsymbol{u}_1 \left(1 - \frac{h}{R_2}\right) - \boldsymbol{j}_s \cdot \boldsymbol{u}_1. \tag{A.20}$$

In the direction $\boldsymbol{u}_2$, analogously,

$$\boldsymbol{j}_{st} \cdot \boldsymbol{u}_2 = \boldsymbol{j}'_s \cdot \boldsymbol{u}_2 \left(1 - \frac{h}{R_1}\right) - \boldsymbol{j}_s \cdot \boldsymbol{u}_2. \tag{A.21}$$

Combining (A.20) and (A.21) yields

$$\boldsymbol{j}_{st} = \boldsymbol{j}'_s - h\boldsymbol{j}'_s \left(\frac{1}{R_2} \boldsymbol{u}_1 \boldsymbol{u}_1 + \frac{1}{R_1} \boldsymbol{u}_2 \boldsymbol{u}_2\right) - \boldsymbol{j}_s. \tag{A.22}$$

It is convenient to introduce the (symmetric) dyadic

$$\boldsymbol{\Gamma} = \frac{1}{R_2} \boldsymbol{u}_1 \boldsymbol{u}_1 + \frac{1}{R_1} \boldsymbol{u}_2 \boldsymbol{u}_2. \tag{A.23}$$

In terms of $\boldsymbol{\Gamma}$, the net surface density (A.22) becomes

$$\boldsymbol{j}_{st} = \boldsymbol{j}'_s - h\boldsymbol{j}'_s \cdot \boldsymbol{\Gamma} - \boldsymbol{j}_s. \tag{A.24}$$

Inserting the value (A.14) for $\boldsymbol{j}'_s$ into (A.24) yields

$$(\boldsymbol{j}_{st})_{do} = J\boldsymbol{c}_s - \boldsymbol{\Gamma} \cdot \boldsymbol{c}_s = \boldsymbol{\pi} \cdot \boldsymbol{c}_s, \tag{A.25}$$

where $\boldsymbol{\pi}$ is the dyadic

$$\boldsymbol{\pi} = J\mathbf{I}_t - \boldsymbol{\Gamma} = \frac{1}{R_1}\boldsymbol{u}_1\boldsymbol{u}_1 + \frac{1}{R_2}\boldsymbol{u}_2\boldsymbol{u}_2. \tag{A.26}$$

The symbol $\mathbf{I}_t = \boldsymbol{u}_1\boldsymbol{u}_1 + \boldsymbol{u}_2\boldsymbol{u}_2$ denotes the unit dyadic in the tangent plane. It is clear, from (A.25), that the net surface current of a doublet layer does *not* vanish when the surface is curved. The layer of zero residual current is not the *doublet*, but a layer formed with a $\boldsymbol{j}'_s$ equal to

$$(\boldsymbol{j}'_s)_{\text{zero}} = \boldsymbol{j}_s + \boldsymbol{\Gamma}\cdot\boldsymbol{c}_s. \tag{A.27}$$

Finally, the equidensity layer is characterized by a net current

$$(\boldsymbol{j}'_{st})_{\text{equi}} = -\boldsymbol{\Gamma}\cdot\boldsymbol{c}_s. \tag{A.28}$$

The volume densities corresponding to the various double layers follow easily from the representation (A.17) which holds for the doublet. The only difference between the three layers resides in the different values of the residual surface currents. Keeping (A.12) in mind, we find (Van Bladel 1989*d*)

$$(\boldsymbol{j})_{\text{do}} = -\boldsymbol{c}_s\frac{\partial\delta_s}{\partial n} = J\boldsymbol{c}_s\delta(n) - \boldsymbol{c}_s\delta'(n), \tag{A.29}$$

$$(\boldsymbol{j})_{\text{zero}} = -\boldsymbol{\pi}\cdot\boldsymbol{c}_s\delta_s - \boldsymbol{c}_s\frac{\partial\delta_s}{\partial n} = \boldsymbol{\Gamma}\cdot\boldsymbol{c}_s\delta(n) - \boldsymbol{c}_s\delta'(n), \tag{A.30}$$

$$(\boldsymbol{j})_{\text{equi}} = -J\boldsymbol{c}_s\delta_s - \boldsymbol{c}_s\frac{\partial\delta_s}{\partial n} = -\boldsymbol{c}_s\delta'(n). \tag{A.31}$$

A.3 Boundary conditions at double current sheets

The value of the curl in the (v_1, v_2, n) coordinates is given in (2.104). On the basis of this formula, Maxwell's equations can be written as

$$\operatorname{grad}_s E_n \times \boldsymbol{u}_n + \boldsymbol{\Gamma}\cdot(\boldsymbol{E}_t \times \boldsymbol{u}_n) + \boldsymbol{u}_n \times \frac{\partial \boldsymbol{E}_t}{\partial n} = -\mathrm{j}\omega\mu_0\boldsymbol{H}_t, \tag{A.32}$$

$$\operatorname{div}_s(\boldsymbol{E}_t \times \boldsymbol{u}_n) = -\mathrm{j}\omega\mu_0 H_n, \tag{A.33}$$

$$\operatorname{grad}_s H_n \times \boldsymbol{u}_n + \boldsymbol{\Gamma}\cdot(\boldsymbol{H}_t \times \boldsymbol{u}_n) + \boldsymbol{u}_n \times \frac{\partial \boldsymbol{H}_t}{\partial n} = \mathrm{j}\omega\varepsilon_0\boldsymbol{E}_t + \boldsymbol{J}_t, \tag{A.34}$$

$$\operatorname{div}_s(\boldsymbol{H}_t \times \boldsymbol{u}_n) = \mathrm{j}\omega\varepsilon_0 E_n + J_n. \tag{A.35}$$

The surface divergence is defined in (1.132), and the surface gradient in (2.105). The current densities $\boldsymbol{J}$ relevant to our problem are those shown in (A.29)–(A.31). We use the form in $\delta(n)$ because it is appropriate for a solution by separation of variables. To determine field behaviour in the layer, we write each field component in the general form

$$E_k = E_{k0}Y(n) + E_{k1}\delta(n) + E_{k2}\delta'(n) + \cdots. \tag{A.36}$$

These expressions are inserted into (A.32)–(A.35), and terms in $\delta(n)$ and $\delta'(n)$ are equated on both sides of the equations. Detailed calculations show that, for the *zero current layer*,

$$j\omega\varepsilon_0 \boldsymbol{E}_t = (\text{grad}_s \, \text{div}_s \, \boldsymbol{C}_s + k^2 \boldsymbol{C}_s) Y(n), \tag{A.37}$$

$$j\omega\varepsilon_0 E_n = \text{div}_s \, \boldsymbol{C}_s \, \delta(n), \tag{A.38}$$

$$\boldsymbol{H}_t = \boldsymbol{u}_n \times \boldsymbol{C}_s \delta(n), \tag{A.39}$$

$$H_n = \text{div}_s (\boldsymbol{C}_s \times \boldsymbol{u}_n) \, Y(n). \tag{A.40}$$

The fields for the other double layers follow by adding, to these expressions, the contribution from the residual surface current $\boldsymbol{J}_{st}$, viz.

$$j\omega\varepsilon_0 E_n = - \text{div}_s \boldsymbol{J}_{st} Y(n), \tag{A.41}$$

$$\boldsymbol{H}_t = (\boldsymbol{J}_{st} \times \boldsymbol{u}_n) Y(n). \tag{A.42}$$

Translated into boundary conditions these equations generate, through their terms in $Y(n)$, the jump conditions

$$j\omega\varepsilon_0 (\boldsymbol{E}_t^+ - \boldsymbol{E}_t^-) = \text{grad}_s \, \text{div}_s \, \boldsymbol{C}_s + k^2 \boldsymbol{C}_s, \tag{A.43}$$

$$j\omega\varepsilon_0 (E_n^+ - E_n^-) = -\text{div}_s \boldsymbol{J}_{st}, \tag{A.44}$$

$$\boldsymbol{H}_t^+ - \boldsymbol{H}_t^- = \boldsymbol{J}_{st} \times \boldsymbol{u}_n, \tag{A.45}$$

$$H_n^+ - H_n^- = \text{div}_s (\boldsymbol{C}_s \times \boldsymbol{u}_n). \tag{A.46}$$

We observe that the boundary conditions on $\boldsymbol{E}_t$ and H_n are the same for all double layers (Panicali 1977; Van Bladel 1989*a*), but that those on E_n and $\boldsymbol{H}_t$ differ. Across the 'zero' layer, for example, both E_n and $\boldsymbol{H}_t$ are continuous; but, across the 'doublet' layer,

$$j\omega\varepsilon_0 (E_n^+ - E_n^-) = -\text{div}_s (\boldsymbol{\pi} \cdot \boldsymbol{C}_s), \tag{A.47}$$

$$\boldsymbol{H}_t^+ - \boldsymbol{H}_t^- = (\boldsymbol{\pi} \cdot \boldsymbol{C}_s) \times \boldsymbol{u}_n. \tag{A.48}$$

Finally, across the 'equidensity' layer (Van Bladel 1989*b*),

$$j\omega\varepsilon_0 (E_n^+ - E_n^-) = \text{div}_s (\boldsymbol{\Gamma} \cdot \boldsymbol{C}_s), \tag{A.49}$$

$$\boldsymbol{H}_t^+ - \boldsymbol{H}_t^- = \boldsymbol{u}_s \times (\boldsymbol{\Gamma} \cdot \boldsymbol{C}_s). \tag{A.50}$$

The reader may have noticed that boundary condition (A.46) is not in agreement with the classical condition 'B_n continuous'. The reason for this discrepancy is easy to understand. In the usual proof, the divergence theorem is applied to $\boldsymbol{B}$ in a very flat cylindrical pillbox (Fig. A.3). The height of the

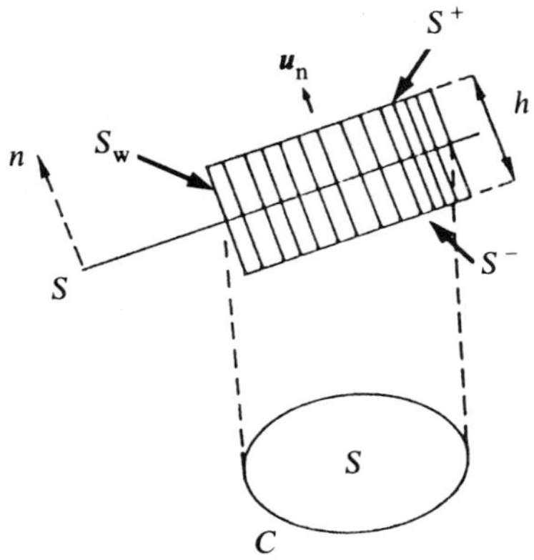

Fig. A.3. A cylindrical pillbox used for the investigation of the boundary condition on B_n.

box is then allowed to shrink to zero, and it is assumed that the flux through the sidewall S_w vanishes in the process. This step leads to the equality of the fluxes through S^+ and S^-, and therefore to the condition $B_n^+ = B_n^-$. Implicit in the argument, however, is the assumption that the normal component of $\boldsymbol{B}$ on S_w remains bounded. In the case of the double sheet, $\boldsymbol{B}$ does *not* remain bounded, but becomes infinite on S_w according to (A.39). Thus,

$$\boldsymbol{B}_t/\mu_0 = \boldsymbol{H}_t = \boldsymbol{u}_n \times \boldsymbol{C}_s \delta(n). \qquad \text{(A.51)}$$

It follows that the flux through S_w keeps a nonzero value, even in the limit $h \to 0$. This property, properly exploited, leads to the correct jump condition for B_n.

References

Panicali, A.R. (1977). On the boundary conditions at surface current distributions described by the first derivative of Dirac's impulse. *IEEE Transactions on Antennas and Propagation*, **25**, 901–903.

Van Bladel, J. (1989*a*). Boundary conditions at double current sheet. *Electronics Letters*, **25**, 98–9.

Van Bladel, J. (1989*b*). More boundary conditions at double current sheet. *Electronics Letters*, **25**, 836–7.

Van Bladel, J. (1989*c*). *On the δ-function representation of double layers of charge*, pp. 1–14. Internal Report 89–3. Laboratory of Electromagnetism and Acoustics, University of Ghent.

Van Bladel, J. (1989*d*). *On the δ-function representation of double layers of current*, pp. 1–15. Internal Report 89–4. Laboratory of Electromagnetism and Acoustics, University of Ghent.

APPENDIX B

Dyadic Analysis

B.1 The dyadic notation

A three-dimensional vector $\boldsymbol{v}'$ is a linear function of another vector $\boldsymbol{v}$ when the following relationships hold:

$$\begin{aligned} v'_x &= a_{xx}v_x + a_{xy}v_y + a_{xz}v_z \\ v'_y &= a_{yx}v_x + a_{yy}v_y + a_{yz}v_z \\ v'_z &= a_{zx}v_x + a_{zy}v_y + a_{zz}v_z. \end{aligned} \tag{B.1}$$

A particularly compact way of writing (B.1) is

$$\boldsymbol{v}' = \mathbf{a} \cdot \boldsymbol{v}, \tag{B.2}$$

where $\mathbf{a}$ is the dyadic

$$\begin{aligned} \mathbf{a} = {} & a_{xx}\boldsymbol{u}_x\boldsymbol{u}_x + a_{xy}\boldsymbol{u}_x\boldsymbol{u}_y + a_{xz}\boldsymbol{u}_x\boldsymbol{u}_z \\ & + a_{yx}\boldsymbol{u}_y\boldsymbol{u}_x + a_{yy}\boldsymbol{u}_y\boldsymbol{u}_y + a_{yz}\boldsymbol{u}_y\boldsymbol{u}_z \\ & + a_{zx}\boldsymbol{u}_z\boldsymbol{u}_x + a_{zy}\boldsymbol{u}_z\boldsymbol{u}_y + a_{zz}\boldsymbol{u}_x\boldsymbol{u}_z. \end{aligned} \tag{B.3}$$

This dyadic is the sum of nine partial dyadics of the type $a_{ik}\boldsymbol{u}_i\boldsymbol{u}_k$. To reproduce the linear relationship (B.1), we must introduce suitable operating rules for the scalar product of a dyadic $\boldsymbol{bc}$ with a vector $\boldsymbol{d}$. These are

$$\begin{aligned} (\boldsymbol{bc}) \cdot \boldsymbol{d} &\stackrel{\text{def}}{=} \boldsymbol{b}(\boldsymbol{c} \cdot \boldsymbol{d}), \\ \boldsymbol{d} \cdot (\boldsymbol{bc}) &\stackrel{\text{def}}{=} (\boldsymbol{d} \cdot \boldsymbol{b})\boldsymbol{c}. \end{aligned} \tag{B.4}$$

A more compact notation for $\mathbf{a}$ is

$$\begin{aligned} \mathbf{a} &= \boldsymbol{a}'_x\boldsymbol{u}_x + \boldsymbol{a}'_y\boldsymbol{u}_y + \boldsymbol{a}'_z\boldsymbol{u}_z \\ &= \boldsymbol{u}_x\boldsymbol{a}_x + \boldsymbol{u}_y\boldsymbol{a}_y + \boldsymbol{u}_z\boldsymbol{a}_z, \end{aligned} \tag{B.5}$$

where

$$\boldsymbol{a}'_x = a_{xx}\boldsymbol{u}_x + a_{yx}\boldsymbol{u}_y + a_{yz}\boldsymbol{u}_z \tag{B.6}$$

is the vector whose components form the first column of the a_{ik} matrix, and

$$\boldsymbol{a}_x = a_{xx}\boldsymbol{u}_x + a_{xy}\boldsymbol{u}_y + a_{xy}\boldsymbol{u}_z \tag{B.7}$$

is the vector whose components form the first row of the same matrix. Similar definitions hold for $\boldsymbol{a}'_y$, $\boldsymbol{a}'_z$, $\boldsymbol{a}_y$, and $\boldsymbol{a}_z$.

B.2 The transpose dyadic

It is obvious that $\mathbf{a} \cdot \boldsymbol{d}$ is generally different from $\boldsymbol{d} \cdot \mathbf{a}$. In other words, the order in which $\mathbf{a}$ and $\boldsymbol{d}$ appear should be carefully respected. The transpose of $\mathbf{a}$ is a dyadic $\mathbf{a}^{\mathrm{T}}$ such that

$$\mathbf{a} \cdot \boldsymbol{d} = \boldsymbol{d} \cdot \mathbf{a}^{\mathrm{T}}. \tag{B.8}$$

One can easily check that the transpose is obtained by interchanging rows and columns. More precisely,

$$\mathbf{a}^{\mathrm{T}} = \boldsymbol{a}_x \boldsymbol{u}_x + \boldsymbol{a}_y \boldsymbol{u}_y + \boldsymbol{a}_z \boldsymbol{u}_z = \boldsymbol{u}_x \boldsymbol{a}'_x + \boldsymbol{u}_y \boldsymbol{a}'_y + \boldsymbol{u}_z \boldsymbol{a}'_z. \tag{B.9}$$

A symmetric dyadic is equal to its own transpose. Thus,

$$\mathbf{a} \cdot \boldsymbol{b} = \boldsymbol{b} \cdot \mathbf{a}. \tag{B.10}$$

The condition for this occurring is $a_{ik} = a_{ki}$. The antisymmetric dyadic is equal to minus its transpose. We write

$$\mathbf{a} \cdot \boldsymbol{b} = -\boldsymbol{b} \cdot \mathbf{a}. \tag{B.11}$$

The condition is $a_{ik} = -a_{ki}$.

B.3 Products

Products which have proved useful are:

(a) the direct product

$$(\boldsymbol{ab}) \cdot (\boldsymbol{cd}) \stackrel{\text{def}}{=} \boldsymbol{a}(\boldsymbol{b} \cdot \boldsymbol{c})\boldsymbol{d} \qquad \text{(a dyadic);} \tag{B.12}$$

(b) the skew products

$$\left.\begin{aligned} (\boldsymbol{ab}) \times \boldsymbol{c} &\stackrel{\text{def}}{=} \boldsymbol{a}(\boldsymbol{b} \times \boldsymbol{c}) \qquad \text{(a dyadic),} \\ \boldsymbol{c} \times (\boldsymbol{ab}) &\stackrel{\text{def}}{=} (\boldsymbol{c} \times \boldsymbol{a})\boldsymbol{b} \qquad \text{(a dyadic);} \end{aligned}\right. \tag{B.13}$$

(c) the double product

$$(\boldsymbol{ab}) : (\boldsymbol{cd}) \stackrel{\text{def}}{=} (\boldsymbol{a} \cdot \boldsymbol{d})(\boldsymbol{b} \cdot \boldsymbol{c}) \qquad \text{(a scalar);} \tag{B.14}$$

(d) the double cross product

$$(\boldsymbol{ab}) \overset{\times}{\times} (\boldsymbol{cd}) \stackrel{\text{def}}{=} (\boldsymbol{b} \times \boldsymbol{c})(\boldsymbol{a} \times \boldsymbol{d}) \qquad \text{(a dyadic).} \tag{B.15}$$

B.4 The identity dyadic

The identity dyadic $\mathbf{I}$ is characterized by the property

$$\mathbf{I}\cdot\boldsymbol{b} = \boldsymbol{b}\cdot\mathbf{I} = \boldsymbol{b}. \tag{B.16}$$

This dyadic (the idemfactor) can be written, in Cartesian coordinates, as

$$\mathbf{I} = \boldsymbol{u}_x\boldsymbol{u}_x + \boldsymbol{u}_y\boldsymbol{u}_y + \boldsymbol{u}_z\boldsymbol{u}_z. \tag{B.17}$$

The trace of a dyadic is the sum of its diagonal elements. Thus,

$$\operatorname{tr}\mathbf{a} = a_{xx} + a_{yy} + a_{zz}. \tag{B.18}$$

The trace is a scalar, i.e. it is invariant with respect to orthogonal transformations of the base vectors. It is a simple matter to show that

$$\operatorname{tr}\mathbf{a} = \mathbf{I}:\mathbf{a}. \tag{B.19}$$

Every antisymmetric dyadic can be written in terms of $\mathbf{I}$ and a suitable vector $\boldsymbol{b}$ as

$$\begin{aligned}\mathbf{a} &= -b_z\boldsymbol{u}_x\boldsymbol{u}_y + b_y\boldsymbol{u}_x\boldsymbol{u}_z + b_z\boldsymbol{u}_y\boldsymbol{u}_x - b_x\boldsymbol{u}_y\boldsymbol{u}_z - b_y\boldsymbol{u}_z\boldsymbol{u}_x + b_x\boldsymbol{u}_z\boldsymbol{u}_y \\ &= \boldsymbol{b}\times\mathbf{I} = \mathbf{I}\times\boldsymbol{b} = -(\boldsymbol{b}\times\mathbf{I})^{\mathrm{T}}.\end{aligned} \tag{B.20}$$

Also

$$(\boldsymbol{a}\times\mathbf{I})\cdot\boldsymbol{b} = \boldsymbol{a}\times\boldsymbol{b}. \tag{B.21}$$

B.5 Operators

The quantity $\operatorname{grad}\boldsymbol{a}$ is the dyadic

$$\begin{aligned}\operatorname{grad}\boldsymbol{a} = \nabla\boldsymbol{a} &= \boldsymbol{u}_x\frac{\partial\boldsymbol{a}}{\partial x} + \boldsymbol{u}_y\frac{\partial\boldsymbol{a}}{\partial y} + \boldsymbol{u}_z\frac{\partial\boldsymbol{a}}{\partial z} \\ &= \operatorname{grad}a_x\boldsymbol{u}_x + \operatorname{grad}a_y\boldsymbol{u}_y + \operatorname{grad}a_z\boldsymbol{u}_z.\end{aligned} \tag{B.22}$$

Its expression in spherical coordinates is

$$\begin{aligned}\operatorname{grad}\boldsymbol{a} &= \left(\operatorname{grad}a_R - \frac{a_\varphi\boldsymbol{u}_\varphi}{R} - \frac{a_\theta\boldsymbol{u}_\theta}{R}\right)\boldsymbol{u}_R \\ &+ \left(\operatorname{grad}a_\theta + \frac{a_R\boldsymbol{u}_\theta}{R} - \frac{a_\varphi\boldsymbol{u}_\varphi}{R\tan\theta}\right)\boldsymbol{u}_\theta \\ &+ \left[\operatorname{grad}a_\varphi + \left(\frac{a_R}{R} + \frac{a_\theta}{R\tan\theta}\right)\boldsymbol{u}_\varphi\right]\boldsymbol{u}_\varphi.\end{aligned} \tag{B.23}$$

The action of a linear operator $\mathscr{L}$ on a dyadic $\mathbf{a}$ is defined by the formula

$$\mathscr{L}\,\mathbf{a} = (\mathscr{L}\boldsymbol{a}'_x)\boldsymbol{u}_x + (\mathscr{L}\boldsymbol{a}'_y)\boldsymbol{u}_y + (\mathscr{L}\boldsymbol{a}'_z)\boldsymbol{u}_z. \tag{B.24}$$

Examples are

$$\operatorname{div}\mathbf{a} = (\operatorname{div}\boldsymbol{a}'_x)\boldsymbol{u}_x + (\operatorname{div}\boldsymbol{a}'_y)\boldsymbol{u}_y + (\operatorname{div}\boldsymbol{a}'_z)\boldsymbol{u}_z \quad \text{(a vector)}, \tag{B.25}$$

$$\operatorname{curl}\mathbf{a} = (\operatorname{curl}\boldsymbol{a}'_x)\boldsymbol{u}_x + (\operatorname{curl}\boldsymbol{a}'_y)\boldsymbol{u}_y + (\operatorname{curl}\boldsymbol{a}'_z)\boldsymbol{u}_z \quad \text{(a dyadic)}. \tag{B.26}$$

Relationships involving the identity dyadic are

$$\operatorname{div}(\phi\mathbf{I}) = \operatorname{grad}\phi, \tag{B.27}$$

$$\operatorname{div}(\mathbf{I}\times\boldsymbol{a}) = \operatorname{curl}\boldsymbol{a}, \tag{B.28}$$

$$\operatorname{curl}(\phi\,\mathbf{I}) = \operatorname{grad}\phi\times\mathbf{I}, \tag{B.29}$$

$$\operatorname{curl}(\mathbf{I}\times\boldsymbol{a}) = (\operatorname{grad}\boldsymbol{a})^{\mathrm{T}} - \mathbf{I}\operatorname{div}\boldsymbol{a}, \tag{B.30}$$

$$\operatorname{curl}\operatorname{curl}(\phi\mathbf{I}) = \operatorname{grad}\operatorname{grad}\phi - \mathbf{I}\,\nabla^2\phi. \tag{B.31}$$

B.6 Integral formulas

$$\iiint_V \operatorname{curl}\mathbf{a}\,\mathrm{d}V = \iint_S \boldsymbol{u}_{\mathrm{n}}\times\mathbf{a}\,\mathrm{d}S, \tag{B.32}$$

$$\iiint_V \operatorname{grad}\boldsymbol{a}\,\mathrm{d}V = \iint_S \boldsymbol{u}_{\mathrm{n}}\boldsymbol{a}\,\mathrm{d}S, \tag{B.33}$$

$$\iiint_V [-\boldsymbol{b}\cdot\operatorname{curl}\operatorname{curl}\mathbf{a} + (\operatorname{curl}\operatorname{curl}\boldsymbol{b})\cdot\mathbf{a}]\,\mathrm{d}V = \iint_S [(\boldsymbol{u}_{\mathrm{n}}\times\boldsymbol{b})\cdot\operatorname{curl}\mathbf{a} + (\boldsymbol{u}_{\mathrm{n}}\times\operatorname{curl}\boldsymbol{b})\cdot\mathbf{a}]\,\mathrm{d}S. \tag{B.34}$$

The symbol $\boldsymbol{u}_{\mathrm{n}}$ denotes the outward-oriented normal unit vector.

Additional relationships can be found in the References (Tai, 1987).

B.7 Green's dyadics

The Green's dyadic makes it possible to solve the general linear differential problem

$$\mathscr{L}\boldsymbol{f} = \boldsymbol{g} \tag{B.35}$$

by means of an integration. Thus,

$$\begin{aligned}
f_x(\boldsymbol{r}_0) &= \iiint_V [G_x^x(\boldsymbol{r}_0|\boldsymbol{r})g_x(\boldsymbol{r}) + G_x^y(\boldsymbol{r}_0|\boldsymbol{r})g_y(\boldsymbol{r}) + G_x^z(\boldsymbol{r}_0|\boldsymbol{r})g_z(\boldsymbol{r})]\,\mathrm{d}V,\\
f_y(\boldsymbol{r}_0) &= \iiint_V [G_y^x(\boldsymbol{r}_0|\boldsymbol{r})g_x(\boldsymbol{r}) + G_y^y(\boldsymbol{r}_0|\boldsymbol{r})g_y(\boldsymbol{r}) + G_y^z(\boldsymbol{r}_0|\boldsymbol{r})g_z(\boldsymbol{r})]\,\mathrm{d}V,\\
f_z(\boldsymbol{r}_0) &= \iiint_V [G_z^x(\boldsymbol{r}_0|\boldsymbol{r})g_x(\boldsymbol{r}) + G_z^y(\boldsymbol{r}_0|\boldsymbol{r})g_y(\boldsymbol{r}) + G_z^z(\boldsymbol{r}_0|\boldsymbol{r})g_z(\boldsymbol{r})]\,\mathrm{d}V.
\end{aligned} \tag{B.36}$$

The dyadic contains nine scalar Green's functions. A function such as $G_x^y(\boldsymbol{r}_0|\boldsymbol{r})$ clearly represents the contribution from a y-oriented unit source at $\boldsymbol{r}$, to the x component of f at $\boldsymbol{r}_0$. Relationship (B.36) may be written in the more compact form

$$f(\boldsymbol{r}_0)=\int_V [g_x(\boldsymbol{r})\boldsymbol{G}'_x(\boldsymbol{r}_0|\boldsymbol{r})+g_y(\boldsymbol{r})\boldsymbol{G}'_y(\boldsymbol{r}_0|\boldsymbol{r})+g_z(\boldsymbol{r})\boldsymbol{G}'_z(\boldsymbol{r}_0|\boldsymbol{r})]\,\mathrm{d}V. \tag{B.37}$$

The $\boldsymbol{G}'$ vectors may be assimilated to the column vectors of a Green's dyadic $\mathbf{G}(\boldsymbol{r}_0|\boldsymbol{r})$. Still more compactly, therefore,

$$\begin{aligned} f(\boldsymbol{r}_0) &= \iiint_V (\boldsymbol{G}'_x\boldsymbol{u}_x\cdot\boldsymbol{g}+\boldsymbol{G}'_y\boldsymbol{u}_y\cdot\boldsymbol{g}+\boldsymbol{G}'_z\boldsymbol{u}_z\cdot\boldsymbol{g})\,\mathrm{d}V \\ &= \iiint_V \mathbf{G}(\boldsymbol{r}_0|\boldsymbol{r})\cdot\boldsymbol{g}(\boldsymbol{r})\,\mathrm{d}V. \end{aligned} \tag{B.38}$$

It is evident that this formulation holds only when problem (B.35) is well-posed, i.e. when it has one and only one solution (see e.g. Van Bladel 1985).

References

LINDELL, I.V. (1988). *Complex vectors and dyadics for electromagneticists.* pp. 1–50, Report 36, Electromagnetics Laboratory, Helsinki University of Technology.

MORSE, P.M., and FESHBACH, H. (1953). *Methods of Theoretical Physics.* McGraw-Hill, New York.

TAI, C.T. (1987). Some essential formulas in dyadic analysis and their applications, *Radio Science*, **22**, 1283–8.

VAN BLADEL, J. (1985). *Electromagnetic fields*, p. 12, Appendix 3. Hemisphere, Washington. Reprinted, with corrections, from a text published in 1964 by McGraw-Hill.

APPENDIX C
Modal expansion in a cavity

The modal representation of the fields in a cavity is based on the use of two complete sets of orthonormal eigenvectors. The E field, for example, is conveniently expanded in the *electric* (or short-circuit) eigenvectors, which satisfy

$$\nabla^2 a_m = \lambda_m a_m,$$

$$u_n \times a_m = 0 \quad \text{and} \quad \operatorname{div} a_m = 0 \quad \text{on } S. \tag{C.1}$$

The subscript m stands for an integer triple (i, j, k). The eigenvectors are real, and satisfy the orthonormal property

$$\iiint_V a_m \cdot a_n \, dV = \delta_{mn}. \tag{C.2}$$

The transformation embodied in (C.1) is self-adjoint.

The electric eigenvectors are of either the irrotational or the solenoidal type. The irrotational eigenvectors are derived from the Dirichlet eigenfunctions by means of the formula

$$f_m = \operatorname{grad} \phi_m, \tag{C.3}$$

where

$$\nabla^2 \phi_m + \mu_m^2 \phi_m = 0 \quad \text{in } V, \qquad \phi_m = 0 \quad \text{on } S. \tag{C.4}$$

The norm of the eigenvectors is

$$\iiint_V |f_m|^2 \, dV = \mu_m^2 \iiint_V \phi_m^2 \, dV = 1 \tag{C.5}$$

The solenoidal eigenvectors, denoted by e_m, are defined by

if $k_m \neq o$ then $\nabla \cdot e_m = 0$ $\quad -\operatorname{curl} \operatorname{curl} e_m + k_m^2 e_m = 0 \quad \text{in } V,$ $\quad k_m \neq o$

if $k_m = o$, then $e_m = \nabla \phi$, so included above $\cdots$

$$u_n \times e_m = 0 \quad \text{on } S, \quad \text{and} \quad \iiint_V |e_m|^2 \, dV = 1. \tag{C.6}$$

For the magnetic field the expansions are in terms of the *magnetic* (or open circuit) eigenvectors, defined by

$$\nabla^2 b_m = \rho_m b_m,$$

$$u_n \cdot b_m = 0 \quad \text{and} \quad u_n \times \operatorname{curl} b_m = 0 \qquad \text{on } S. \tag{C.7}$$

The transformation embodied in (C.7) is again self-adjoint. The b_m satisfy the orthonormal property (C.2), and they are either solenoidal or irrotational.

The irrotational family is derived from the *Neumann* eigenfunctions according to the relationships

$$\boldsymbol{g}_m = \operatorname{grad} \psi_m,$$

$$\nabla^2 \psi_m + \nu_m^2 \psi_m = 0 \quad \text{in } V, \qquad \frac{\partial \psi_m}{\partial n} = 0 \quad \text{on } S, \tag{C.8}$$

$$\iiint_V |\boldsymbol{g}_m|^2 \, \mathrm{d}V = \nu^2{}_m \iiint_V \psi_m^2 \, \mathrm{d}V = 1.$$

One of the Neumann eigenfunctions is $\psi_0 = 1$, corresponding to the index triple (0, 0, 0). It does not generate a $\boldsymbol{g}_m$, because its gradient is zero.

The solenoidal eigenvectors $\boldsymbol{h}_m$ are directly related to their electric counterpart $\boldsymbol{e}_m$ through the equations

$$\boldsymbol{h}_m = (1/k_m) \operatorname{curl} \boldsymbol{e}_m, \qquad \boldsymbol{e}_m = (1/k_m) \operatorname{curl} \boldsymbol{h}_m,$$

$$\iiint_V |\boldsymbol{e}_m|^2 \, \mathrm{d}V = \iiint_V |\boldsymbol{h}_m|^2 \, \mathrm{d}V = 1. \tag{C.9}$$

The electric eigenvectors form a complete set, in terms of which the expansion of a vector function $\boldsymbol{a}(\boldsymbol{r})$ takes the form (Van Bladel 1960)

$$\boldsymbol{a}(\boldsymbol{r}) = \operatorname{grad} \phi + \operatorname{curl} \boldsymbol{v}, \tag{C.10a}$$

where

$$\operatorname{grad} \phi = \sum_{m=0}^{\infty} \boldsymbol{f}_m \left(- \iiint_V \phi_m \operatorname{div} \boldsymbol{a} \, \mathrm{d}V \right),$$

$$\operatorname{curl} \boldsymbol{v} = \sum_{m=0}^{\infty} \frac{\boldsymbol{e}_m}{k_m^2} \left(\iiint_V \operatorname{curl} \boldsymbol{a} \cdot \operatorname{curl} \boldsymbol{e}_m \, \mathrm{d}V - \iint_S (\boldsymbol{u}_{\mathrm{n}} \times \boldsymbol{a}) \cdot \operatorname{curl} \boldsymbol{e}_m \, \mathrm{d}S \right). \tag{C.10b}$$

The scalar potential ϕ is constant on S. The magnetic eigenvectors also form a complete set, in terms of which $\boldsymbol{a}(\boldsymbol{r})$ takes the form

$$\boldsymbol{a}(\boldsymbol{r}) = \operatorname{grad} \psi + \operatorname{curl} \boldsymbol{v}, \tag{C.11a}$$

where

$$\operatorname{grad} \psi = \sum_{m=0}^{\infty} \boldsymbol{g}_m \left(- \iiint_V \psi_m \operatorname{div} \boldsymbol{a} \, \mathrm{d}V + \iint_S (\boldsymbol{u}_{\mathrm{n}} \cdot \boldsymbol{a}) \psi_m \, \mathrm{d}S \right),$$

$$\operatorname{curl} \boldsymbol{w} = \sum_{m=0}^{\infty} \frac{\boldsymbol{h}_m}{k_m^2} \iiint_V \operatorname{curl} \boldsymbol{a} \cdot \operatorname{curl} \boldsymbol{h}_m \, \mathrm{d}V. \tag{C.11b}$$

Here $\partial \psi / \partial n = 0$ on S, and $\boldsymbol{w}$ is perpendicular to S. Equations (C.10a, b) and (C.11a, b) clearly show which terms disappear when $\boldsymbol{a}$ is either irrotational or solenoidal.

The expansion formulas can be applied, in particular, to the fields in a cavity excited by time-harmonic currents. These currents can be of the electric or the magnetic type, with respective volume densities $\boldsymbol{J}$ and $\boldsymbol{K}$. Let us assume that the cavity contains a homogeneous medium of parameters ε and μ, and is bounded by a perfectly conducting wall S_w. The wall can be interrupted by an aperture S_a, in which a tangential electric field exists (Fig. C.1). The field expansions are (see e.g. Van Bladel 1985):

$$\boldsymbol{E}(\boldsymbol{r}) = -\frac{1}{j\omega\varepsilon}\sum_{m=0}^{\infty} \boldsymbol{f}_m(\boldsymbol{r}) \iiint_V \boldsymbol{J}\cdot\boldsymbol{f}_m \,\mathrm{d}V + \sum_{m=0}^{\infty} \boldsymbol{e}_m(\boldsymbol{r})\left(\frac{j\omega\mu}{k^2-k_m^2}\iiint_V \boldsymbol{J}\cdot\boldsymbol{e}_m \,\mathrm{d}V \right.$$
$$\left. + \frac{k_m}{k^2-k_m^2}\iiint_V \boldsymbol{K}\cdot\boldsymbol{h}_m \,\mathrm{d}V + \frac{k_m}{k^2-k_m^2}\iint_{S_a} (\boldsymbol{u}_n \times \boldsymbol{E})\cdot\boldsymbol{h}_m \,\mathrm{d}S\right),$$

$$\boldsymbol{H}(\boldsymbol{r}) = -\frac{1}{j\omega\mu}\sum_{m=0}^{\infty} \boldsymbol{g}_m(\boldsymbol{r})\left(\iiint_V \boldsymbol{K}\cdot\boldsymbol{g}_m \,\mathrm{d}V + \iint_{S_a} (\boldsymbol{u}_n \times \boldsymbol{E})\cdot\boldsymbol{g}_m \,\mathrm{d}S\right) \tag{C.12}$$
$$+ \sum_{m=0}^{\infty} \boldsymbol{h}_m(\boldsymbol{r})\left(-\frac{k_m}{k^2-k_m^2}\iiint_V \boldsymbol{J}\cdot\boldsymbol{e}_m \,\mathrm{d}V + \frac{j\omega\varepsilon}{k^2-k_m^2}\iiint_V \boldsymbol{K}\cdot\boldsymbol{h}_m \,\mathrm{d}V \right.$$
$$\left. + \frac{j\omega\varepsilon}{k^2-k_m^2}\iint_{S_a} (\boldsymbol{u}_n \times \boldsymbol{E})\cdot\boldsymbol{h}_m \,\mathrm{d}S\right).$$

In these expressions, $k^2 = \omega^2\varepsilon\mu$. We observe that resonances occur for $k = k_m$, and that they are restricted to the solenoidal eigenvectors.

In the previous analysis we have tacitly assumed that the volume was simply connected and simply bounded. When V is doubly bounded or doubly connected, the set of eigenvectors must include a suitable harmonic vector (Van Bladel 1960).

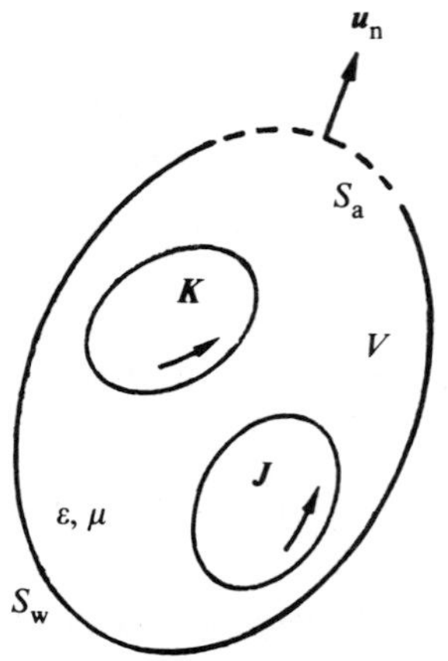

Fig. C.1. A cavity with electric and magnetic sources.

References

HARRINGTON, R.F. (1961). *Time-harmonic Electromagnetic Fields*, Chapter 8. McGraw-Hill, New York.

KUROKAWA, K. (1958). The expansions of electromagnetic fields in cavities. *IEEE Transactions on Microwave Theory and Techniques*, **6**, 178–87.

SLATER, J.C. (1950). *Microwave Electronics*, Chapter 4. Van Nostrand, New York.

VAN BLADEL, J. (1960). On Helmholtz's theorem in multiply-bounded and multiply-connected regions. *Journal of the Franklin Institute*, **269**, 445–62.

VAN BLADEL, J. (1985). *Electromagnetic Fields*, Chapter 10. Hemisphere, Washington. Reprinted, with corrections, from a text published in 1964 by McGraw-Hill, New York.

APPENDIX D
Modal expansion in a waveguide

The time-harmonic fields in a waveguide with constant cross-section (Fig. D.1) can be expanded as follows:

$$\mathbf{E}(\mathbf{r}) = \sum_{\mathbf{m}=0}^{\infty} V_{\mathbf{m}}(z)\,\mathrm{grad}\,\phi_{\mathbf{m}} + \sum_{\mathbf{n}\neq 0}^{\infty} V_{\mathbf{n}}(z)\,\mathrm{grad}\,\psi_{\mathbf{n}} \times \mathbf{u}_z + \sum_{\mathbf{m}=0}^{\infty} A_{\mathbf{m}}(z)\phi_{\mathbf{m}}\mathbf{u}_z, \tag{D.1}$$

$$\mathbf{H}(\mathbf{r}) = \sum_{\mathbf{m}=0}^{\infty} I_{\mathbf{m}}(z)\mathbf{u}_z \times \mathrm{grad}\,\phi_{\mathbf{m}} + \sum_{\mathbf{n}\neq 0}^{\infty} I_{\mathbf{n}}(z)\,\mathrm{grad}\,\psi_{\mathbf{n}} + \sum_{\mathbf{n}=0}^{\infty} B_{\mathbf{n}}(z)\psi_{\mathbf{n}}\mathbf{u}_z.$$

In these expressions, $\mathbf{m}$ and $\mathbf{n}$ are integer pairs, the sums are doubly infinite, and $\phi_{\mathbf{m}}$ and $\psi_{\mathbf{n}}$ are respectively the (two-dimensional) Dirichlet and Neumann eigenfunctions for the cross-section S. They are defined as in (C.4) and (C.8). Their normalization integrals, in particular, are

$$\iint_S |\mathrm{grad}\,\phi_{\mathbf{m}}|^2\,\mathrm{d}S = \mu_{\mathbf{m}}^2 \iint_S \phi_{\mathbf{m}}^2\,\mathrm{d}S, \tag{D.2}$$
$$\iint_S |\mathrm{grad}\,\psi_{\mathbf{n}}|^2\,\mathrm{d}S = \nu_{\mathbf{n}}^2 \iint_S \psi_{\mathbf{n}}^2\,\mathrm{d}S.$$

The functions $\phi_{\mathbf{m}}$ and $\psi_{\mathbf{n}}$ depend only on x and y. One of the Neumann eigenfunctions is $\psi_0 = 1$. Its gradient is zero, which means that grad ψ_0 does not appear in the series for $\mathbf{E}$ and $\mathbf{H}$. We express this fact by writing $\mathbf{n} \neq 0$ under the appropriate summation signs.

The form of expansion (D.1) shows that the '$\mathbf{m}$' eigenfunctions correspond to transverse magnetic (or E) modes, whereas the '$\mathbf{n}$' eigenfunctions give rise to transverse electric (or H) modes.

Assume that the waveguide is excited by electric and magnetic volume currents ($\mathbf{J}$, $\mathbf{K}$). Then, for the E (or TM) modes:

$$\frac{\mathrm{d}V_{\mathbf{m}}}{\mathrm{d}z} + \mathrm{j}\omega\mu_0 I_{\mathbf{m}} - A_{\mathbf{m}} = -\iint_S \mathbf{K}\cdot(\mathbf{u}_z \times \mathrm{grad}\,\phi_{\mathbf{m}})\,\mathrm{d}S$$
$$- \int_C (\mathbf{u}_\mathrm{n} \times \mathbf{E})\cdot(\mathbf{u}_z \times \mathrm{grad}\,\phi_{\mathbf{m}})\,\mathrm{d}C = -g_{\mathbf{m}}(z), \tag{D.3}$$
$$\frac{\mathrm{d}I_{\mathbf{m}}}{\mathrm{d}z} + \mathrm{j}\omega\varepsilon_0 V_{\mathbf{m}} = -\iint_S \mathbf{J}\cdot\mathrm{grad}\,\phi_{\mathbf{m}}\,\mathrm{d}S = -f_{\mathbf{m}1}(z),$$
$$I_{\mathbf{m}} + \frac{\mathrm{j}\omega\varepsilon_0}{\mu_{\mathbf{m}}^2} A_{\mathbf{m}} = -\iint_S (\mathbf{J}\cdot\mathbf{u}_z)\,\phi_{\mathbf{m}}\,\mathrm{d}S = -f_{\mathbf{m}2}(z).$$

Here $\boldsymbol{u}_n$ is a unit vector perpendicular to the waveguide wall. The contour integrals containing $\boldsymbol{u}_n \times \boldsymbol{E}$ represent the excitation through an aperture. They differ from zero when the contour C 'cuts' the aperture, i.e. when the z coordinate corresponds to a point 'under the aperture'.

Similar equations hold for the H (or TE) modes, viz.

$$\frac{dV_n}{dz} + j\omega\mu_0 I_n = -\iint_S \boldsymbol{K}\cdot\operatorname{grad}\psi_n\,dS - \int_C (\boldsymbol{u}_n \times \boldsymbol{E})\cdot\operatorname{grad}\psi_n\,dC = -g_{n1}(z),$$

$$\frac{dI_n}{dz} + j\omega\varepsilon_0 V_n - B_n = -\iint_S \boldsymbol{J}\cdot(\operatorname{grad}\psi_n \times \boldsymbol{u}_z)\,dS = -f_n(z), \tag{D.4}$$

$$V_n + \frac{j\omega\mu_0}{v_n^2} B_n = -\iint_S (\boldsymbol{K}\cdot\boldsymbol{u}_z)\psi_n\,dS - \int_C (\boldsymbol{u}_n \times \boldsymbol{E})\cdot\boldsymbol{u}_z\,\psi_n\,dC = -g_{n2}(z).$$

To solve (D.3) and (D.4), it pays to eliminate two coefficients (say A_m and I_m) to obtain a second-order differential equation for the third one (say V_m). Some of the resulting equations are

$$\frac{d^2V_m}{dz^2} + (k^2 - \mu_m^2)V_m = \frac{\gamma_m^2}{j\omega\varepsilon_0} f_{m1} - \frac{\mu_m^2}{j\omega\varepsilon_0}\frac{df_{m2}}{dz} - \frac{dg_m}{dz}, \tag{D.5}$$

$$\frac{d^2V_n}{dz^2} + (k^2 - v_n^2)V_n = j\omega\mu_0 f_n - \frac{dg_{n1}}{dz} + v_n^2 g_{n2}.$$

They are of the general type

$$\frac{d^2f}{dz^2} - \gamma^2 f = h(z). \tag{D.6}$$

Their solution is most conveniently effected by means of a Green's function. In an infinite guide (i.e. in the interval $-\infty < z < \infty$), application of the radiation condition (or the damping condition) at large distances requires the Green's function to be of the type

$$G_1(z\,|\,z'\,|\,\gamma) = -(1/2\gamma)e^{-\gamma|z-z'|} = G_1(z'\,|\,z\,|\,\gamma), \tag{D.7}$$

with

$$\gamma = \begin{cases} \sqrt{(\mu^2 - k^2)} & \text{for } k < \mu \text{ (damped mode)},\\ j\sqrt{(k^2 - \mu^2)} & \text{for } k \geqslant \mu \text{ (propagated mode)}. \end{cases} \tag{D.8}$$

The solution of (D.6) is explicitly

$$\begin{aligned} f(z) &= \int_{-\infty}^{\infty} G_1(z\,|\,z'\,|\,\gamma)h(z')\,dz' \\ &= -\frac{1}{2\gamma}e^{-\gamma z}\int_{-\infty}^{z} e^{\gamma z'}h(z')\,dz' - \frac{1}{2\gamma}e^{\gamma z}\int_{z}^{\infty} e^{-\gamma z'}h(z')\,dz'. \end{aligned} \tag{D.9}$$

When the second member of (D.6) is a derivative, the equation becomes

$$\frac{d^2 f}{dz^2} - \gamma^2 f = \frac{dq}{dz}. \tag{D.10}$$

In the infinite guide, the solution is

$$f(z) = \int_{-\infty}^{\infty} G_2(z \mid z' \mid \gamma)\, q(z')\, dz' = \int_{-\infty}^{\infty} \frac{dG_1(z \mid z' \mid \gamma)}{dz}\, q(z')\, dz', \tag{D.11}$$

where the Green's function is

$$G_2(z \mid z' \mid \gamma) = \tfrac{1}{2}\, \mathrm{sgn}\,(z - z')\, e^{-\gamma |z - z'|}. \tag{D.12}$$

The detailed form of (D.11) is therefore

$$f(z) = \tfrac{1}{2} e^{-\gamma z} \int_{-\infty}^{z} e^{\gamma z'}\, q(z')\, dz' - \tfrac{1}{2} e^{\gamma z} \int_{z}^{\infty} e^{-\gamma z'}\, q(z')\, dz'. \tag{D.13}$$

The Neumann eigenfunction $\psi_0 = 1$ contributes the term $B_0(z)\boldsymbol{u}_z$ in (D.1). Because the ψ_n are orthogonal, expansion coefficient B_0 is given by

$$B_0(z) = \frac{1}{s} \iint_S H_z\, dS = [H_z]_{\mathrm{ave}}, \tag{D.14}$$

where the average is over the cross-section S, of area s. Maxwell's equation curl $\boldsymbol{E} = -\mathrm{j}\omega\mu_0 \boldsymbol{H}$, projected on the z axis, yields

$$\mathrm{div}_{xy}(\boldsymbol{E} \times \boldsymbol{u}_z) = -\mathrm{j}\omega\mu_0 H_z - K_z. \tag{D.15}$$

Integrating both members over the cross-section S gives

$$\begin{aligned} B_0(z) &= -\frac{1}{\mathrm{j}\omega\mu_0}\left(\frac{1}{s}\iint_S \boldsymbol{K}\cdot\boldsymbol{u}_z\, dS + \frac{1}{s}\int_C (\boldsymbol{u}_n \times \boldsymbol{E})\cdot\boldsymbol{u}_z\, dC\right) \\ &= -\frac{1}{\mathrm{j}\omega\mu_0}\left((K_z)_{\mathrm{ave}} + \frac{1}{s}\int_C (\boldsymbol{u}_c \cdot \boldsymbol{E})\, dC\right). \end{aligned} \tag{D.16}$$

The term in $B_0(z)$ can therefore only exist at those points of the z axis where there is either a volume magnetic current $\boldsymbol{K}$ or an (aperture) surface current $\boldsymbol{u}_n \times \boldsymbol{E}$. We have shown, in Section 3.28, that the omission of $B_0(z)$ can lead to important errors.

To conclude, we present a few formulas of frequent use in waveguide analysis:

$$\operatorname{curl}[f(z)A(x,y)\boldsymbol{u}_z] = f(z)\operatorname{grad}A \times \boldsymbol{u}_z, \tag{D.17}$$

$$\operatorname{curl}[f(z)\operatorname{grad}A(x,y)] = \frac{\mathrm{d}f}{\mathrm{d}z}\boldsymbol{u}_z \times \operatorname{grad}A, \tag{D.18}$$

$$\begin{aligned}\operatorname{curl}[f(z)\operatorname{grad}A \times \boldsymbol{u}_z] &= \operatorname{curl}\operatorname{curl}[f(z)A(x,y)\boldsymbol{u}_z],\\ &= \frac{\mathrm{d}f}{\mathrm{d}z}\operatorname{grad}A - f(z)\nabla^2 A\boldsymbol{u}_z.\end{aligned} \tag{D.19}$$

References

Collin, R.E. (1973). On the incompleteness of E and H modes in waveguides. *Canadian Journal of Physics*, **51**, 1135–40.

Collin, R.E. (1990). *Field theory of guided waves*, IEEE Press, New York. Revised and expanded version of a text first published by McGraw-Hill, New York, in 1960.

Van Bladel, J. (1981). Contribution of the ψ = constant mode to the modal expansion in a waveguide. *Proceedings of the IEE*, **128**, 247–51.

Van Bladel, J. (1985). *Electromagnetic Fields*, Chapter 13. Hemisphere, Washington. Reprinted, with corrections, from a text published in 1964 by McGraw-Hill, New York.

Vu Khac, T., and Carson, C.T. (1973). $m = 0, n = 0$ mode and rectangular-waveguide slot discontinuity. *Electronics Letters*, **9**, 431–2.

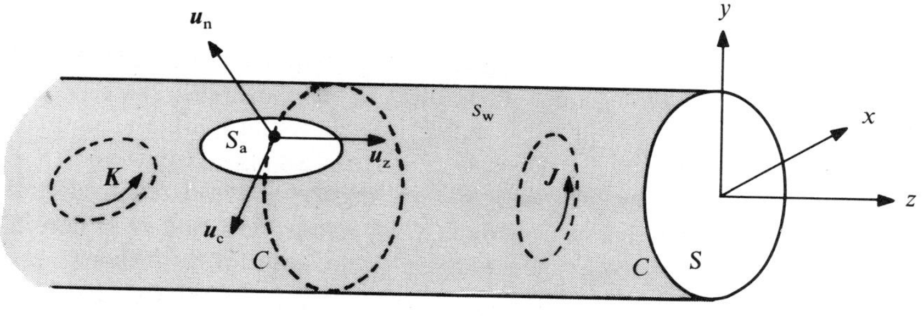

Fig. D.1. A waveguide with electric and magnetic sources

APPENDIX E
Basis functions

In the late 1950s, the availability of digital computers of progressively growing power started a fundamental evolution in the techniques used to solve electromagnetic problems. This evolution is still with us, thirty years later. The computer requires the originally 'continuous' problem to be transformed into a 'discrete' one; differential operators, for example, are replaced by difference operators, and integral equations by sets of linear equations (for an early example, see Cristal *et al.* 1961). In its most general form, the discretization process is termed 'method of moments' (Harrington 1968). The method has now reached a high degree of sophistication (Umashankar 1988; Wang 1990). To describe its main features, consider the problem

$$\mathscr{L}f(\mathbf{r}) = g(\mathbf{r}) \quad (f \in \mathscr{D}), \tag{E.1}$$

where $\mathscr{L}$ is a linear operator (integral or differential), g a known forcing function, and f—the unknown of the problem—a function which must belong to a given domain $\mathscr{D}$ (e.g. the family of functions which are continuous in a volume $V, \ldots$).

A solution for f is often sought in the form

$$f = \sum_{n=1}^{\infty} A_n f_n. \tag{E.2}$$

The f_n are the *basis functions*, and the A_n are the expansion coefficients yet to be determined. When the f_n form a complete set, expansion (E.2) converges to the exact solution. In numerical practice, however, one limits the sum to its first N terms. Such a truncation clearly yields only an approximation to f. The coefficients A_1 to A_N may now be determined by methods such as 'point matching'. In that particular approach, (E.1) is only satisfied at N carefully chosen points. The problem then reduces to the solution of N linear equations with N unknowns, a discretized form which is eminently suitable for programming on a digital computer.

A more general way to determine the A_n consists in forming the scalar product of both members of (E.1) with N *testing functions* w_i. This procedure yields N equations of the type

$$\sum_n \langle w_i, Lf_n \rangle A_n = \langle w_i, g \rangle = g_i. \tag{E.3}$$

In matrix form:

$$\mathbf{Z} \cdot A = g, \tag{E.4}$$

where $\mathbf{Z}_{in} = \langle w_i, \mathscr{L} f_n \rangle.$

Here A and g are the N-dimensional vectors formed by the A_n and the g_n. The choice of the scalar product (real, complex, ...) depends on the problem in hand.

The support of the basis functions $f_n(\boldsymbol{r})$ may be the whole domain V of variation of $\boldsymbol{r}$, in which case the functions are termed 'entire domain functions'. Obvious examples are $\sin nx$ and $\cos nx$ in the interval $(0, 2\pi)$. It is often advantageous, however, to choose 'subdomain' functions, i.e. functions

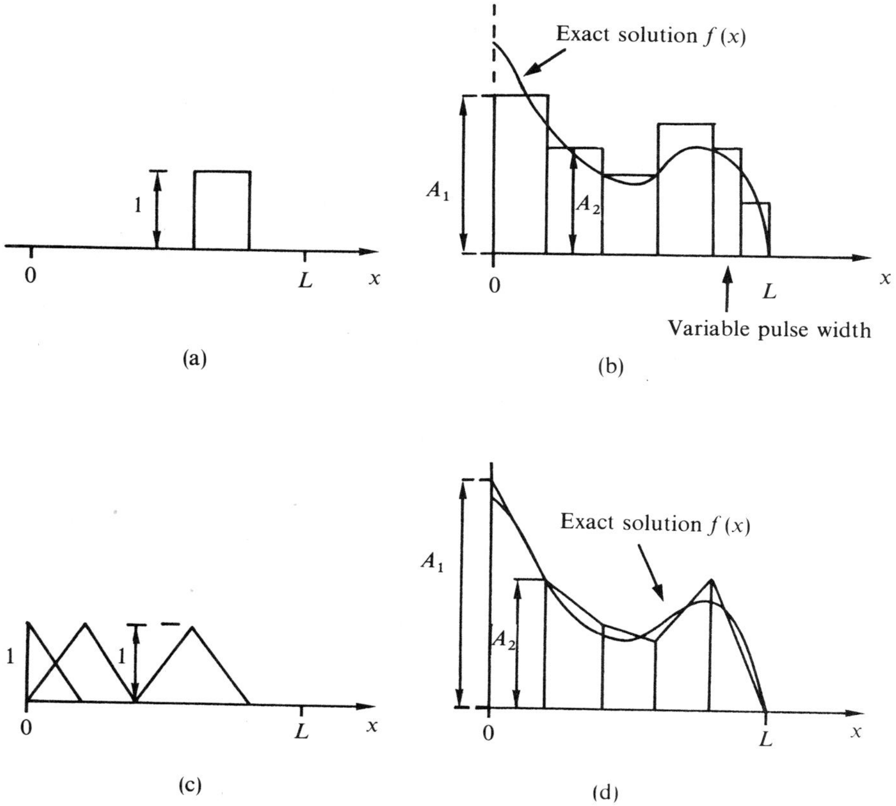

Fig. E.1. (a) A rectangular pulse. (b) An approximation of a function by rectangular pulses. (c) Triangular pulses. (d) An approximation of a function by triangular pulses.

with a support V^* which covers only part of V. These functions are identically zero in $V-V^*$. Two frequently used one-dimensional subdomain functions are as follows.

(a) The rectangular pulse (Fig. E.1 a, b): The unknowns of the problem are the heights $A_1, \ldots, A_n$ of the pulses. The function is approximated by a succession of horizontal plateaus. The width of the pulses may be chosen at will. It is advantageous, for example, to introduce a high density of pulses in those regions where $f(\mathbf{r})$ varies rapidly.

(b) The triangle and the half-triangle (Fig. E.1c, d): The function is now represented by a succession of linear segments.

Other subdomain functions, such as the half-sinusoid, have also been used.

It is intuitively evident that low values of N do not produce good accuracy, and hence that high values of N may be needed to obtain a satisfactory approximation to the unknown function f. The increase in N has the obvious disadvantage of requiring more computer time. The choice of an adequate value of N depends on the problem in hand.

References

Cristal, E.G., and Van Bladel, J. (1961). Fields in cavity excited accelerators. *Journal of Applied Physics*, **32**, 1715–24.

Harrington, R.F. (1968). *Field computation by moment methods.* Macmillan, New York.

Umashankar, K.R. (1988). Numerical analysis of electromagnetic wave scattering and interaction based on frequency-domain integral equation and method of moments techniques. *Wave motion*, **10**, 493–525.

Wang, J.J.H. (1990). *Generalized Moment Methods in Electromagnetics—Formulation and Computer Solution of Integral Equations.* J. Wiley, 1990.

APPENDIX F
Variational principles

Let $J(f)$ be a functional of f, i.e. an expression which takes a well-defined value for each f belonging to a given ensemble (the 'admissible' functions). This value varies by an amount δJ when f varies by δf. The functional is said to be stationary about f_0 when the variation δJ is of the order of $\| \delta f \|^2$ or higher. To explore the vicinity of f_0, we set

$$f(\mathbf{r}) = f_0(\mathbf{r}) + \varepsilon \eta(\mathbf{r}), \tag{F.1}$$

where ε is a small parameter, and the ηs are admissible functions, i.e. functions which belong to the basket admitted for comparison. When the $f(\mathbf{r})$ are inserted in $J(f)$, the resulting δJ must be of the order of ε^2 or higher. This 'stationarity' condition is not satisfied unless f_0 satisfies an equation of the type $\mathscr{M} f_0 = g$, where $\mathscr{M}$ is an operator which depends on the nature of the functional. The equation $\mathscr{M} f_0 = g$ is the Euler equation of the functional. In frequent cases, the Euler equation is insufficient to ensure stationarity, and f_0 is required to satisfy additionally some boundary conditions, which are then termed 'natural'. To illustrate the method, consider the functional encountered in (4.50), rewritten here as

$$J(f) = \int_0^1 f(x) \left(\int_0^1 K(x \mid x') f(x') \, \mathrm{d}x' \right) \mathrm{d}x - 2 \int_0^1 f(x) \, \mathrm{d}x. \tag{F.2}$$

The functions are real, and the kernel is real and symmetric (i.e. $K(x \mid x') = K(x' \mid x)$). Inserting (F.1) into (F.2) yields, because of the symmetry of the kernel,

$$\begin{aligned} J(f) = J(f_0) &+ 2\varepsilon \int_0^1 \eta(x) \left[\int_0^1 K(x \mid x') f_0(x') \, \mathrm{d}x' - 1 \right] \mathrm{d}x \\ &+ \varepsilon^2 \int_0^1 \eta(x) \left(\int_0^1 K(x \mid x') \, \eta(x') \, \mathrm{d}x' \right) \mathrm{d}x. \end{aligned} \tag{F.3}$$

For the term in ε to be zero for all admissible $\eta(x)$, the expression between brackets must vanish. As a result f_0 must satisfy the Euler equation

$$\int_0^1 K(x \mid x') f_0(x') \, \mathrm{d}x' = 1 \quad \text{for } 0 < x < 1. \tag{F.4}$$

This intuitive description of the variational procedure should be backed by more rigorous considerations; these are discussed at length in the specialized literature. In any event the reader will notice that (F.4), in the case of functional (4.50), is precisely the original integral equation (4.48).

The stationary value of the functional often has a physical meaning; it may represent, for example, a capacitance, an input impedance, or an energy. It should be observed, in that respect, that a rough approximation for f_0 leads to a much better approximation for the stationary functional $J(f_0)$. The reason is evident: the error on J is of the order of ε^2, while that on f_0 is of the order of ε.

Two approaches are obviously possible to exploit the variational principle:

(1) Solve $\mathscr{L}f = g$ to a good approximation, and insert this solution into $J(f)$ to obtain an (even better) approximation for the functional.

(2) Obtain an approximation to f_0 by starting from the functional instead of from $\mathscr{L}f = g$. This is done by inserting a parameter-laden trial function $f(C_1, \ldots, C_N)$ into $J(f)$. The best approximation to f_0 is obtained by satisfying the requirement

$$\frac{\partial J(C_1, \ldots, C_N)}{\partial C_i} = 0 \quad (i = 1, \ldots, N). \tag{F.5}$$

Good results are obtained if a sufficient number of parameters is included in the trial function.

The way to find Euler's equation in the case of functional (5.25) again consists in setting $\phi = \phi_0 + \varepsilon\eta$, and inserting this expression into (5.25). The resulting integral can be transformed by means of Green's theorem:

$$\iint_S (A\,\nabla_s^2 B + \operatorname{grad}_s A \cdot \operatorname{grad}_s B)\,\mathrm{d}S = \int_C A\,\boldsymbol{u}_m \cdot \operatorname{grad}_s B\,\mathrm{d}C. \tag{F.6}$$

This theorem holds on a surface S bounded by a curve C. The unit vector $\boldsymbol{u}_m$ is perpendicular to C, and lies in the tangent plane of S (Fig. F.1). An

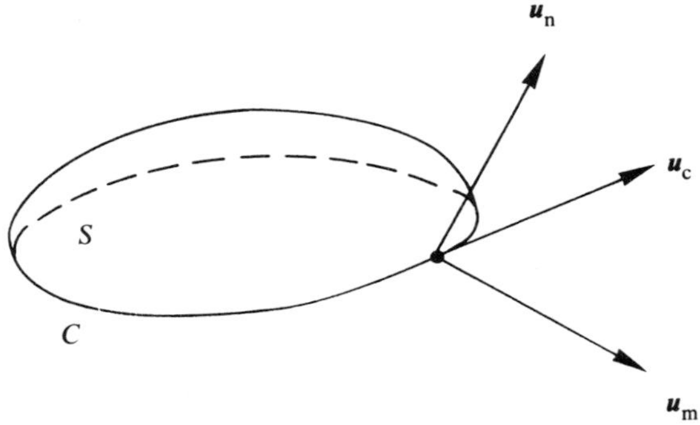

Fig. F.1. An open surface.

application of (F.6) yields

$$J_1(\phi) = J_1(\phi_0) + 2\varepsilon\left(-\iint_S \eta[\nabla_s^2\phi_0 + \mu(\mu+1)\phi_0]\,dS + \int_C \eta \boldsymbol{u}_m \cdot \operatorname{grad}\phi_0\,dC\right) + \text{term in } \varepsilon^2. \tag{F.7}$$

Stationarity requires the coefficient of ε to vanish for all admissible η. This requirement leads immediately to the differential equation in (5.21), and to the boundary condition in the same equation, which is therefore a 'natural' one. We conclude that this boundary condition is automatically satisfied by ϕ_0, and hence that $J_1(\phi)$ is stationary with respect to the magnetic eigenfunctions Z_{mn}. If, on the other hand, the supply of admissible functions is restricted by the requirement $\phi = 0$ on C, η vanishes on C, and the line integral in (F.5) is automatically zero. For such case, $J_1(\phi)$ is stationary with respect to the solutions of (5.16), i.e. with respect to the electric eigenfunctions Y_{mn}.

Applied to functional $J_2(\phi)$ in (5.27), the technique described above yields

$$J_2(\phi) = \frac{\iint_S |\operatorname{grad}_s\phi_0|^2\,dS + 2\varepsilon\iint_S \operatorname{grad}_s\phi_0 \cdot \operatorname{grad}_s\eta\,dS + \varepsilon^2\iint_S |\operatorname{grad}_s\eta|^2\,dS}{\iint_S \phi_0^2\,dS + 2\varepsilon\iint_S \eta\phi_0\,dS + \varepsilon^2\iint_S \eta^2\,dS_0}$$

$$= \frac{\iint_S |\operatorname{grad}_s\phi_0^2|\,dS}{\iint_S \phi_0^2\,dS} \tag{F.8}$$

$$\left[1 + 2\varepsilon\left(\frac{\iint_S \operatorname{grad}_s\phi_0 \cdot \operatorname{grad}_s\eta\,dS}{\iint_S |\operatorname{grad}_s\phi_0|^2\,dS} - \frac{\iint_S \eta\phi_0\,dS}{\iint_S \phi_0^2\,dS}\right) + \text{term in } \varepsilon^2\right].$$

The coefficient of ε again vanishes for the Y_{mn} and the Z_{mn}. The stationary value $J_2(\phi_0)$ is

$$J_2(\phi_0) = \frac{\iint_S |\operatorname{grad}_s\phi_0|^2\,dS}{\iint_S \phi_0^2\,dS} = -\frac{\iint_S \phi_0\,\nabla_s^2\phi_0\,dS}{\iint_S \phi_0^2\,dS} = \nu(\nu+1). \tag{F.9}$$

References

The general application of variational methods to electromagnetic problems is discussed in textbooks such as

Collin, R.E. (1990). *Field Theory of Guided Waves.* IEEE Press, New York. Revised and expanded version of a text first published by McGraw-Hill, New York, in 1960.

Harrington, R.F. (1961). *Time-harmonic Electromagnetic Fields,* Chapter 7. McGraw-Hill, New York.

Van Bladel, J. (1985). *Electromagnetic Fields,* Chapter 2. Hemisphere, Washington. Reprinted, with corrections, from a text published in 1989 by McGraw-Hill.

AUTHOR INDEX

SUBJECT INDEX